Dynamics in Small
Confining Systems V

MATERIALS RESEARCH SOCIETY
SYMPOSIUM PROCEEDINGS VOLUME 651

Dynamics in Small Confining Systems V

Symposium held November 27–30, 2000, Boston, Massachusetts, U.S.A.

EDITORS:

J.M. Drake
Exxon Research and Engineering
Annandale, New Jersey, U.S.A.

J. Klafter
Tel Aviv University
Tel Aviv, Israel

Pierre E. Levitz
CNRS
Orléans, France

René M. Overney
University of Washington
Seattle, Washington, U.S.A.

M. Urbakh
Tel Aviv University
Tel Aviv, Israel

Materials Research Society
Warrendale, Pennsylvania

CAMBRIDGE
UNIVERSITY PRESS

University Printing House, Cambridge CB2 8BS, United Kingdom

One Liberty Plaza, 20th Floor, New York, NY 10006, USA

477 Williamstown Road, Port Melbourne, VIC 3207, Australia

314-321, 3rd Floor, Plot 3, Splendor Forum, Jasola District Centre, New Delhi - 110025, India

79 Anson Road, #06-04/06, Singapore 079906

Cambridge University Press is part of the University of Cambridge.

It furthers the University's mission by disseminating knowledge in the pursuit of education, learning and research at the highest international levels of excellence.

www.cambridge.org
Information on this title: www.cambridge.org/9781558995611

Materials Research Society
506 Keystone Drive, Warrendale, PA 15086
http://www.mrs.org

© Materials Research Society 2001

First published 2001
First paperback edition 2013

Single article reprints from this publication are available through University Microfilms Inc., 300 North Zeeb Road, Ann Arbor, MI 48106

CODEN: MRSPDH

A catalogue record for this publication is available from the British Library

ISBN 978-1-558-99561-1 Hardback
ISBN 978-1-107-41233-0 Paperback

This work was supported in part by the Office of Naval Research under Grant Number N00014-01-1-0233. The United States Government has a royalty-free license throughout the world in all copyrightable material contained herein.

CONTENTS

*Invited Paper

*Invited Paper

*Invited Paper

Author Index

Subject Index

*Invited Paper

PREFACE

The sixth symposium on "Dynamics in Small Confining Systems", held November 27–30 at the 2000 MRS Fall Meeting in Boston, Massachusetts, celebrated a decade of this series. The program of the symposium covered a broad range of topics related to static and dynamic properties of confining systems: probing of confined systems, structure and dynamics of liquids at interfaces, nanorheology and tribology, adsorption, diffusion in pores, polymers and membranes, dielectric relaxation and biological aspects. Participants from various disciplines shared different points of view on the questions of how ultrasmall geometries can force a system to behave in ways significantly different from its behavior in the bulk, how this difference affects molecular properties, and how it is probed.

There appears to be a continuing interest in the dynamics and thermodynamics of confined molecular systems. This symposium was an effective way to bring together different disciplines interested in common problems.

The contributions in the proceedings reflect the broad range of topics discussed at the Meeting, all related to dynamics in small confining systems. We hope that the interdisciplinary nature of the proceedings and the Meeting will help to bridge the gap among the different approaches and methods presented. The papers appear in the book in the order of their presentation during the symposium.

We would like to thank everyone who helped in the organization and execution of the 2000 MRS Fall Meeting, especially the authors and presenters whose work has made this proceedings possible and the symposium successful.

Special thanks are due to ExxonMobil, ONR, and Tel Aviv University, which sponsored the symposium.

J.M. Drake
J. Klafter
Pierre E. Levitz
René M. Overney
M. Urbakh

May, 2001

MATERIALS RESEARCH SOCIETY SYMPOSIUM PROCEEDINGS

MATERIALS RESEARCH SOCIETY SYMPOSIUM PROCEEDINGS

Prior Materials Research Society Symposium Proceedings available by contacting Materials Research Society

Dynamics in Small
Confining Systems V

Mat. Res. Soc. Symp. Proc. Vol. 651 © 2001 Materials Research Society

Anomalous Diffusion in Active Intracellular Transport

Avi Caspi, Rony Granek, Michael Elbaum
Department of Materials and Interfaces
Weizmann Institute of Science
Rehovot 76100, Israel

ABSTRACT

The dynamic movements of tracer particles have been used to characterize their local environment in dilute networks of microtubules, and within living cells. In the former case, 300 nm diameter beads are fixed to individual microtubules, so that the movements of the bead reveal undulatory modes of the polymer. The mean square displacement shows a scaling of $t^{\frac{3}{4}}$ in keeping with mode analysis arguments. Inside a cell, beads show a more complicated behavior that reflects internal dynamics of the cytoskeleton and associated motors.

When placed near the cell edge, 3 micron diameter beads coated by proteins that mediate membrane adhesion are engulfed underneath the membrane and drawn toward the center by a contracting flow of actin. On reaching the region surrounding the nucleus, the beads continue to move but lose directionality, so that they wander within a restricted space. Measurement of the mean square displacement now shows a scaling of $t^{\frac{3}{2}}$ up to times of ~1 sec. At longer times the scaling varies between t^1 and $t^{\frac{1}{2}}$ in the various runs. The data do not fit a crossover between ballistic (t^2) and diffusive (t^1) behavior. The movement is clearly driven by non-thermal interactions, as it cannot be stopped by an optical trap. Treatment of the cell to depolymerize microtubules restores ordinary diffusion, while actin depolymerization has no effect, indicating that microtubule-based motor proteins are responsible for the motion. Immunofluorescence microscopy shows that the mesh size of the microtubules is smaller than the bead diameter.

We propose that the observations are related, and that the non-trivial scaling in the polymer system leads to time-dependent friction in a network, which in turn leads to a generalized Einstein relation operative in the intracellular environment. This results, in the driven system, in sub-ballistic motion at short times and sub-diffusive motion at longer times.

INTRODUCTION

In a symposium on dynamics in small confining systems, the living animal cell is almost an ideal laboratory. Enveloped by a single lipid bilayer, it contains structural and functional elements in all dimensions: a bulk solvent phase, a distinct liquid phase confined to vesicles and membrane-bound reticuli, numerous two-dimensional internal membranes, an extended network of filaments that collectively make up the cytoskeleton, and isolated molecular machines that function as individual entities. Coordination of the biochemistry involves a complex transport machinery that relies on a balance between thermal diffusion and deliberate delivery. In the latter case an important role is played by mechanoenzymes, commonly called "motor proteins" that carry cargo in one or the other direction along oriented tracks formed by cytoskeletal filaments. This discussion will focus on the special micro-rheological environment formed by such a system of tracks along which relatively large objects may be driven by molecular motors.

A few words of introduction to the cytoskeleton are in order. Extended filaments are formed by non-covalent polymerization of specific proteins. There are in fact three cytoskeletons of absolutely distinct composition, which nonetheless maintain a degree of interdependence and even mechanical linkage in the cell. These are the networks of filamentous actin (F-actin), microtubules, and intermediate filaments.

F-actin is most closely associated with maintenance of cell shape, cell-cell and cell-substrate adhesions, and in combination with the associated family of myosin motors, with force generation and cell contractility. The polymerized filaments are built of monomeric globular actin (G-actin) and have a chiral double-helix form reminiscent of DNA. The diameter of the actin filament is approximately 8 nm. The persistence length (the ratio of rigidity κ to thermal energy $k_B T$) is on the order of 10 μm, which coincides with the typical filament length. In the cell, F-actin is typically crosslinked by other proteins into sheets or bundles.

Microtubules, the second major cytoskeletal component, are the central mediators of intracellular spatial organization. The various organelles of the cell, including the nucleus, mitochondria, and endoplasmic reticulum, are carried to or maintained in their appropriate positions by interaction with a radially-polarized array of microtubules nucleating from the microtubule-organizing center, or centrosome, that sits near the nucleus. Microtubules are also essential in separating the chromosomes during cell division. They are formed from heterodimers of α- and β- tubulin, so that each monomer of the microtubule has dimensions roughly 4 by 8 nm. Polymerized, they self-assemble into hollow cylinders built of 13 parallel but staggered protofilaments, the total structure having outer diameter of 25 nm. The persistence length is roughly 6 mm [1], far larger than any relevant physical length. The oriented heterodimers give the microtubule a well-defined structural polarization. With respect to the centrosome, the "minus" ends are at the center of the cell while the "plus" are directed outward. (In spite of the nomenclature, this polarity has nothing to do with electric charge.) In some cell types a fraction of the microtubules may detach from the centrosome, but the polar orientation is maintained. Microtubule-associated motor proteins move directionally along the microtubule substrates; cytoplasmic dynenin moves from plus to minus end, i.e. inward in the cell, while most members of the kinesin superfamily move from the minus toward the plus end of the microtubule. These motors carry various types of cargo, of which small vesicles for import to (endocytosis) or export from (exocytosis) the cell are the most prominent.

F-actin and microtubules are universal structures, appearing in almost all types of animal, plant, and yeast cells. The third category of cytoskeletal elements, called intermediate filaments because in electron microscopy their diameters appear intermediate between those of F-actin and microtubules, appear specifically in certain cell types. Their biological function is less well understood, and there are no known motor proteins associated with them.

From the materials perspective, all of the cytoskeletal filaments are in the class of "semi-flexible" polymers. While there is no strict definition for the term, it is applied to polymers for which the persistence length is much longer than the chemical repeat unit, but for which thermal energy (at room temperature) is still sufficient to generate undulations of significant amplitude. Microtubules are thus at the rigid limit of the semi-flexible category. A corollary to the definition, related to the present work, is that semi-flexible polymers display significant differences in dynamics transverse and longitudinal to the local tangent along the contour.

The following discussion treats two experiments that show anomalous diffusion related to microtubules, a monomer subdiffusion driven by thermal energy *in vitro* [2], and an enhanced superdiffusion driven actively by motor proteins *in vivo,* within the cytoplasm of living cells [3]. We then propose that both effects are related to the same local rheological response of microtubules as semi-flexible polymers.

MONOMER DIFFUSION ON MICROTUBULES IN VITRO

What is the motion of a point on a polymer, when the polymer is embedded in a network? This question, related to conventional rheology in a somewhat inverse manner, is particularly relevant to the biological polymers where the long persistence lengths result in very substantial spatial inhomogeneities. The high cost of the materials, either for purchase or in effort to prepare them, weighs against conventional bulk rheological measurements. The emerging field of micro-rheology attempts to circumvent this problem by use of micron-sized particles embedded in a gel of the relevant polymer [4]. These studies have focused largely on F-actin as a model biological polymer. The particle motion may be detected to varying degrees of sensitivity by optical methods, including image-based single particle tracking and differential interferometric displacement analysis, or by scattering methods such as diffusing-wave spectroscopy. The particle must be significantly larger than the network mesh size in order that its displacement will be dominated by filament undulations rather than Brownian motion in the solvent. A two-particle correlation method is required in order to reconcile particle-based local microrheological with conventional rheological measurements [5]. Of course these considerations preclude the measurement of rheological properties in sparse networks. An alternate approach would be to label a single point on a polymer and to track its motion in time. Microtubules then have the technical advantage that the longer persistence length expands the range of time scales on which interesting behavior may be observed, relieving somewhat the disadvantages of convenient but relatively slow video-based methods. With a point-label method there is no lower bound on the required network density.

Experimental

Tubulin protein was purified from bovine brain by cycles of polymerization and depolymerization, followed by phosphocellulose chromatography to remove electrostatically-bound microtubule-associated proteins [6]. Tubulin was stored and used in a Pipes buffer solution:100 mM Na-Pipes pH 6.9, 2 mM $MgSO_4$, 1 mM EGTA, and 1 mM GTP. Tubulin polymerizes into microtubules upon an increase in temperature. Near the physiological range of temperature and concentration, microtubules display a dynamic instability where steady growth is interrupted stochastically by rapid depolymerization events. A phase diagram of the polymerization conditions has been described [7]. For the purpose of experiments, a tubulin solution at a given concentration was introduced to an observation chamber formed by a standard microscope slide to which a long (22 by 40 mm) coverslip was mounted crosswise using strips of Parafilm as spacers and sealant. This was placed on an inverted microscope (Zeiss model 35) equipped with an oil-coupled objective (Zeiss Fluar 100x/NA 1.3) and differential interference contrast (DIC) optics. The temperature of the objective was maintained at 30 ºC by a regulated electrical heater strip (Bioptechs Inc.). Samples were incubated for at least 30 minutes to produce a network of microtubules. Observations were made using a digitally-enhanced CCD camera (iSight model iSC 2050LL) with 1/2" sensor, for a final magnification of 80 nm/pixel. Individual microtubules could be observed with this setup.

Silica beads of 300 nm diameter and carboxylate surface functionalization (Bangs Labs, Inc.) were incubated in a Tris buffer solution (tris [hydroxymethyl] aminomethane), followed by washing in distilled water and resuspension in tubulin buffer. The buffer salt should form an ionic bond to the carboxylate groups on the bead surface, leaving the hydroxymethyl groups available for hydrogen bonding to microtubules. Treated beads were introduced to networks of polymerized microtubules. Beads stuck to microtubules spontaneously, or they could be directed to them deliberately using optical tweezers. The latter were used to verify adhesion, and microtubules could be stretched or even buckled without detaching the beads. (Other methods of bead attachment, particularly using coupled antibodies, proved less successful. Presumably the bead surface was blocked by unpolymerized tubulin, as they gradually detached.) Following introduction of the beads, the chamber was sealed and observations begun.

A small portion of the video field surrounding a bead was recorded directly into the hard disk of a personal computer at the rate of 25 frames/sec using a Matrox Meteor framegrabber, typically for 2-3 minutes. A single frame in which both bead and microtubule are visible is shown in Figure 1a. A cross-correlation algorithm [8], available in the Matrox Inspector or MIL software packages, was used to locate the center of the bead to sub-pixel accuracy frame-by-frame.

Results and Discussion

The raw data consists of an ordered table of frame numbers i and positions (x_i, y_i). A scatter plot, as in Figure 1b, clearly shows the difference between transverse and longitudinal movements. The required coordinate transformations were applied so as to separate out only the transverse undulatory modes. The mean square displacement was then calculated:

$$\left\langle \Delta h^2(x,t) \right\rangle = \left\langle \left(h(x,t) - h(x,0) \right)^2 \right\rangle , \qquad (1)$$

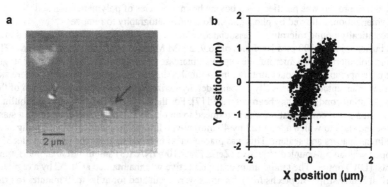

Figure 1. a) The arrow indicates a bead attached to a single microtubule. b) Measured positions of a bead attached to a microtubule are plotted, showing a clear separation into transverse (undulatory, large amplitude) and longitudinal (small amplitude) modes.

where h is the amplitude of the undulation and x the position along the filament. Figure 2a shows three examples. In one case, the microtubules were too short to be entangled or otherwise constrained by other filaments in the network. The result is ordinary diffusion, with a prefactor comparable to that expected for a thin rod of length 40 μm. The second shows a long microtubule in an extremely dilute network, where filament crossings may be separated by hundreds of microns. In this case the undulation amplitude builds up subdiffusively, with a scaling exponent of 0.77. Third, a stressed network may be prepared by polymerizing the microtubules in a test tube and then injecting them already formed into the observation chamber. Individual microtubules then appear bent, and the exponent is further depressed to 0.42. When dense, unstressed networks are produced as described in "Experimental", the mean square displacement appears as in Figure 2b. Within the first 2-3 sec the undulations grow with an exponent of 0.75 (20 series, std. dev. 0.04), while at longer times individual cases roll over to plateaus at different amplitudes.

The subdiffusion can be interpreted by direct mode analysis of the semi-flexible filament to which the small bead is attached. The undulation amplitude at a single point results from the summation of contributions from individual bending modes with energy density κq^4, each one driven by thermal energy $k_B T$:

$$\left\langle \Delta h^2(x,t) \right\rangle = \frac{4}{L} \sum_n \frac{k_B T}{\kappa q_n^4} \left(1 - e^{-w(q_n)t}\right) \sin^2(q_n x), \tag{2}$$

where $q_n = \frac{n\pi}{L}$ is the wavenumber. Each mode is associated with a different relaxation rate:

$$\omega(q_n) = \frac{\kappa q_n^4}{4\pi\eta} \ln\left(\frac{1}{q_n a}\right). \tag{3}$$

For a given time t, those modes with relaxation rates greater (i.e. faster) than t^{-1} will

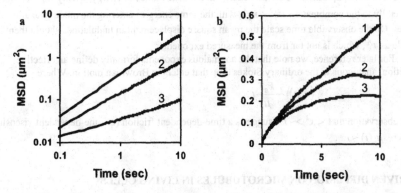

Figure 2. a) Measured mean square displacements for 3 cases: (1) a free microtubule where $<x^2> \sim t^{1.00}$, (2) a weakly entangled microtubule where $<x^2> \sim t^{0.77}$, (3) a microtubule in a pre-stressed network where $<x^2> \sim t^{0.42}$. b) Case (1) does not show a saturation regime on the experimentally accessible time scale, while cases (2) and (3) do. The saturation time is more apparent on a log scale (not shown).

contribute to the mean square displacement of the undulations. The relaxation rate corresponding to the longest wavelength mode sets an important cutoff, beyond which we can expect saturation of the undulation amplitude.

At short times the number of contributing modes grows with time, from the shortest toward longer wavelengths. Thus the response of the filament is more local the shorter the time span on which it is observed. The scaling in Equation 2 is apparent if we consider only terms that carry units, and carry out the sum up to a time-dependent long wavelength cutoff, where $\omega(q^*)t = 1$. This immediately yields $q^* \sim t^{1/4}$. Then

$$\left\langle \Delta h^2(t) \right\rangle = \frac{1}{L} \sum_{q^*(t)}^{1/a} \frac{k_B T}{\kappa q^4} \rightarrow \frac{1}{2\pi} \int_{q^*}^{1/a} dq \frac{k_B T}{\kappa q^4} \sim \frac{k_B T}{\kappa^{1/4}} t^{1/4},$$ (4)

and the short-time $t^{1/4}$ scaling is apparent. At long times all modes are saturated and the mean square displacement of the undulations reaches constant amplitude, where

$$\left\langle \Delta h^2(t) \right\rangle \rightarrow \frac{1}{2\pi} \int_{1/L}^{1/a} dq \frac{k_B T}{\kappa q^4} \sim \frac{k_B T}{\kappa} L^3.$$ (5)

Numerically, a saturation time of 1 sec corresponds to a longest wavelength mode of about 30 μm, which is far shorter than the physical length of the microtubules that lead to the observed saturation in Figure 2b. Thus we conclude that physical entanglements, or crossings, between the filaments effectively impose nodes in the undulations and restrict the mode spectrum relevant to the mean square displacement observations. In a sparse network such as seen in Figure 2a, the time corresponding to the longest wavelength mode is beyond that for which the measurements can yield a reliable exponent.

When the network is sheared, as in the third example of Figure 2a, we can expect that individual microtubules will be placed in tension. The derivations of Equations 2-3 can then be reworked for equipartition of thermal energy into tensile modes of energy density σq^2, or more generally to the combination $\sigma q^2 + \kappa q^4$, with the same energies entering the mode relaxation rates. On the observable time scale the mean square displacement in undulations should then scale as $t^{1/2}$, which is not far from the measured exponent.

For later reference, we note that the anomalous exponents formally define an effective friction, analogous to the ordinary Stokes drag that enters in Brownian motion. Where

$$\left\langle x^2(t) \right\rangle \sim \frac{k_B T}{\mu_e} t,$$ (6)

the observation that $<x^2> \sim t^{1/4}$ suggests a time-dependent friction (or time-dependent viscosity) where $\mu_e(t) \sim t^{1/4}$.

DRIVEN DIFFUSION ON MICROTUBULES IN LIVING CELLS

A classic assay in cell biophysics involves decoration of the cell membrane with small particles, such as beads, whose kinetic paths are meant to reveal the motion of membrane lipids or embedded proteins. We found that under certain conditions, large beads coated with the lectin (sugar-binding protein) Concanavalin A are engulfed into the cells rather than adhering to the membrane externally. This gives us the opportunity to study dynamics of active structures within the cell using beads as reporters. Experiments focused particularly on multinuclear giants formed from fibroblasts of the SV80 cell line, which afford very large, radially symmetric cells. Where the individual progenitors are motile, moving directionally along a flat substrate, the giant cells direct the locomotory apparatus, as it were, in all directions at once. Thus they remain stationary and provide an unambiguous frame of reference. Figure 3a shows a typical example viewed by differential interference contrast imaging.

Experimental

Cells were maintained in Dulbecco's Modified Eagle Medium (DMEM) with 10% bovine calf serum at 37 °C and 7.5% CO_2 in a standard culture incubator. Fibroblasts of the human-origin SV80 and murine NIH 3T3 cell lines were used for the experiments reported here. Multinuclear giants [9] of the SV80 cells were prepared by treatment with low doses of the actin filament disrupting drug cytochalasin D. Briefly, a sparse culture was exposed to the drug at a concentration of 0.2 µg/ml for 3-5 days. During this time the cells are unable to divide, but continue replication of their internal components, including centrosomes and nuclei. The drug is then washed out with fresh medium and the cells are allowed to respread for 1 day. For experiments the cells were replated onto Petri dishes with holes cut in the bottoms and sealed by Parafilm to glass cover slips, in order to permit high-resolution imaging. Prior to observation the CO_2-dependent DMEM medium was exchanged for Leibowitz L-15 medium whose pH is maintained without bicarbonate buffer.

Polystyrene beads of 3 µm diameter and amine surface functionalization (Polysciences, Inc.) were coated by Concanavalin A (Sigma Israel Chemical Co., C-7275) at saturating coverage using glutaraldehyde as a crosslinking reagent, following the bead manufacturer's protocol. Simple physical adsorption of the protein gave identical results, though bare amine beads did not adhere to the cells at all. In the case of non-giant cells, beads were mixed with cells during the normal process of suspension and replating onto the glass-bottomed observation dishes. For giant cells, beads were placed directly on the membrane near the cell edge using optical tweezers.

Observations were made using the same optical system described above, with the addition of a 0.5× demagnifying lens. Brightfield imaging was used for tracking of the large particles, and an improved cross-correlation algorithm in the updated Matrox MIL software applied. The experimental uncertainty in positioning was estimated at 20 nm by running the tracking algorithm on beads that adhered to the cover slip. Further analysis was made using Matlab and Microsoft Excel software. The cell shown in Figure 3b was fixed by sudden exposure to methanol at −20 °C, followed by rinsing with gradual dilutions of phosphate-buffered saline at room temperature. Immunofluorescence staining to α-tubulin was done using the monoclonal antibody DM1A (Sigma Israel Chemical Co., T 9026), followed by a Cy2-labelled fluorescent secondary antibody.

Results and Discussion

When placed near the perimeter on the giant cells, beads were quickly engulfed and then carried centripetally toward the nuclei. The motion was quite reminiscent of the now canonical

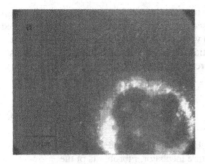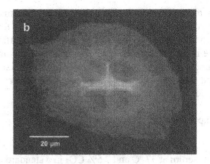

Figure 3. a) a section of a typical multinuclear giant cell as seen live, by optical differential interference contrast (DIC) imaging. The nuclei, in the lower left-hand corner of the image, are at the center of the radially symmetric cell. (Two images were merged to produce the extended focus effect.) b) Immunofluorescence microscopy shows a dense network of microtubules radiating from the cell center, where each nucleus provides a single centrosome.

"surface" movement of attached beads. We verified by confocal fluorescence microscopy and by scanning electron microscopy of fixed cells on which bead movement had been observed optically that this movement occurred, in our system, clearly underneath the plane of the dorsal cell membrane. Details of this movement form part of another study and will be reported separately [10]. For the present purposes, the interesting regime is that when the beads reach the end of their paths. They then "wander" aimlessly, or diffuse within a restricted region comparable to their diameter. The same regime is observed in the individual cells where beads were loaded in suspension. Close visual inspection suggests that this diffusion is not Brownian motion, however. The beads make long, slow excursions rather than the more jittery thermal motion. Moreover, they cannot be stopped or otherwise hindered by an optical tweezer that can easily trap and drag the same beads through a viscous medium at far greater speeds.

Tracking of the particles yielded a list of positions and coordinates as in the *in vitro* case above. A typical example is shown in Figure 4a. Calculation of the mean square displacement immediately gave a surprising result, that $< x^2(t) > \sim t^{3/2}$. From 19 observations on as many cells, we obtained a mean exponent of 1.47, with std. dev. 0.07, calculated by linear regression over the first 1 sec of data. At longer times the curve rolled over toward slower rates of increase. Figure 4b (curve 1) shows an example of the mean square displacement.

The superdiffusive exponent again indicates a non-thermal element driving the bead. Within the cell this could reasonably be attributed either to actin or microtubule-associated motor protein activity. The two possibilities were tested by use of drugs, cytochalasin D and nocodazole, to specifically depolymerize either F-actin or microtubules respectively, leaving the motors without a track along which to move. As seen in Figure 4b, the loss of F-actin had no effect on the dynamics, while the plots in Figures 4a and b show that microtubule depolymerization results in ordinary Brownian motion. Finally, lipid granules of roughly 1 μm diameter, naturally present in the SV80 cells, were found to show the passive scaling characteristic of the *in vitro* observations. Unlike the beads, these granules could be trapped and dragged using the optical tweezers with pN forces, suggesting that they did not directly engage motor proteins.

We could suspect that the exponent of $\frac{3}{2}$ represents a crossover between ballistic movement at constant velocity v, where $<x> \sim vt$ or $<x^2> \sim v^2 t^2$, to diffusive motion where $<x^2> \sim Dt^1$. Assuming constant friction μ and exponentially time-correlated force $<F(t_1)F(t_2)> = \mu^2 <v^2> e^{-|t_1 - t_2|/\tau}$, one obtains:

$$\langle x^2(t) \rangle = 2v^2 \tau \left[t - \tau \left(1 - e^{-\frac{t}{\tau}} \right) \right]. \tag{7}$$

Attempts to model the movement with a variety of time constants τ and even non-exponential force correlation functions were not successful in recovering the observed behavior. The measured roll-off is too sharp, and at the same time the $\frac{3}{2}$ power always appears cleanly and independently of the sample-specific time on which the roll-off takes place.

Immunofluorescent staining of the microtubules in these cells shows that near the nuclei, and therefore near the centrosomes, the beads will be embedded within a dense array. The bead diameter is much larger than the separation between filaments. This suggests an alternate explanation for the anomalous enhanced diffusion scaling. In being pushed along the microtubules by active motors, the bead must push the surrounding microtubules out of the way. How will they respond to this applied force? There is no *a priori* reason to expect them to respond as a Newtonian fluid with a simple scalar viscosity.

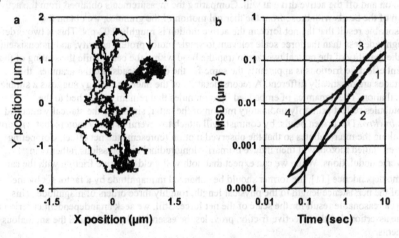

Figure 4. a) a trace of the bead position within the cell. The long trace is from a normal cell, while the curve marked by the arrow shows a case where microtubules are depolymerized by nocodazole. b) Mean square displacement curves for 4 cases: (1) a normal cell, where $<x^2> \sim t^{1.50}$ within the first second, (2) a cell with microtubules depolymerized as above, where $<x^2> \sim t^{1.06}$, (3) a cell with F-actin depolymerized by cytochalasin D, where $<x^2> \sim t^{1.50}$ as in the untreated cell, and (4) mean square displacement of passively-driven lipid granules naturally present in the cells, showing $<x^2> \sim t^{0.75}$.

We propose that the micro-rheological environment felt by the driven bead is precisely that of a network of semi-flexible filaments, i.e. one that responds to a force with time-dependent friction:

$$x \sim \frac{F}{\mu_e(t)} t; \qquad \langle x^2 \rangle \sim t^{3/2} \to \mu_e(t) \sim t^{1/4} . \tag{8}$$

Indeed the observed mean square displacement indicates the same scaling in the effective friction $\mu_e(t) \sim t^{1/4}$ observed for the lipid granules in the cells, and for the microtubule undulations in the *in vitro* experiment.

This proposal may be formulated in the context of a generalized Einstein relation:

$$\frac{4k_B T}{\mu_e(t)} = \frac{d}{dt} \langle \bar{x}^2(t) \rangle_{TH}; \qquad \frac{dx}{dt} = \frac{F}{\mu_e(t)} . \tag{9}$$

The most obvious question is then whether we can extract the effective force acting in the driven motion. For a randomly directed force with zero mean, Equation 9 can be reformulated to:

$$\langle \bar{x}^2(t) \rangle_F = \frac{\langle F^2 \rangle}{(4k_B T)^2} \langle \bar{x}^2(t) \rangle_{TH}^2 , \tag{10}$$

and in principle the thermal and driven motions can be compared. Unfortunately it is not possible to turn on and off the active drive at will. Comparing the measurements obtained from the active motion of the beads with those from the thermal motion of the granules, we obtain the unreasonable result that the net force in the active motion is roughly 0.02 pN. This is two orders of magnitude less than the force scale relevant to single motor protein activity, and inconsistent with the inability of the optical tweezer to trap the bead within the cell. While the scaling of the relevant effective frictions is apparently the same for the driven beads and the granules, the amplitudes are drastically different. A reconsideration of the phenomenology suggests a simple reconciliation. The dynamics of embedded lipid granules that remain unattached to the microtubules reflect primarily undulatory motion of the latter, similar to the statically attached beads *in vitro*. The driven beads, by contrast, will attach to several or many motors that will drive them along the microtubules so that the observed motion represents the net effect. Competing and even stalled motors will then induce primarily longitudinal compressions rather than transverse undulations. While we can expect that both will yield effective friction with the same $t^{1/4}$ time dependence [11], the former should be enhanced in magnitude by a factor set by the ratio of the persistence length to the physical length, roughly three orders of magnitude. This returns a reasonable result for the scale of the net force. Still, we seek an independent criterion to test the assertion that the effective friction provides the essential physics behind the anomalous exponents.

Such a test is provided by the dynamics of the beads at longer times. If the $3/2$ exponent is simply a crossover between 2 and 1, the long-time behavior should roll over to diffusive at times longer than the correlation time of the motor generated force. If the $3/2$ exponent comes from time-dependent effective friction $\mu_e(t) \sim t^{1/4}$ the prediction is quite different. While properly worked out in Laplace space, the predicted scaling can be seen from simple arguments. For times long enough that the driving force is decorrelated, $< F(t_1)F(t_2) > \sim \delta(t_1 - t_2)$. If this time is shorter than the scale on which the effective friction becomes saturated and essentially rezeroed ("loss of memory" effect), then Equation 9 yields:

$$\left\langle x^2(t) \right\rangle = \int_0^t dt_1 \int_0^t dt_2 \frac{\left\langle F(t_1)F(t_2) \right\rangle}{\mu_e(t-t_1)\mu_e(t-t_2)} \rightarrow \int_0^t dt' \frac{1}{\mu_e^2(t-t')} \sim t^{\frac{1}{2}}. \tag{10}$$

In the experimental measurements, the typical behavior is indeed a crossover from the short-time $t^{\frac{1}{2}}$ scaling to a distinctly subdiffusive scaling at longer times. While the restricted time domain available makes it risky to try to fix an exponent, the data are consistent with $t^{\frac{1}{2}}$ scaling and certainly inconsistent with the proposal of a crossover from t^2 to t^1 behavior. Figure 5 shows a clear example of this crossover. This domain is observed to differing degrees in different samples. It should disappear at times longer than the "loss of memory" time in $\mu_e(t)$, when normal diffusion would be restored.

SUMMARY

In summary, we have observed anomalous sub-diffusion in thermally-driven undulations of microtubules, and sub-ballistic motion of engulfed particles along microtubules in living cells. In the latter case the behavior at times longer than the correlation time of the driving force is strongly sub-diffusive. All of the observed scaling in the mean square displacement measurements can be explained in terms of a time-dependent friction arising from the semi-flexible nature of the microtubule polymers.

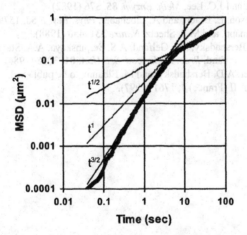

Figure 5. The bold line shows the measured crossover between short-time $t^{\frac{3}{4}}$ and long-time $t^{\frac{1}{2}}$ behavior, as indicated by the thin guide lines. The data are not consistent with a crossover between conventional "ballistic" t^2 and "diffusive" t scaling.

ACKNOWLEDGEMENTS

The authors are grateful to Prof. Alexander Bershadsky for help and advice in all matters relating to cell biology, to Katia Arnold for assistance in immunofluorescent staining, and to Prof. Yossi Klafter for many helpful discussions on anomalous diffusion. This work was supported by the Israel Science Foundation, and by the Gerhardt M.J. Schmidt Center for Supramolecular Architecture. M.E. is incumbent of the Delta Career Development Chair.

REFERENCES

1. M. Elbaum, D. Kuchnir Fygenson, A. Libchaber, *Phys. Rev. Lett.* **76**, 4078 (1996).
2. A. Caspi, M. Elbaum, R. Granek, A. Lachish, and D. Zbaida, *Phys. Rev. Lett.* **80**, 1106 (1998).
3. A. Caspi, R. Granek, and M. Elbaum, *Phys. Rev. Lett.* **85**, 5655 (2000).
4. T.G. Mason and D.A. Weitz, *Phys.Rev. Lett.* **74**, 1250 (1995); F. Amblard, A.C. Maggs, B. Yurke, A.N.Pargellis, and S. Leibler, *Phys. Rev. Lett.* **77**, 4470 (1996); T.G. Mason, K. Ganeson, J.H. van Zanten, D. Wirtz, and S.C. Kuo, *Phys. Rev. Lett.* **79**, 3282 (1997); F. Gittes, B. Schnurr, P.D. Olmstead, F.C. MacKintosh, and C.F. Schmidt, *Phys. Rev. Lett.* **79**, 3286 (1997); F.C. MacKintosh and C.F. Schmidt, *Curr. Opin. Colloid In.* **4**, 300 (1999).
5. J.C. Crocker, M.T. Valentine, E.R. Weeks, T. Gisler, P.D. Kaplan, A.G. Yodh, and D.A. Weitz, *Phys. Rev. Lett.* **85**, 888 (2000).
6. R.C. Williams, Jr. and J.C. Lee, *Meth. Enzym.* **85**, 376 (1982).
7. D. Kuchnir Fygenson, E. Braun, and A. Libchaber, *Phys. Rev. E* **50**, 1579 (1994).
8. J. Gelles, J.P. Schnapp, and M.P. Sheetz, *Nature* **331**, 450 (1988).
9. L.A. Lyass, A.D. Bershadsky, V.I. Gelfand, A.S. Serpinskaya, A.A. Stavrovskaya, J.M. Vasiliev, and I.M. Gelfand. *Proc. Natl. Acad. Sci. USA.* **81**, 3098 (1984).
10. A. Caspi, O. Yeger. A.D. Bershadsky, and M. Elbaum, to be published.
11. R. Granek, J. Phys. II (France) **7**, 1761 (1997).

Mat. Res. Soc. Symp. Proc. Vol. 651 © 2001 Materials Research Society

Structure of Charged Polymer Chains in Confined Geometry

Elliot P. Gilbert [1], Loïc Auvray [2] and Jyotsana Lal [1]

[1] Intense Pulsed Neutron Source, Argonne National Laboratory, Argonne, IL 60439, U.S.A.
[2] Laboratoire Leon Brillouin, Centre d'Etudes de Saclay, Gif-sur-Yvette Cedex, France.

ABSTRACT

The intra- and interchain structure of sodium poly(styrenesulphonate) when free and when confined in contrast matched porous Vycor has been investigated by SANS. When confined, a peak is observed whose intensity increases with molecular weight and the $1/q$ scattering region is extended compared to the bulk. We infer that the chains are sufficiently extended, under the influence of confinement, to highlight the large scale disordered structure of Vycor. The asymptotic behavior of the observed interchain structure factor is $\approx 1/q^2$ and $\approx 1/q$ for free and confined chains respectively.

INTRODUCTION

An understanding of polymer conformation in reduced dimensionality has enormous technological application, such as microfabrication and miniaturization [1]. While neutron scattering has provided invaluable information of chain conformation in bulk [2], there have been relatively few scattering experiments of chains under confinement and these mainly focus on neutral chains. The significantly more complex, and biologically relevant, situation of charged chains (e.g. DNA) in confined geometry currently lacks an experimental framework with which to verify fundamental principles. We report here small-angle neutron scattering (SANS) studies of poly(styrenesulphonate) (PSSNa) when free and when confined in the nanopores of Vycor. By employing solutions composed of either pure deuterated PSSNa or suitable mixtures of protonated and deuterated PSSNa, we have separated the intramolecular and intermolecular contributions to the total scattering, and provide a theoretical understanding of their q-dependencies.

EXPERIMENTAL DETAILS

All polyelectrolytes had a degree of sulphonation > 95% and were dialysed prior to solution preparation; molecular weights, M_W and polydispersities, M_W/M_N, are shown in table 1. Since Vycor is highly susceptible to contamination, disks were cleaned by repeated boiling in H_2O_2 solution followed by heating to 500•C for at least three hours. The cleaned disks were opalescent but, on entering the solution, became translucent with bubbles subsequently being released, indicating the solution-displacement of air. The confined samples were studied after 2, 7 and 9 days for the low, intermediate and high M_W samples respectively corresponding to the time for which no visible bubbles remained. Experiments were performed on the SAD instrument at IPNS, Argonne National Laboratory. Absolute values of intensity were obtained from calibration against a polymer standard.

Table I. Polyelectrolytes studied, R_g and Sharp and Bloomfield model fitting parameters for bulk ZAC solutions. R_g limits given for polystyrene, rod conformation and Benoit and Doty relation

M_W (Da) H/D	31000 H	25000 D	76000 H	77400 D	451000 H	425000 D
M_W/M_N	1.1	1.1	< 1.1	1.1	1.05	1.04
$<N>_{ZAC}$	142		368		2142	
$<M_W>_{ZAC}$	29500		76300		445000	
R_g (neutral)	36 Å		58 Å		143 Å	
R_g (rod)	85 Å		262 Å		1440 Å	
R_g (BD)	40 Å		70 Å		178 Å	
Parameter	Fit	Pred.	Fit	Pred.	Fit	Pred.
l_p (Å)	15(1)		17(1)		18(1)	
Scale factor	0.90(2)	1.20	1.8(1)	2.42	16.2(8)	18.1

The SANS from bare Vycor exhibits an intense peak at q_{max} ca. 0.02 Å^{-1} arising from quasiperiodicity of the, average internal diameter of 70 Å, pore structure [3]. The scattering length density (SLD) of Vycor was determined by contrast variation experiments with mixtures of Millipore H_2O (resistivity equal to 18 MΩcm) and D_2O. The match point was determined to be 3.79 x 10^{10}cm^{-2}; equivalent in SLD to a mixture of 37.6% H_2O and 62.4% D_2O. To confirm this, Vycor was filled with such a water mixture and gave rise to a SANS signal in which the correlation peak is absent and the scattered intensity at q_{max} was reduced by three orders of magnitude. This water composition was used to prepare all polyelectrolyte solutions. The same Vycor disk was used for both the sample and background scattering runs to avoid effects from differences between individual disks.

THEORY

The scattering from a deuterated/ protonated mixture of a single polymer species, with index of polymerisation, N, in solution may be written [4]:

$$I(q) = (\bar{\rho}_D - \bar{\rho}_H)^2 x(1-x)v\Phi NP(q) + (x\bar{\rho}_D + (1-x)\bar{\rho}_H - \bar{\rho}_0)^2 [v\Phi NP(q) + V\Phi^2 Q(q)] \quad (1)$$

where $I(q)$ is the total scattering intensity per unit volume (cm^{-1}), $\bar{\rho}_D, \bar{\rho}_H, \bar{\rho}_0$ are the SLD of the deuterated, protonated polymer, solvent respectively, x is the mole fraction of deuterated chains, v is the monomer molecular volume, V is the sample volume and Φ is the total volume fraction of polymer in solution. $P(q)$ and $Q(q)$ are the (intrachain) form factor and interchain structure factors respectively as normalised in [4]. When the average contrast between the polymer and solvent is adjusted to zero (ZAC, zero average contrast condition), the scattered intensity becomes:

$$I(q) = (\bar{\rho}_D - \bar{\rho}_H)^2 x(1-x)v\Phi NP(q) \quad (2)$$

enabling direct determination of $P(q)$. To maximise the scattering intensity, the composition of the polymer mixture is typically chosen with x equal to 0.5 with the solvent matched accordingly but, for confined studies, both the solute and solvent must match Vycor. With SLDs for $PSS^D Na$ and $PSS^H Na$ of 6.8879 and 2.8321 x 10^{-10} cm^{-2} respectively [5], this is satisfied for a mixture of 23.6:76.4 mole% but necessarily results in a further reduction in scattering intensity in these already weakly scattering systems.

For a medium with neutral pores, a concentration is typically chosen such that the Debye length, κ^{-1}, is less than the pore diameter. Initial experiments were performed with solutions with $\kappa^{-1} \approx 7$ Å but resulted in negligible chain penetration indicating the charged nature of the pore surface. The concentration used for all ZAC solutions here was 0.306 gcm^{-3} (ca. four times the initial study) except for the intermediate weight solution which was 0.238 gcm^{-3}. Similar deuterated-only solutions were also prepared for the low and high M_W chains with a concentration of 0.314 gcm^{-3} so as to maintain a constant monomer concentration.

The radius of gyration, R_g, for a fully extended chain is $L/\sqrt{12}$, where L is the contour length (equal to Na where a is the monomer size equal to 2.5 Å). Using this value to calculate the overlap concentration, c*, for the ZAC solutions gives 4.52 x 10^{-2}, 6.79 x 10^{-3} and 2.00 x 10^{-4} gcm^{-3} respectively where:

$$c^* = M_w/N_A R_g^3 \qquad (3)$$

and N_A is Avogadro's number. For comparison, R_g for the neutral parent polystyrene chain in a theta solvent is $0.27 M_w^{0.506}$ [6]. R_g limits for the chains studied are summarised in Table I.

RESULTS AND DISCUSSION

The scattering from the bulk deuterated chain is weak due to the low compressibility of the system. The principal feature is a peak at q ca. 0.25 Å^{-1} associated with an average interchain distance of ca. 25 Å whose position is independent of molecular weight. From the bulk ZAC solutions, we are sensitive only to the chain form factor. Kratky plots, I(q) versus q^{-2}, exhibit a plateau region centred at q ca. 0.05 Å^{-1} and, at higher q, vary as q^{-1}. The scattering vector associated with this crossover, q*, corresponds to the transition between the asymptotic behaviour of a Gaussian chain to that of a rod. The persistence length, l_p, can thus be extracted from fitting the scattering to a Sharp and Bloomfield model for wormlike chains if L > 10l_p and ql_p < 2 (Figure 1)[7]. Setting L equal to 356, 919 and 5356 Å for the low, intermediate and high M_W chains based on their average degree of polymerisation, $<N>_{ZAC}$, the parameters obtained are summarised in Table I. One may express R_g as a function of L and l_p by [8]:

$$\left\langle R_g^2 \right\rangle = l_p^2 \left[\frac{L}{3l_p} - 1 + \frac{2l_p}{L} - \frac{2l_p^2}{L^2} \left(1 - \exp(-\frac{L}{l_p}) \right) \right] \qquad (4)$$

R_g values obtained from (4) are much closer to those for the Gaussian relative to rod conformation as expected for the chain concentrations » c* investigated (Table I). l_p is also independent of M_W at this high monomer concentration. A comparison of the free and confined scattering at large scattering vectors (ql_p>>1) indicates that the intensity associated with the confined chain is approximately one quarter that of the bulk solution. Since Vycor has a porosity of 28%, such a reduction in high q scattering is in line with expectations and suggests that chains have fully entered. The principal feature in the confined scattering is a peak at approximately the same position as observed from bare Vycor whose intensity increases with M_W. In previous studies of the neutral parent polymer under confinement, the SANS from chains of $M_W \geq 137$ kDa showed the slight presence of this peak despite being at ZAC [4]. This was interpreted in terms of chains spanning pores over a range comparable to or longer than the Vycor pore correlations. We may exclude the possibility that this peak arises from a contrast mismatch since not only has the contrast match point of Vycor been verified but the bulk ZAC scattering gives rise to I(q), l_p and R_g in good agreement with the existing literature [5]. Figure 1 shows scattering data plotted in the form of a qP(q) versus q form to emphasise the asymptotic rod-like behaviour and shows that this conformation region extends to much lower q than the corresponding bulk indicating an increase in the persistence length under confinement. The low q cut-off for q^{-1} scattering occurs at q ca. 0.05 Å^{-1}, independent of M_W; below this q, the scattering around the peak masks the polymer scattering. We also note that the observed increase in peak intensity with M_W is as would be expected if an increasing number of pores were connected by an extended chain. In common with the ZAC chains, the confined deuterated polyelectrolytes exhibit a correlation peak with the higher M_W chain associated with a greater intensity [9]. The deuterated sample gives rise to lower scattering for fixed M_W due to the presence of intermolecular effects which are absent under the ZAC condition [9].

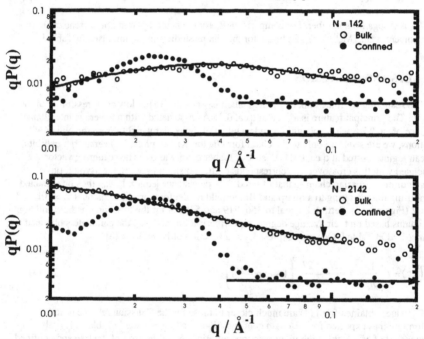

Figure 1. *qP(q) from free and confined, zero average contrast polyelectrolytes for low and high M_W. Free chains have been fitted to the Sharp and Bloomfield model for $ql_p < 2$. Also shown for high M_W is the crossover region at q^* between q^{-2} and q^{-1} scattering.*

There have been few studies in which Q(q) has been measured for free polyelectrolyte chains and none in which a quantitative analysis of its variation with q has been discussed; however, the combination of ZAC and deuterated chain scattering ($x = 1$ in (1)) here enables the separation of intrachain, P(q) and interchain contributions, Q(q). Using the measured P(q) from ZAC and inserting the known values for the prefactors into (1), we may obtain Q(q) by an appropriate weighted subtraction (Figure 2). One of the striking features of our data is that Q(q) of the free and confined chains have well-defined asymptotic behaviour; Q(q) decays as $\approx -q^{-2}$ for the bulk chains whereas the confined chain function decays only as $\approx -q^{-1}$.

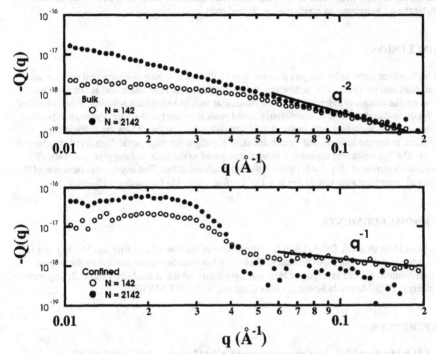

Figure 2. -Q(q) for a) bulk and b) confined chains for low and high molecular weights.

We note that the effect of confined charged chains in water (dielectric constant ≈ 80), surrounded by low dielectric constant ($= 3.85$) walls of Vycor glass would be to concentrate electric field lines along the axial direction of the pores [10]. This has the effect of enhancing the electrostatic interaction along the pore axis and may also be responsible for the stretched conformation as evidenced by the scattering peak. Further, since we are studying the semi-dilute regime, we may expect Q(q) to be independent of M_W in the bulk, at large q, where the scattering corresponds to dimensions $\approx l_p$ (Figure 2a). We thus believe that the observed behaviour is a simple consequence of the local rod-like structure of the chains. It has been suggested that, for bulk polyelectrolytes, $Q(q)$ can be approximated by $q^{-(2/\nu)}$ where $\nu=1$ for a locally rigid rod in contrast to the experimentally verified q^{-3} decay for neutral chains [11]. This is indeed verified by our experimental data for the bulk systems (Figure 2a). It may also be easily shown the asymptotic limit of $Q(q)$ for a solution of rod-like particles is $\pi^2/4q^2L^2$ and is proportional to the amplitude scattered by a portion of rod of size $\approx q^{-1}$ [16]. On the contrary, if we assume that the confined polyelectrolyte chains are locally aligned in the pores, as for a single rod, the main contribution to $Q(q)$ will vary as q^{-1} [16]. We find for the bulk that the $Q(q)$ functions not only overlay at high q but also illustrate the predicted q^{-2} power law. While statistics limit our conclusions at high M_W, in the case of the confined chains, the low M_W system does indeed exhibit the q^{-1} behaviour as predicted for aligned rods.

CONCLUSIONS

SANS has been used to investigate the structure of PSSNa in a bulk concentrated solution and when confined in Vycor. The persistence length obtained from the Sharp and Bloomfield model for wormlike chains is independent of the molecular weight in the bulk solution. When confined in Vycor, and under the ZAC condition, a broad peak is present at the bare Vycor peak position whose intensity increases with molecular weight. We infer that the chains are sufficiently extended, under the influence of confinement, to highlight the large scale disordered structure of Vycor. The 1/q scattering region is extended compared to the bulk and supports this view. We have also determined $Q(q)$ for both the bulk and confined states. The asymptotic behavior of the observed interchain structure factor is $\approx 1/q^2$ for free and $\approx 1/q$ for confined chains.

ACKNOWLEDGMENTS

We would like to thank Dr. A. Lapp for preparation of the low M_W chains and Mr. E. Lang for assistance in performing SANS studies. This work has benefited from the use of the Intense Pulsed Neutron Source at Argonne National Laboratory which is funded by the U.S. Department of Energy, BES-Materials Science, under Contract W-31-109-ENG-38.

REFERENCES

[1] S.H.J. Idziak and Y.L. Li, *Current Opinion Colloid Interface Sci.*, **3**, 293 (1998).
[2] P.-G. de Gennes in *Scaling Concepts in Polymer Physics*, (Cornell University Press, 1979).
[3] P. Levitz et al., *J. Phys. Chem.*, **95**, 6151 (1991).
[4] J. Lal, S.K. Sinha and L. Auvray, *J. Phys. II (France)*, **7** 1597 (1997).
[5] M.N. Spiteri et al., *Phys. Rev. Lett.*, **77**, 5218 (1996).
[6] J.-P. Cotton, *J. Phys. Lett.*, **41**, L231 (1980).
[7] P. Sharp and V.A. Bloomfield, *Biopolymers*, **6**, 1201 (1968).
[8] H. Benoît and P. Doty, *J. Phys. Chem.*, **57**, 958 (1953).
[9] E.P. Gilbert, L. Auvray and J. Lal, submitted.
[10] A. Parsegian, *Nature*, **221**, 844 (1969).
[11] G. Jannink et al., *Europhys. Lett.*, **27**, 47 (1994).

Mat. Res. Soc. Symp. Proc. Vol. 651 © 2001 Materials Research Society

Enzymes and Cells Confined in Silica Nanopores

Jacques Livage, Cécile Roux, Thibaud Coradin, Souad Fennouh, Stéphanie Guyon, Laurie Bergogné, Anne Coiffier and Odile Bouvet[1]
Laboratoire de Chimie de la Matière Condensée, Université Pierre et Marie Curie, 4 place Jussieu, 75252 Paris, France
[1]Unité des Entérobactéries, Institut Pasteur, 28 rue du Dr Roux, 75724 Paris, France

ABSTRACT

The sol-gel process opens new possibilities in the field of biotechnologies. Sol-gel glasses are formed at room temperature *via* the polymerization of molecular precursors. Enzymes can be added to the solution of precursors and trapped within the growing silica network. Small substrate molecules can diffuse through the pores allowing reactions to be performed in-situ, within the silica gels. Enzyme are encased by the hydrated silica in a cage tailored to their size, they retain their biocatalytic activity and may even be stabilized within the sol-gel matrix.

Whole cell bacteria have also been immobilized within sol-gel glasses. They behave as a "bag of enzymes" and their membrane protects enzymes against denaturation and leaching. The cellular organization of bacteria cells is preserved upon encapsulation. Experiments performed with *Escherichia coli* induced to β-galactosidase show that they still exhibit noticeable enzymatic activity. Some degradation of the cell walls may even occur increasing the "measured" activity. However silica gels made from aqueous precursors seem to prevent bacteria from natural degradation upon ageing.

Antibody-antigen recognition has been shown to be feasible within sol-gel matrices. Trapped antibodies bind specifically the corresponding haptens and can be used for the detection of traces of chemicals. Even whole cell protozoa have been encapsulated without any alteration of their cellular organization. For medical applications, trapped parasitic protozoa have been used as antigens for blood tests with human sera. Antigen-antibody interactions were followed by the so-called Enzyme Linked ImmunoSorbent Assays (ELISA).

INTRODUCTION

Entrapment in crosslinked organic polymers is a well known method for the immobilization of enzymes and whole cells. Entrapped biomolecules are physically confined within the polymer matrix and can be reused several times. Organic polymers such as polyacrylamide gels are currently used in biotechnology but silica glasses could offer some advantages such as improved mechanical strength and chemical stability. Moreover they don't swell in aqueous or organic solvents preventing leaching of entrapped biomolecules. However glasses are made at high temperature and, up to now, enzyme immobilization can only be performed *via* adsorption or covalent binding onto the surface of porous glasses [1].

The so-called sol-gel process opens new possibilities in the field of biotechnology [2-4]. Sol-gel glasses are formed at room temperature *via* the polymerization of molecular precursors such as metal alkoxides. Proteins can be added to the solution of precursors. Hydrolysis and condensation then lead to the formation of an oxide network in which biomolecules remain trapped. Small analytes can diffuse through the pores allowing bioreactions to be performed inside the sol-gel glass. Trapped enzymes still retain their biocatalytic activity and may even be stabilized within the sol-gel cage. A wide range of biological species such as antibodies and whole cells have been trapped within sol-gel matrices. They usually retain their activity but weak interactions with the silica cage actually occur that can change their behavior.

SOL-GEL CONFINEMENT IN SILICA MATRICES

Sol-gel silica can be synthesized at room temperature via the hydrolysis and condensation of TetraMethyl OrthoSilicate (TMOS), $Si(OCH_3)_4$. Hydrolysis gives reactive silanol groups whereas condensation leads to the formation of bridging oxygen as follows:

$$\equiv Si\text{-}OCH_3 + H_2O \Rightarrow \equiv Si\text{-}OH + CH_3OH \qquad \text{(hydrolysis)}$$

$$\equiv Si\text{-}OH + HO\text{-}Si\equiv \Rightarrow \equiv Si\text{-}O\text{-}Si\equiv + H_2O \qquad \text{(condensation)}$$

The overall reaction is then

$$Si(OCH_3)_4 + 2H_2O \Rightarrow SiO_2 + 4 CH_3OH$$

Silicon alkoxides are not miscible with water so that a common solvent has to be used. The parent alcohol is currently chosen but a large variety of other organic solvents can also be used [5]. Silicon alkoxides are not very sensitive to hydrolysis. Gelation takes place within several days and even weeks, when pure water is added. Therefore acids and bases have to be added in order to increase the gelation rate. Acid catalysis, below the Point of Zero Charge of silica (pH≈3) mainly increases hydrolysis rates. Chain polymers are formed leading to microporous monolithic gels (average pore diameter < 20 Å). Basic catalysis (pH > 3) enhances condensation, giving dense spherical colloids and mesoporous gels (average pore diameter in the range 50-100 Å) [6].

The encapsulation of proteins within sol-gel matrices is usually performed in two steps in order to avoid denaturation. The first step is the acid hydrolysis of pure TMOS, without alcohol as a co-solvent. Most alkoxy groups are removed giving silisic acid $Si(OH)_4$. A suspension of enzymes in a buffered solution (pH≈7) is then added. Basic condensation takes place rapidly and a mesoporous oxide is formed around the biomolecule. Enzymes are immobilized within the silica gel but their active site may remain accessible and they can react with small chemical analytes through the porous matrix.

The chemical control of the pore size is one of the major challenges of sol-gel encapsulation. Pores have to be small enough to avoid the leaching of entrapped proteins but large enough to allow the diffusion of analytes. The pore size has then to be tailored according to the nature of biospecies. Figure 1 shows the pore size and specific area variation as a function of the amount of water, h=[H_2O]/[TMOS], added during the first hydrolysis step.

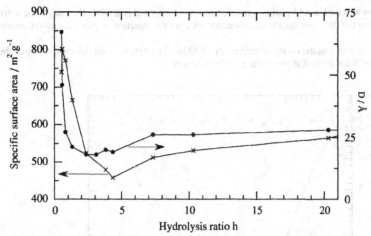

Figure 1. *Mean pore size as a function of the hydrolysis ratio h=[H$_2$O]/[TMOS].
The pore size is measured on xerogels dried at room temperature.*

They go through a minimum around h≈3. Below this value, an emulsion is formed that disappears just before gelation when the buffered aqueous solution is added in order to adjust the pH around 7.

CATALYTIC ACTIVITY OF ENZYMES

Enzymes are biological catalysts which are responsible for the chemical reactions of living organisms. Chemically these proteins are made of amino acids linked together *via* covalent peptide bonds. Their high specificity and huge catalytic power is due to the fact that the geometry of the active site can fit exactly that of the substrate. Therefore even small changes in the enzyme conformation can reduce drastically their catalytic activity.

A large number of enzymes have been trapped within sol-gel glasses since the pioneering work of D. Avnir in Jerusalem [7]. Most publications study glucose oxidase (GOD) as a model for enzyme encapsulation. GOD catalyzes the oxidation of D-glucose by molecular oxygen into D-gluconolactone and D-gluconic acid as follow:

$$C_6H_{12}O_6 + O_2 \Rightarrow C_6H_{10}O_6 + H_2O_2$$

The enzymatic activity can be followed in a number of ways. The formation of hydrogen peroxide can be detected by optical measurements using a peroxidase to oxidize an organic dye [8]. The redox reaction at the active site of GOD can be followed with an electrochemical mediator [9]. In our group oxygen consumption was measured directly with an oxygen sensitive electrode. In these experiments the sol-gel solution containing GOD was deposited onto the Pt cathode of a Clark electrode. Oxygen concentration was measured by amperometric titration at imposed potential. The enzymatic activity is determined *via* the decrease in oxygen concentration after the injection of glucose into the aqueous solution.

Figure 2 shows that the catalytic activity of GOD in silica gels is about the same as when the enzyme is just immobilized at the surface of the electrode

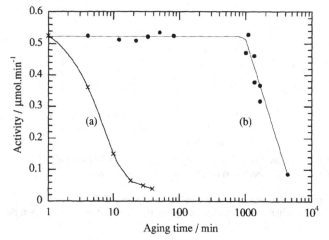

Figure 2. *GOD activity as a function of time measured onto a Clark electrode :(a) enzyme immobilized at the electrode surface, (b) encapsulated enzyme*

Therefore sol-gel encapsulation does not lead to denaturation during the formation of the silica network. Even after drying at room temperature, silica xerogels still contain enough water to provide a mainly aqueous environment to avoid denaturation of the enzymes. Figure 2 also shows that the catalytic activity of the trapped enzyme is constant during *ca.* 500 min whereas it progressively decreases in the case of enzymes immobilized at the electrode surface. The half-time $t_{1/2}$ (corresponding to a 50% decrease of catalytic activity) is close to 300 min for encapsulated GOD instead of 10 min when the enzyme is only immobilized. Its thermal behavior is also improved. Entrapped GOD retains 60% of its activity at 63°C after 20 h, whereas the half-life of free GOD is only 6.5 min. [10].

These experiments suggest that the silica matrix protects encapsulated enzymes against leaching and denaturation. Leaching is prevented by the small size of silica pores whereas denaturation could be prevented by some restriction of molecular motions. Upon encapsulation the enzyme is encased by the hydrated silica in a cage tailored to its size. The silica matrix constrains the motions of the encapsulated protein molecules and may prevent irreversible structural deformations [11]. Electrostatic interactions may also occur between anionic silicate sites and specific positively charged residues on the protein surface. If these residues are essential for either preserving the protein structure or catalysis, such interactions will decrease the enzymatic activity. Silica surfaces are negatively charged above the Point of Zero Charge (pH≈3) and electrostatic interactions mainly depend on the IsoElectric Point (IEP) of the protein.

Experiments performed with three different oxidases, glucose oxidase (IEP=3.8), glycolate oxidase (IEP=4.6) and lactate oxidase (IEP=9.6) show that only GOD retain its activity upon encapsulation [10]. The detrimental electrostatic interactions can be overcome by complexing the enzyme with a polyelectrolyte that shield the critical charged sites and lactate oxidase can be stabilized by complexing with the weak base PVI (poly(N-vinylimidazole) [12].

Confinement within silica gels does not only protect enzymes against denaturation. It can also provide a chemical surrounding that favors the enzymatic activity. Lipases provide a nice example of such interactions between enzymes and the sol-gel matrix. These enzymes act on ester bonds. In aqueous media they are able to hydrolyze fats and oils into fatty acids and glycerol. In organic media esterification or transesterification reactions occur. However lipases are not soluble in organic solvents and freeze-dried powders are currently used. A least one hydration layer is required for enzymes to retain their bio-activity. Therefore the activity of lipases in organic media is not very good. Actually lipases are interfacial activated enzymes. In an aqueous solution, an amphiphilic peptidic loop covers the active site just like a lid. At a lipid/water interface, this lid undergoes a conformational rearrangement which renders the active site accessible to the substrate [13].

Transesterification reactions performed with lipases trapped within silica gels show that their activity strongly depend on the water content of the gels. Aged gels that contain a large amount of water (> 60% in weight) exhibit a good enzymatic activity. They provide a mainly aqueous environment similar to that observed in biological media. However the enzymatic activity decreases drastically upon drying. This mainly depends on the chemical nature of the silica matrix. Gels hydrolyzed with a very small amount of water (h=[H_2O]/[TMOS] $\approx$ 0.6) still exhibit a good activity even after freeze-drying, whereas the activity of fully hydrolyzed (h$\approx$20) silica xerogels is very low. This might be due to the fact that the methyl groups of the alkoxide precursor, $Si(OMe)_4$, have not been completely removed during the formation of the silica gel. Lipases are then encapsulated in pores that contain both hydrophilic Si-OH and hydrophobic Si-OMe surface groups. The amphiphilic peptidic loop may then be in the "open" position. The gel shrinks upon drying preventing further motion of the loop and the active site remains open when the gel is freeze-dried. This is no longer the case with fully hydrolyzed silica gels. Nanopores are then covered only with hydrophilic Si-OH groups and the peptidic loop is in the "close" position. In wet gels there is enough space to allow the motion of the loop, but not in dry gels.

Very nice experiments have been published by M. Reetz who showed that lipases can be almost 100 time more active when trapped within a hybrid silica matrix [14]. The co-hydrolysis of $Si(OMe)_4$ and $RSi(OMe)_3$ precursors provides alkyl groups that offer a lipophilic environment that could interact with the active site of lipases and increase their catalytic activity. Such entrapped lipases are now commercially available and offer new possibilities for organic chemistry, food industry and oil processing.

BACTERIA IN SILICA GELS

Many intracellular microbial enzymes are produced in large enough quantities to be used in industrial processes. However the cost for their isolation and purification may be quite high. Therefore it might be interesting to be able to directly immobilize micro-organisms containing these enzymes. Moreover, retaining the enzyme within its natural surrounding preserves its stability. Whole-cell immobilization with organic polymers has been extensively studied. It is a powerful tool for a precise control of the micro-environment of the cell in a wide variety of reactor configurations and external conditions [15].

The first sol-gel experiments were published more than ten years ago by G. Carturan et al. with *"Saccharomyces cerevisiae"* cells [16]. These yeast are involved in the conversion of sugar and carbohydrates into ethyl alcohol and CO_2 and are currently employed in the fermentation of beers and the raising of bread. Encapsulated yeast cells follow the well-known Michaelis-Menten law and exhibit almost the same activity as in a solution for more than a year. They even retain their viability and the budding of yeast cells can still be observed after immobilization at the surface of sol-gel films [17].

The sol-gel encapsulation of whole cell bacteria *Escherichia coli* was reported much later [18]. These bacteria were trapped within two different sol-gel matrices, synthesized from alkoxide or aqueous precursors, in order to study the role of the micro-environment.
Sol-gel entrapment with alkoxide precursors was performed via the same two-step procedure as for enzymes. The pH of the acid solution of hydrolyzed TMOS is first increased with a phosphate buffer in order to prevent the denaturation of bacteria. *E. coli* cells, extracted by centrifugation from their culture medium, are suspended in a phosphate buffer solution (10^9 cells/ml) and then mixed with the sol-gel solution. A silica gel is formed within a few minutes and left for aging several days at room temperature.
A two steps procedure was also followed for the aqueous route using a solution of sodium silicate (0.4 M, 2ml) and colloidal silica (Ludox HS 40, 1 ml). The pH of such solutions is close to 12. It is first acidified by adding HCl (4 M, 180 µl) in order to decrease the pH down to 7. The bacteria suspension is then added with a phosphate buffer. Gelation takes places within a few minutes, encapsulating the bacteria cells.

Ultra structural observations by transmission electron microscopy of thin slices of wet gels show that the integrity of *Escherichia coli* cells is preserved after sol-gel encapsulation (Figure 3). The capsule, cell walls and plasma membrane are not destroyed and there is a clear separation between the cell and the porous silica matrix.

Figure 3. *Transmission electron microscopy of an Escherichia coli cell*
confined within a wet silica gel made from TMOS (bar = 0.5 µm)

E. coli bacteria were induced, in their culture medium, with IPTG (IsoPropylThioGalactoside) in order to express β-galactosidase enzyme. The β-galactosidase activity of trapped bacteria was measured using p-NPG (p-nitrophenyl-β-D-galactopyranoside) as a substrate. p-NPG can enter the cell without a specific permease and once inside the cell it is cleaved by β-galactosidase into galactose and nitrophenol :

p-NPG + H_2O ⇒ p-nitrophenol + β-D-galactose

The saturating concentration of the substrate [p-NPG] has been chosen in order to be much larger than the Michaelis constant of β-galactosidase ($K_M \approx 0.1$ mM for free bacteria in solution) [18]. The rate of the reaction (V_{max}) is then nearly independent of [p-NPG] and proportional to the enzyme concentration. The formation of the yellow p-nitrophenol is followed by optical absorption of the solution at $\lambda = 400$ nm. The enzymatic activity was deduced from the initial velocity of the reaction and compared with the activity of free bacteria suspended in a phosphate buffer solution (Table I).

Table I. Evolution with time of β-galactosidase activity of free and encapsulated *E. coli*

	β-galactosidase activity (μmol/min) of *E. coli*		
	free bacteria	in TMOS silica	in aqueous silica
ageing 1 day	0.030	0.045	0.035
ageing 7 days	0.060	0.071	0.050

These experiments show that after one day, the β-galactosidase activity of bacteria in silica gels is slightly higher thant that of free bacteria. This might be due to some denaturation of the cell walls that become more porous allowing easier diffusion of the substrate within the cell. Silica gels prepared from TMOS give the best activity, presumably because of the presence of CH_3OH in the hydrolyzed solution. Results obtained after one week ageing in the buffered solution without nutrients are rather interesting. They show that the activity of bacteria trapped in aqueous gels is now smaller than that of free bacteria suggesting that the silica matrix partially prevents the natural degradation of the cell walls.

It might be interesting to point out that the enzymatic activity of *E. coli* trapped in aqueous gels containing a nutrient such as gelatin does not change significantly after one week of ageing suggesting that these bacteria still exhibit some metabolic activity.

IMMUNOASSAYS IN SILICA GELS

Antigen-antibody reactions have been performed within sol-gel matrices extending the field of sol-gel chemistry toward immunosensors [19]. However antibodies are large biomolecules. IgG the major immunoglobulin in normal human serum is a monomeric protein with a molecular weight close to 150 kDa and dimensions of about 19 x 56 x 240 $Å^3$. Therefore most immunoassays have been performed with antibodies either bonded to the surface [20] or trapped within the sol-gel matrix. Specific haptens are then used, they are much smaller and can diffuse easily through the pores of the sol-gel matrix. These reactions have been developed for the biodetection of chemicals such as atrazine, a widely used herbicide. Anti-atrazine antibodies are trapped within the sol-gel matrix and nanograms of atrazine are applied on SiO_2 sol-gel columns doped with this antibody. Titration of eluted solutions shows that high amounts of atrazine remain bound to anti-atrazine antibodies inside the silica gel [21]. TNT titration using anti-TNT antibodies has also been reported recently [22].

For medical applications, whole cell parasitic protozoa (*Leishmania donovani infantum*) have been trapped within sol-gel matrices and used as antigens for blood tests with human or dog sera [23]. As for bacteria, transmission electron microscopy shows that the cellular organization of the parasites is well preserved and that the plasma membrane is unaltered. This is very important as antigenic determinants are situated at the outside surface of the membrane.

The recognition of trapped *Leishmania* cells by antibodies in a serum was followed by the so-called Enzyme Linked ImmunoSorbent Assays (ELISA), a widely used test in parasitology [24]. Specific antibodies bound to Leishmania cells in the silica gel are detected via an enzyme conjugate, horseradish peroxidase (HRP) which, after binding to the antibody-antigen complex will catalyze the oxidation of an organic dye by H_2O_2. Optical density measurements show a clear difference between negative and positive sera (Figure 4).

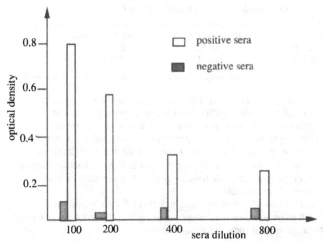

Figure 4. *Optical detection of antigen-antibody association via ELISA*

The optical density of positive tests decreases regularly with dilution showing that mainly specific antigen-antibody interactions are involved. Only non specifically bound immunoglobulins are washed out giving a very low residual coloration.

These experiments show that large antibodies can diffuse through the pores of the silica matrix and bind specifically to epitopes on the surface of the cell. This of course is no longer possible when pores are not large enough to allow such a diffusion. Assays performed using different silica matrices obtained with different hydrolysis ratios show that pore size is actually a critical parameter (Table II).

Table II. ELISA tests with gels of different porosity

Hydrolysis ration h = [H$_2$O]/[Si]	0.57	3	20
Average pore diameter (Å)	45	18	24
OD+/OD-	5.6	1.8	3.4

CONCLUSION

The biological applications of sol-gel chemistry appear to be very promising. Immobilization within silica gels offer several advantages compared to organic polymeric matrices which are nowadays widely used in biotechnology. Biomolecules are trapped inside hard porous glasses that do not swell in water. They are confined within hydrated silica pores and surrounded by an aqueous medium close to that observed in biological media. Motions are restricted by the silica cage avoiding irreversible denaturation. The silica matrix protects biomaterials against external aggression (pH, temperature, solvents...). It may even enhance their activity by providing a controlled micro-environment. Sol-gel encapsulation does not destroy the cellular organization of micro-organisms. This might be one of the major advance of sol-gel chemistry and some promising examples show that living cells can be immobilized within sol-gel matrices. Mammalian tissue such as pancreatic islets have been recently been encapsulated in porous silica gels and transplanted into a diabetic mouse where they retain their activity [25, 26]. The fine porosity of silica gels protects transplanted cells against antibody aggression but permits nutrients to reach the cell and byproducts to escape. Such transplants, if viable for extended lengths of time, could emerge as a new treatment for diabetes.

REFERENCES

1. H. Weetall, *Trends in Biotechnology*, **3**, 276 (1985).
2. D. Avnir, S. Braun, O. Lev, M. Ottolenghi, *Chem. Mater.*, **6**, 1605 (1994).
3. C. Dave, B. Dunn, J.S. Valentine, J.I. Zink, *Anal. Chem.*, **66**, 1120A (1994).
4. J. Livage, *C.R Acad. Sci. Paris*, **322**, 417 (1996).
5. C.J. Brinker, G.W. Scherer, *Sol-Gel Science*, (Academic press, 1990).
6. R. K. Iler, the *Chemistry of Silica*, (John Wiley, 1979).
7. S. Braun, S. Rappoport, R. Zusman, D. Avnir, M. Ottolenghi, *Mater. Lett.*, **10**, 1 (1990).
8. S.A. Yamanaka, F. Nishida, L.M. Ellerby, C.R. Nishida, B. Dunn, J.S. Valentine, J.I. Zink, *Chem. Mater.*, **4**, 495 (1992).
9. P. Audebert, C. Demaille, C. Sanchez, *Chem. Mater.*, **5**, 911 (1993).
10. Q. Chen, G.L. Kenausis, A. Heller, *J. Am. Chem. Soc.*, **120**, 4582 (1998).
11 J.D. Badjic, N.M. Kostic, *Chem. Mater.*, **11**, 3671 (1999).

12. J. Heller, A. Heller, *J. Am. Chem. Soc.*, **120**, 4586 (1998)
13. R.D. Schmid, R. Verger, *Angew. Chem. Int. Ed.*, **37**, 1608 (1998).
14. M.T. Reetz, *Adv. Mater.*, **9**, 943 (1997).
15. R.G. Willaert, G.V. Baron, L. de Baker in *Immobilized Living Cell Systems*, ed. R.G. Willaert, G.V. Baron, L. de Baker (John Wiley, 1996) pp.21-95.
16. G. Carturan, R. Campostrini, S. Dire, V. Scardi, E. de Alteriis, *J. Mol. Catal.*, **57**, L13 (1989).
17. L. Inama, S. Dire, G. Carturan, A. Carazza, *J. Biotechnol.*, **30**, 197 (1993).
18. S. Fennouh, S. Guyon, C. Jourdat, J.Livage, C. Roux, *C.R. Acad. Sci. Paris, IIc*, 625 (1999).
19. J. Livage, C. Roux, J.M. Da Costa, I. Desportes, J.Y. Quinson, *J. Sol-Gel Sci. Tech.*, **7**, 45 (1996).
20. R. Collino, J. Therasse, P. Binder, F. Chaput, J.P. Boilot, Y. Levy, *J. Sol-Gel Sci. Tech.*, **2**, 823 (1994).
21. A. Turniansky, D. Avnir, A. Bronshtein, N. Aharonson, M. Altstein, *J. Sol-Gel Sci. Tech.*, **7**, 135 (1996).
22. E.H. Lan, B. Dunn, J.I. Zink, *Chem. Mater.*, **12**, 1874 (2000).
23. J.Y. Barreau, J.M. da Costa, I. Desportes, J. Livage, L. Monjour, M. Gentilini, *C.R. Acad. Sci. Paris*, **317**, 653 (1994).
24. P. Venkatesan, D. Wakelin, *Parasitology Today*, **9**, 228 (1993).
25. G. Carturan, G. Dellagiacoma, M. Rossi, R. Dal Monte, M. Muraca, *Sol-Gel Optics IV*, *SPIE Proc.*, **3136**, 366 (1997).
26. E.J.A. Pope, K. Braun, C.M. Peterson, *J. Sol-Gel Sci. Tech.*, **8**, 635 (1997).

Mat. Res. Soc. Symp. Proc. Vol. 651 © 2001 Materials Research Society

Dynamics of Sol-Gel Clusters and Branched Polymers

A. G. Zilman and R. Granek
Department of Materials and Interfaces
Weizmann Institute of Science
Rehovot 76100, Israel

Abstract

We study the dynamics of sol-gel clusters. We find that the dynamic structure factor decays in time as a stretched exponential, and the viscoelastic modulus is an algebraic function of time. The difference beween screened and non-screened systems in the context of cluster dynamics is discussed.

Polymers in solution form variety of self-similar structures. For example, a linear polymer molecule forms a coil with a radius of gyration R_g scaling as $R_g \sim N^\nu$ where N is the number of monomers and the exponent ν depends on the solvent quality [1, 2]. Another example are sol-gel clusters which are formed in a reaction bath where monomers can polymerize and cross-link. The fractal aggregates formed this way are closely related to percolation clusters[3, 6] At a critical concentration of the monomers, an infinite cluster is formed, a process known as gelation. Dilution of sol-gel clusters is the most convenient way to produce branched polymers[3]. Sol-gel clusters are polydisperse in their molecular weight, the average molecular weight N_z being a function of the monomer concentration.

Sol-gel clusters are fractal objects consisting of randomly cross-linked polymeric segments[3]. The fractal dimension of a cluster is defined as follows. Consider a sphere of a radius l around a given point belonging to a cluster. The distance l is measured along the polymer segments of the cluster (i.e., it is the 'chemical length'[17]). The number of monomers inside the sphere scales as $n \sim l^D$, where D is the intrinsic fractal dimension. When the cluster folds in the d-dimensional Euclidean space, the distance r between two points of the clusters defines the fractal dimension in the embedding space, d_f so that $n \sim r^{d_f}$; $d_f \leq d$. The dimension of the embedding space $d = 3$ for the bulk solutions and $d = 2$ for thin films. The ratio $D/d_f = \nu$ (the Flory exponent) describes via the scaling $r \sim l$, how the D-dimensional fractal is folded in the embedding space. For example, a linear self-avoiding chain has $D = 1$, $d_f = 5/3$. There is an additional characteristic of a fractal: the spectral dimension d_s which is related to the fractal's connectivity. The mean square displacement of a random walker on a fractal is given by $l^2(t) \sim t^{d_s/D}$. For a usual Euclidean space of integer dimensionality we have $d_s = D$ and the diffusion becomes regular linear diffusion. Since $\nu \leq 1$, we have $d_s \leq D \leq d_f \leq d$. While D and d_s are intrinsic cluster characteristics, d_f and ν reflect the way in which the cluster is crumpled in the embedding space and might depend on the details of the intracluster interactions. For percolation clusters, $D \simeq 1.8$, $d_s \simeq 1.3$ and $d_f \simeq 2.5$, while for dilute clusters $d_f = 2$ [3, 17].

Dynamics of sol-gel clusters is of a great theoretical and practical interest. There are quite different mechanisms governing the dynamics of sol-gel systems at different timescales. For times $t < \frac{\eta \xi^3}{T}$, the dynamics is governed mainly by the behavior of the individual polymer chain segments between the cross-links (ξ is the mean distance between the cross-links, η is the viscosity of the solvent and T is the temperature). For times $\frac{\eta\xi^3}{T} < t < \frac{\eta R_g^3}{T}$, where $R_g \simeq N_z^{1/d_f}$ is the radius of gyration of an average cluster, the dynamics of the individual clusters is dominant, and for times $\frac{\eta R_g^3}{T} < t$ the lateral diffusion of interacting clusters in the environment of other clusters becomes important.[8]. Above the gelation transition, the non-ergodic dynamics of the infinite cluster must be considered. In addition, in dense systems (e.g., the reaction bath close to the gel point) the hydrodynamic interaction between the monomers can be screened, giving rise to additional effects. The dynamics of sol-gel systems has been studied extensively both theoretically and experimentally (mainly via scattering and rheological measurements), producing a wealth of experimental and theoretical results [5, 8, 4, 7, 19, 20]. The results vary greatly depending on the particular dynamical regime or a cross-over area under study. Therefore it is interesting to study the effects arising from the dynamics of a *single* cluster, as a first step towards understanding more complicated effects.

The clusters under discussion consist of branched cross-linked polymer chains, which fluctuate in space due to thermal fluctuations. This motion should be distinguished from the diffusion of a particle *on* a 'frozen' fractal [12]. We start by generalizing the Zimm model for linear self-avoiding polymers [2], to arbitrary *flexible* fractal branched clusters. The internal position of a monomer in a cluster is described, even if just symbolically, by the "vector" $\vec{l}$. The vector $\vec{R}(\vec{l})$ denotes the position of this monomer in the d-dimensional embedding space. The free-energy of the cluster of the molecular weight N is described by a generalized Edwards Hamiltonian ($k_B = 1$ in our units) [9]

$$H\left[\{\vec{R}(\vec{l})\}\right] = \frac{T}{b^D} \int d^D l \left(\nabla_D \vec{R}(\vec{l})\right)^2$$
$$+ v \int d^D l \int d^D l' \delta^d \left(\vec{R}(\vec{l}) - \vec{R}(\vec{l}')\right) \tag{1}$$

where b is the momomer linear size and v is the excluded volume parameter. The first term in Eq.(1) describes the entropic elasticity. The self-avoidance, described by the second term, makes an exact treatment of the dynamics impossible, but it becomes tractable in the so-called linearization approximation which we now describe.

Bearing in mind the special features of fractals, mathematical operations can be defined in a fashion similar to Euclidean spaces. In particular, one can define the Laplace operator, $\nabla_D^2 \phi(\vec{l}) \equiv \lim_{b \to 0} b^{-2D/d_s} \sum_{|\vec{l}' - \vec{l}| = b} [\phi(\vec{l}') - \phi(\vec{l})]$. We shall make use of the orthogonal set of eigenstates $\Psi_E(\vec{l})$ of the Laplacian in the fractal space [11, 4, 6], defined by

$$\nabla_D^2 \Psi_E(\vec{l}) = -E \Psi_E(\vec{l}) \tag{2}$$

The $\Psi_E(\vec{l})$'s are normalized as $\int d^D l \Psi_E^*(\vec{l}) \Psi_{E'}(\vec{l}) = N \delta_{E,E'}$, where N is the fractal molecular weight, and $\Psi_E^*(\vec{l}) = \Psi_E(-\vec{l})$. These eigenstates have been extensively studied in the context of fractons[11]. In particular, it has been shown that the density of eigenstates $N(E)$ scales with E as $N(E) \sim E^{d_s/2 - 1}$. This allows us to construct an *effective Hamiltonian* H_{eff} which is Gaussian and diagonal in the manifold eigenstate space

$$H_{\text{eff}} = \frac{1}{2} T \sum_E \frac{\vec{R}(E)^2}{\langle |\vec{R}(E)|^2 \rangle} \tag{3}$$

where $\vec{R}(\vec{l}) = \sum_E \vec{R}_E \Psi_E(\vec{l})$. The mean value $\langle |\vec{R}(E)|^2 \rangle$ are determined self-consistently from the knowledge the real space correlation function

$$\langle (\vec{R}(\vec{l}) - \vec{R}(\vec{l}'))^2 \rangle = 2 \int dE N(E) \langle |\vec{R}(E)|^2 \rangle \left(1 - \bar{\Psi}_E(l)(|\vec{l} - \vec{l}'|)\right) \tag{4}$$

where $\bar{\Psi}_E(|\vec{l} - \vec{l'}|) = \langle \Psi_E(\vec{l}) \Psi_E^*(\vec{l'}) \rangle_{\text{dis}}$ is a disorder ensemble average eigenstate correlator. Following the general considerations described in Refs. [11] and [4], the latter should have the following scaling form

$$\bar{\Psi}_E(l) = f\left(E^{d_s/2D}l\right)$$

(Note that the dimensions of E are $[E] = [l]^{-2D/d_s}$). The static correlation function $\langle (\vec{R}(\vec{l}) - \vec{R}(\vec{l'}))^2 \rangle$ can, in principle, be calculated from the original Hamiltonian (1) to give the exponent ν as a function of D, d_s and d[9]. The correlator in Eq. (4) must be equal to $\sim b^{2-2\nu}|\vec{l} - \vec{l'}|^{2\nu}$ which implies that

$$\langle |\vec{R}(E)|^2 \rangle \simeq b^{2-2\nu} E^{-d_s(1/d_f+1/2)} \tag{5}$$

We now turn to the dynamics of the polymeric fractal clusters. The Langevin equations of motion, in the creeping flow approximation, are [2, 21]

$$\frac{d\vec{R}(\vec{l})}{dt} = -\int d^{d_s}l' \mathbf{L}\left(\vec{R}(\vec{l}) - \vec{R}(\vec{l'})\right) \cdot \frac{\delta H_{\text{eff}}}{\delta \vec{R}(\vec{l'})} + \vec{f}(t,\vec{l}) \tag{6}$$

where $\vec{f}(t,\vec{l})$ is thermal white noise. Here $\mathbf{L}(\vec{r})$ is the hydrodynamic interaction between different monomers [13] which in d-dimensions is given by $\mathbf{L}(\vec{r}) \propto (\hat{r}\hat{r} + 1)/\eta r^{d-2}$, with η the solvent viscosity. (Note that the viscosity dimensions are $[\eta] = [l]^{2-d}[M]/[t]$.) We now perform the preaveraging approximation [2], where $\mathbf{L}$ is replaced by its equilibrium average: $\langle \mathbf{L}(\vec{R}(\vec{l}) - \vec{R}(\vec{l'})) \rangle_{\text{eq}} = \Lambda(\langle |\vec{R}(\vec{l}) - \vec{R}(\vec{l'})| \rangle)\mathbf{1}$ where $\Lambda(r) \sim 1/\eta r^{d-2}$. Therefore Eq. (6) becomes linear and in the eigenstate space it is

$$\frac{d\vec{R}(E)}{dt} = -\omega(E)\vec{R}(E) + \vec{f}(E,t) \tag{7}$$

where

$$\omega(E) = \frac{T\Lambda(E)}{\langle |\vec{R}(E)|^2 \rangle} \tag{8}$$

Here $\Lambda(E)$ is the manifold eigenstate transform of the hydrodynamic interaction kernel $\Lambda(R_{ll'})$ averaged over fractal disorder.

$$\Lambda(E) \simeq \frac{1}{\eta} \int \frac{d^D l \, \bar{\Psi}_E(l)}{\langle |\vec{R}(\vec{l}) - \vec{R}(0)|^{d-2} \rangle} \simeq \frac{E^{\frac{d_s}{2}(d-2)/d_f - \frac{d_s}{2}}}{\eta b^{(d-2)(1-\nu)}} \tag{9}$$

Consequently, the mode relaxation rate becomes

$$\omega(E) \simeq \frac{T E^{dd_s/2d_f}}{\eta b^{d(1-\nu)}} \tag{10}$$

Eqs. (5) and (10) are analogous to the Zimm model results for linear polymers [2]. It is important to emphasize that the eigeinfrequencies $\omega(E)$ of the cluster dynamics depend on the intrinsic characteristics of the cluster such as d_s.

We are now able to calculate the time-dependent mean-square displacement $\langle \Delta(t)^2 \rangle$ of a single monomer:

$$\langle \Delta(t)^2 \rangle \equiv \langle (\vec{R}(\vec{l},t) - \vec{R}(\vec{l},0))^2 \rangle$$
$$= 2\int dE N(E)\langle |\vec{R}(E)|^2 \rangle \left(1 - e^{-\omega(E)t}\right) \tag{11}$$

Using Eqs. (5) and (10) for $|\vec{R}(E)|^2$ and $\omega(E)$ we obtain (for times $\eta b^d/T \ll t \ll \eta R_g^d/T$, where R_g is the radius of gyration)

$$\langle \Delta(t)^2 \rangle \simeq \left(\frac{T}{\eta} t \right)^{2/d} \tag{12}$$

It is seen that a monomer performs *anomalous subdiffusion* $\langle \Delta(t)^2 \rangle \sim t^\alpha$ (with $\alpha < 1$) in the embedding space. The anomalous exponent $\alpha = 2/d$ is *independent* of all fractal character-istics (i.e., independent of the exponents D, d_s and d_f). This can be attributed to the long range hydrodynamic interaction which strongly couples the motion of monomers through the Euclidean space, even when the chemical distance between them is large. As a consequence, the fractal characteristics may become irrelevant. This becomes more evident when we con-sider below the Rouse model. It also agrees with the scaling hypothesis $\langle \Delta(t)^2 \rangle = R_g^2 f(t/\tau)$ with $\tau = \eta R_g^d/T$ ($f(x)$ is a scaling function), if we assume that for $t \ll \tau$ the result should not depend on R_g, which implies that $f(x) \sim x^{2/d}$ for $x \ll 1$.

One conventional technique to study the dynamics of polymers is the neutron scattering. The quantity measured in scattering experiments is the dynamic structure factor $S(q,t)$ equal to the Fourier transform of the density-density correlation function. It can be conveniently expressed as [14]

$$S(q,t) \simeq \frac{1}{b^D} \int d^D l \exp \left[-\frac{1}{6} q^2 \langle (\vec{R}(\vec{l},t) - \vec{R}(0,0))^2 \rangle \right] \tag{13}$$

The correlation function in the exponent can be written as a sum of static and dynamic parts

$$\langle (\vec{R}(\vec{l},t) - \vec{R}(0,0))^2 \rangle = \langle (\vec{R}(\vec{l}) - \vec{R}(0))^2 \rangle + \langle \Delta(\vec{l},t)^2 \rangle \tag{14}$$

where

$$\langle \Delta(\vec{l},t)^2 \rangle = 2 \int dE N(E) \, g_E(l) \langle |\vec{R}(E)|^2 \rangle \left(1 - e^{-\omega(E)t} \right) \tag{15}$$

The latter reduces to the single monomer mean-square displacement $\langle \Delta(t)^2 \rangle$ of Eq. (11) for $\vec{l} = 0$. In can be shown that for $\eta/q^d T \ll t \ll \eta R_g^d/T$ the *decay* of $S(q,t)$ is dominated by $\langle \Delta(0,t)^2 \rangle$, the mean-square displacement of a single monomer. Thus this decay becomes nearly a *stretched exponential*

$$S(q,t) \simeq S(q) \, e^{-\text{const.} \left(\frac{Tq^d}{\eta} t \right)^{2/d}} \tag{16}$$

where $S(q) \sim q^{-d_f}$ is the static structure factor. The stretching exponent $2/d$ is therefore *independent* of all fractal characteristics.

Second, we consider the contribution of a single cluster to the shear viscoelastic modulus $G(\omega)$ of the solution, which can be measured in rheological experiments[2] . In the time domain it is given by $G(t) = \frac{T}{v_c} G_R(t)$ where v_c is the volume available per cluster and the intrinsic modulus $G_R(t)$ is (for $t \ll \eta R_g^d/T$)

$$G_R(t) = \int dE N(E) e^{-\omega(E)t} \simeq \left(\frac{\eta R_g^d}{T t} \right)^{\frac{d_f}{d}} \tag{17}$$

In the frequency domain $G_R(\omega) \sim (i\omega)^{d_f/d}$. Thus the viscoelastic spectrum *is* sensitive to d_f, even if not to d_s, in contrast to behavior exhibited by $S(q,t)$. Note that $G_R(t)$ saturates to unity for $t \gtrsim \eta R_g^d/T$, which implies that for the long tail size distribution at the gel point the time dependence of the macroscopic modulus will be modified [5].

So far we have considered self-similar structures with the hydrodynamic interaction influencing the dynamics. In dense systems hydrodynamic backflow effects may become screened above a certain length $\xi_H \sim \xi$ (ξ is the mesh size) and we recover the free draining limit described by the Rouse model [2, 4]. The role of the inter-cluster excluded volume interaction on the cluster distribution in space, i.e., on the exponent ν (or d_f), is less evident. However, even if the actual value of ν in the regime $\xi \ll r \ll R_g$ is not known, we can still use it as a model parameter in our dynamic self-consistent theory. For the case of screened hydrodynamics we can still apply the formalism used above, except that now the friction is local, being simply the Stokes drag on a single monomer [2]. We thus use $\mathbf{L} = \Lambda \mathbf{1}$ with $\Lambda \simeq b^D \delta^D(\vec{l} - \vec{l}')/\eta b^{d-2}$. Accordingly, $\Lambda(E)$ is independent of E, and the relaxation rate becomes

$$\omega(E) \simeq \frac{T E^{d_s(1/d_f + 1/2)}}{\eta b^{d-D-2\nu}} \tag{18}$$

Using Eqs. (5) and (18) in Eq. (11) we now obtain

$$\Delta(t)^2 \simeq b^2 \left(\frac{T}{\eta b^d} t\right)^{\frac{2}{2+d_f}} \tag{19}$$

Thus, when the hydrodynamics is screened, the anomalous diffusion exponent *is* dependent on d_f. For the intrinsic viscoelastic modulus we obtain

$$G_R(t) \simeq N \left(\frac{\eta b^d}{T t}\right)^{\frac{d_f}{2+d_f}} \tag{20}$$

where $N \simeq (L/b)^D \simeq (R_g/b)^{d_f}$ is the fractal "mass". Hence $G_R(\omega) \sim (i\omega)^{d_f/(2+d_f)}$. Expressions for $G(t)$ and $\Delta(t)^2$ have been obtained also in the previous works in different limits using different methods [4, 8, 15]. Our model agrees with the previous studies in appropriate limits and allows one to calculate $G(t)$ and $S(q,t)$ within a single framework without invoking any *ad hoc* hypothesis regarding the relaxations times distribution [8, 20].

Next we consider the lateral diffusion of the clusters. The diffusion coefficient $\mathcal{D}$ can be calculated from Eq.(7), noting that the motion of the cluster center of mass is described by the $E = 0$ mode of the Eq.(7). The result is

$$\mathcal{D} \simeq \frac{T}{\eta R_g} \quad \text{for the non-screened hydrodynamics}$$

$$\mathcal{D} \simeq \frac{T}{\eta N} \quad \text{when the hydrodynamics is screened}$$

This shows that, similar to the case of linear polymers, when the hydrodynamics is not screened, the external flow is excluded from inside the cluster, and it behaves essentially as a rigid sphere of the radius $R_g \sim N^{1/d_f}$. In real systems the interactions between different clusters are important, which can be modeled via introduction of the effective viscosity η_{eff}. The latter may depend on the density of the system [8]. Another effect which has not been taken into account in this study is polydispersity. Both lateral diffusion of the clusters and the polydispersity will be considered in future work, as the next step towards description of real sol-gel systems.

To conclude, we have shown that the dynamic structure factor of a branched fractal cluster decays in time a stretched exponential $S(q,t) \sim e^{-(\Gamma_q t)^\alpha}$. In dilute systems the hydrodynamic interaction is unscreened and the stretching exponent $\alpha = 2/d$ is a universal constant independent of any cluster characteristics. The decay rate is proportional to the third power of the scattering wavevector $\Gamma_q \sim q^3$. These results are in good agreement with light scattering and neutron spin echo experiments [7, 19]. In dense systems the hydrodynamic interaction is screened, resulting with $\alpha = d_f/d$ and the decay rate is a compliacted power of the scattering wavevector $\Gamma_q \sim q^{2d/d_f}$. Comparison of the viscoelastic modulus, which we predict to be an algebraic function of time with the experimental data is less conclusive, due to many effects which play role in experimental systems, unaccounted for in this study.

The authors are grateful to M. Cates and S. Alexander for helpful discussions.

References

[1] P.-G. de Gennes, *Scaling Concepts in Polymer Physics* (Cornell Uni., Ithaca, N.Y., 1979).

[2] M. Doi and S. F. Edwards, *The Theory of Polymer Dynamics* (Clarendon, Oxford, 1986), pp. 97-103.

[3] M. Daoud et al., J. Physique Lett. **45**, L-199 (1984); M. Daoud and L. Leibler, Macromolecules **21**, 1497 (1988).

[4] M. E. Cates, J. Physique **46**, 1059 (1985).

[5] J. E. Martin et al., Phys. Rev. A **39**, 1325 (1989).

[6] D. Stauffer and A. Aharony, *Introduction to Percolation Theory*, Second Edition (Taylor & Francis, London, 1992).

[7] M. Adam et al., Macromolecules **30**, 5920 (1997).

[8] J. E. Martin et al., Phys. Rev. A **43**, 858 (1991), Sec. IV;

[9] J. A. Aronovitz and T. C. Lubensky, Europhys. Lett. **4**, 395 (1987).

[10] Y. Kantor, in *Statistical Mechanics of Membranes and Surfaces*, eds. D. Nelson et al. (World Scientific, Singapore, 1989).

[11] S. Alexander and R. Orbach, J. Physique Lett. **43**, L-625 (1982); R. Rammal and G. Toulouse, J. Physique Lett. **44**, L-13 (1983).

[12] A. Blumen et al., in *Optical Spectroscopy of Glasses*, ed. I. Zschokke (Reidel, Dordrecht, 1986), pp. 199, and references therein.

[13] Ref. [2], pp. 88-89.

[14] Ref. [2], pp. 132-135.

[15] K. J. Wiese, Eur. Phys. J. B **1**, 273 (1998);*ibid*, 269 (1998).

[16] P.-G. de Gennes, Macromolecules **9**, 587 (1986).

[17] S.Havlin, D. Ben-Avraham, Adv. Phys, **36** (1987), 695.

[18] M. Daoud, J. Phys (France) **51** (1990), 2843

[19] A. Lapp *et al.*, J. Phys II **2** (1992), 1495

[20] M. Rubinstein *et al*, in *Space-Time Organization in Macromolecular Fluids* (1990) p.66.

[21] A. G. Zilman, R.Granek, Phys. Rev. E **58** (3) (1998), R2725.

Mat. Res. Soc. Symp. Proc. Vol. 651 © 2001 Materials Research Society

Computational Study of Polymerization in Carbon Nanotubes

Steven J. Stuart, Brad M. Dickson, Bobby G. Sumpter[1], and Donald W. Noid[1]
Department of Chemistry, Clemson University, Clemson, SC 29634-0973, USA.
[1]Chemical and Analytical Science Division, Oak Ridge National Laboratory, Oak Ridge, TN 37830, USA.

ABSTRACT

Molecular dynamics simulations of ethylene polymerization have been performed using a chemically realistic, reactive potential. These simulations have been performed in the bulk liquid and in the interior of both (10,10) and (7,7) nanotubes as a means of investigating the effects of nanoscale confinement on the polymerization reaction. The structure of the resulting polymer was found to be similar in the bulk and in the (10,10) tube at the elevated temperatures investigated, while only very small oligomers were formed in the (7,7) tube. The reaction rate was substantially reduced in the nanotubes, when compared to the bulk, primarily as a result of spatial interference due to reaction products. These simulations have implications for the possible use of nanotubes as synthetic reaction vessels, as well as for the general understanding of association reactions in confined spaces.

INTRODUCTION

Carbon nanotubes are under investigation for a number of their remarkable properties. Among these is their ability to be filled via capillary forces in order to create nanoparticles, nanowires, and other novel structures.[1-5] Although most applications involve filling nanotubes with aqueous solutions, molten salts, or molten metals, previous work using mesoporous silica fibers as a synthetic support for polyethylene[6] suggests that polymerization in nanoscale geometries will also generate unique structures. In addition to the likelihood of forming novel polymeric structures, there is also the interesting possibility of observing fractal kinetics for association reactions in spaces of reduced dimensionality.[7] For these reasons, the polymerization of ethylene was studied in the confined geometry of nanotubes of two different diameters, and compared to comparable simulations in the bulk liquid phase.

MODEL AND COMPUTATIONS

Because the reactivity of the hydrocarbons was of critical importance, it was crucial to perform the simulations a model that is capable of accurately modeling dissociation and formation reactions in hydrocarbons. For this reason, the AIREBO (adaptive intermolecular reactive empirical bond-order) potential[8] was selected. This is a bond-order potential based on Brenner's well-known REBO potential for hydrocarbons[9,10]. The AIREBO potential preserves the treatment of C—C, C—H, and H—H covalent bonding interactions that has been validated in numerous studies with the REBO potential, while introducing terms corresponding to dispersion, torsional, and exchange repulsion interactions in a way that does not interfere with the covalent bonding potential.

The focus in this study is on the geometric effects of confinement within a carbon nanotube, rather than any potential reactions with the nanotube walls. For this reason, and in order to make the simulations more efficient, a cylindrically symmetric confining potential was used to represent the carbon nanotubes. This potential had the form

$$V(r) = A\left(\frac{\sigma}{R-r}\right)^{2p+q} - B\left(\frac{\sigma}{R-r}\right)^{p+q} + C\left(\frac{\sigma}{R-r}\right)^{q} + D \qquad (1)$$

where r is the distance from the center of the nanotube. Equation (1) represents the simplest soft-wall cylindrical confining potential that results in an adsorption minimum near the tube wall, a local maximum at the center of the tube, and an infinite interaction as the particle approaches the tube wall ($r=R$). Taking R as the tube radius from 0 K simulations with the AIREBO potential, and selecting $p=4$ arbitrarily, the remaining parameters were chosen to fit to the cylindrically averaged interaction of C and H atoms with a single-walled nanotube under the AIREBO potential.

Simulations of the polymerization of ethylene were performed in three separate geometries: a bulk system, a (10,10) carbon nanotube (6.6 Å radius), and a (7,7) carbon nanotube (4.7 Å radius). All systems were simulated at a density of approximately 0.75 g/cm^3. The bulk system consisted of 128 ethylene monomers in a cubic box of side 19.9 Å, with periodic boundary conditions. The (10,10) nanotube system contained 78 ethylene monomers in a nanotube of length 60 Å, and the (7,7) nanotube system contained 100 monomers in a nanotube of length 557 Å. One-dimensional periodic boundary conditions were applied to these systems. (Note that the volume of the soft-walled carbon nanotube is not uniquely defined. The ½ kT radius of approximately 3.9 Å for the (10,10) tube and 1.9 Å for the (7,7) tube at 3000 K was used to determine the tube volume.) All simulations were performed under the canonical ensemble. The temperature was controlled using a generalized Langevin thermostat.[11] At the constant density studied, the pressure of the liquid phase before polymerization was approximately 75-80 kbar (depending on the temperature examined).

RESULTS

To simulate the polymerization process, the systems were heated to elevated temperatures of between 2800 K and 3400 K, causing thermal generation of radicals and subsequent polymerization. Note that at these temperatures the polymerization does not result in strictly saturated polyethylene, but also polyacetylene and other unsaturated species. Indeed, the equilibrium product at long times would be glassy or graphitic carbon, if the hydrogen byproducts were allowed to escape from the closed system.

For each of the three system geometries (bulk; (10,10) tube; (7,7) tube), polymerization was simulated at each of four temperatures, and the change in molecular structure was monitored. At each temperature the polymerization was allowed to proceed for at least 20 ns in the tube geometries and 5 ns in the bulk, for a total of over 180 ns of dynamics.

The progress of the polymerization was monitored via average molecular weights. Figure 1 illustrates the change in weight-averaged molecular weight, M_w, with time for the three different systems at 3200 K. The growth rate of the polymer is considerably greater in the bulk system than it is in either confined geometry, and is larger in the (10,10) nanotube than the (7,7) nanotube. Another difference that is apparent from Figure 1 is that the polymerization reaction reaches a plateau at 3200 K in both of the confined systems, but not in the bulk. The polymer growth saturates at an average molecular weight of about twice the monomer weight (28 g/mol) in the (7,7) tube and about four times the monomer weight in the (10,10) tube. The bulk system,

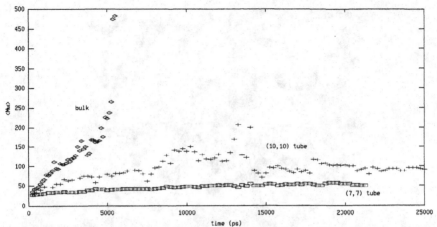

Figure 1. *Weight-averaged molecular weight, M_w, during polymerization for three different system geometries. The bulk system is represented by diamonds (red); the (10,10) nanotube by crosses (green), and the (7,7) nanotube by squares (blue). The unreacted initial system has M_w=28 g/mol.*

on the other hand, shows no sign of saturation at an average molecular weight of over 15 times the monomer weight. Indeed, the configuration sampled at 5.5 ns includes a single C_{94} oligomer chain that includes over a third of all of the carbons in the system (not quite at the percolation threshold indicative of an infinite molecular weight in this periodic system). The behavior is qualitatively quite similar at the other temperatures studied.

Figures 2 and 3 illustrate representative geometries of the growing polymer chain at late stages of their growth. While the polymers confined to the nanotubes are generally confined to the linear shape of the nanotube, the morphology is surprisingly similar in the confined and non-confined systems. In particular, the polymer is not in a highly linear (all-trans) configuration. In general, the polymer growth proceeds via thermal initiation and radical chain propagation, as would be expected for ethylene under these high-temperature, non-catalyzed conditions. Because of the high reaction temperatures, the polymer that is formed has a high degree of branching and ring formation. Note that this is true in both the bulk and the (10,10) nanotube; the tube diameter is large enough to support short branches and cyclic groups (see Figure 3). Analysis of pair correlation functions (not shown here) also indicates that the structure in the (10,10) and bulk systems is largely similar, differing mainly in the average chain length and the distribution of dihedral angles. The limited statistics available from these single runs do not permit an in-depth analysis of branching ratios or ring formation. Likewise, the statistics for individual runs do not permit any quantitative comparison of reaction order, preventing any conclusions regarding the existence of fractal kinetics or time-dependent rate constants in these systems. In the smaller, (7,7) tube, there is little branching and negligible ring formation due to both the limited degree of polymerization in that system and the extremely narrow (~4 Å) inner diameter of that tube.

The dependence of reaction rate on system temperature is presented in Figure 4. The rate constant was obtained by fitting the observed M_w behavior for each system and temperature with that expected assuming second-order kinetics. As seen from Figure 4, the temperature

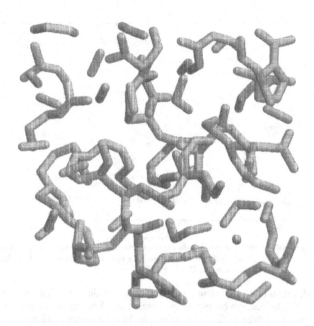

Figure 2. *Conformation of bulk system after 5.5 ns of polymerization at 3400 K. Only carbon-carbon bonds are displayed, for clarity. The figure shows several independently growing oligomer chain, the largest of which contains a C_{143} chain. Note the presence of short branches, as well as small ring structures.*

dependence shows the expected Arrhenius behavior. The slopes for each curve are similar, with activation energies of 65±13 kcal/mol in the bulk, 52±8 kcal/mol in the (10,10) tube, and 54±5 kcal/mol in the (7,7) tube. Any contribution to the rate decrease due to energetic interactions with the confining tube wall is thus minor, if present at all, and the primary reason for the decrease of reaction rate in the confined geometries is entropic in nature.

The magnitude of the rate decrease is quite significant, with the polymerization in the (10,10) tube proceeding at an average rate 18 times slower than in the bulk, and the polymerization in the (7,7) tube proceeding fully 50 times slower than in bulk. It is to be expected that the rate will decrease somewhat upon confinement, due to nonproductive collisions

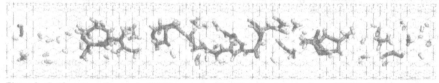

Figure 3. *Conformation of system in (10,10) nanotube after 28 ns of polymerization at 3400 K. Note the presence of short branches, as well as small ring structures. Small bubble-like phases of hydrogen gas and gaseous hydrocarbon fragments are visible to the far left and far right.*

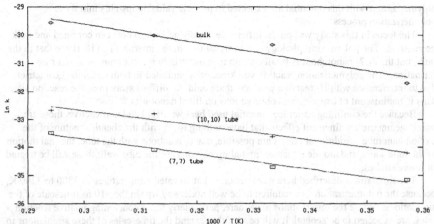

Figure 4. *Arrhenius plot of polymerization rate (assumed 2nd order) in various geometries. The bulk system is represented by diamonds (red); the (10,10) nanotube by crosses (green); and the (7,7) nanotube by squares (blue)*

with the nonreactive wall, but the magnitude of this decrease is surprising. Surface area arguments alone cannot account for the 18-fold decrease in reaction rate in the (10,10) tube, for example, since only ~50% of the surface area of each monomer is occupied by nanotube wall.

An explanation for the large decrease in reaction rate, as well as the observed plateau in average molecular weight, is found upon examining the reaction trajectories in detail. Figure 3 illustrates the conformation of the polymer in a (10,10) nanotube after 28 ns at 3400 K. Notice that the H radicals that dissociated in the thermal initiation step, as well as in subsequent formation of unsaturated polymer units, have associated into H_2 molecules and separated into a separate phase of hydrogen gas. These nanoscale "bubbles" of hydrogen form high-pressure barriers between individual islands of growing oligomer chains, and act to prevent the diffusion and formation of large molecular-weight polymer chains. Thus the reaction is self-limiting, with the H_2 reaction product inhibiting the continued growth of the polymer. Hydrogen molecules also form and associate into bubbles, in the three-dimensional system. This does not substantially inhibit the growth of the polymer, however, as the bubbles are not large enough to span any dimension of the system. Note that the self-limiting nature of the reaction in the carbon nanotubes due to reaction byproducts is very similar to that observed in many experimental studies of filled carbon nanotubes. When the nanotube contents are subsequently reacted, the presence of a gaseous reaction byproduct tends to result in the formation of small nanoparticles, rather than nanowires.[4,5,12]

CONCLUSIONS

Polymerization of ethylene has been observed via computer simulation at elevated temperatures of near 3000 K and pressures of near 75 kbar in both confined and bulk geometries, using a reactive potential. This is significant because it is one of very few all-atom simulations of bulk-scale polymerization with a chemically realistic potential. Future simulations of polymerization with this potential are expected to provide valuable insights into the polymerization process.

The focus in this study was on the difference in polymerization between confined and bulk geometries. The polymer morphology was fairly similar in the interior of a (10,10) to that in the bulk, but the (7,7) nanotube was small enough to substantially restrict branching and ring formation. The polymerization reaction was kinetically inhibited in both nanotube geometries, due to interference with H_2 reaction products that could not diffuse away from the reaction site. This is reminiscent of experimental observations on filled nanotubes.

Because the confining nanotube potential used here was rigid and nonreactive, these are purely geometric confinement effects that have nothing to do with the chemical nature of the carbon nanotube. Additional effects are possible, due to reactions with the tube, thermal motion of the tube walls, and atomic structure (including chirality) of the tube wall; these will be treated in future studies.

The simulations described here were performed at elevated temperatures of 2800 to 3200 K, because the polymerization was required to be well underway within the ~10 ns timescale of the simulations. While the observations made here, particularly those concerning relative reaction rates, are expected to be general, it will be useful to extend the timescales of these simulations to the microsecond regime in order to be able to make use of more moderate temperatures.

ACKNOWLEDGEMENTS

Acknowledgement is made to the donors of the Petroleum Research Fund, administered by the ACS, and to the Research Corporation for partial support of this research. Thomas Zacharia of Oak Ridge National Laboratory is also acknowledged for a generous donation of computer time.

REFERENCES

1. M. R. Pederson and J. Q. Broughton, *Phys. Rev. Lett.*, **69**, 2689 (1992).
2. P. M. Ajayan and S. Iijima, *Nature*, **361**, 333 (1993).
3. E. Dujardin, T. W. Ebbesen, H. Hiura, and K. Tanigaki, *Science*, **265**, 1850 (1994).
4. D. Ugarte, A. Châtelain, and W. A. de Heer, *Science*, **274**, 1897 (1996).
5. M. Terrones, N. Grobert, W. K. Hsu, Y. Q. Zhu, W. B. Hu, H. Terrones, J. P. Hare, H. W. Kroto, and D. R. M. Walton, *MRS Bull.*, **24** (8), 43 (1999).
6. K. Kageyama, J. I. Tamazawa, and T. Aida, *Science*, **285**, 2113 (1999).
7. R. Kopelman, *Science*, **241**, 1620 (1988).
8. S. J. Stuart, A. B. Tutein, and J. A. Harrison, *J. Chem. Phys.*, **112**, 6472 (2000).
9. D. W. Brenner, *Phys. Rev. B*, **42**, 9458 (1990); **46**, 1948, (1992).
10. D. W. Brenner, J. A. Harrison, C. T. White, and R. J. Colton, *Thin Solid Films*, **206**, 220 (1991).
11. S. A. Adelman and J. D. Doll, *J. Chem. Phys.*, **64**, 2375 (1976).
12. A. Chu, J. Cook, R. J. R. Heesom, J. L. Hutchison, M. L. H. Green, and J. Sloan, *Chem. Mater.*, **8**, 2751 (1996).

Mat. Res. Soc. Symp. Proc. Vol. 651 © 2001 Materials Research Society

Synchrotron X-ray Studies of Molecular Ordering in Confined Liquids

Hyunjung Kim, O. H. Seeck[1], D. R. Lee, I. D. Kaendler, D. Shu, J. K. Basu[2], and S. K. Sinha

Advanced Photon Source, Argonne National Laboratory,
Argonne, IL 60439, U.S.A.
[1]IFF, FZ Jülich GmbH, 52425 Jülich, Germany
[2]Materials Research Laboratory, University of Illinois at Urbana-Champaign, Urbana, IL 61801, U. S. A.

ABSTRACT

X-ray specular and off-specular reflectivity studies have been carried out to study the density modulations in liquids confined between two smooth silicon mirrors. The special technique as well as the advantages of using high energy and high brilliance synchrotron x-ray beams for carrying out such experiments will be discussed. Results will be presented on the ordering of octamethyl-cyclotetrasiloxane (OMCTS) as a function of the confining pressure, where we find evidence of layering as the gap is decreased from macroscopic down to a few nanometers.

INTRODUCTION

The structural and other properties of fluids confined between solid surfaces differ considerably from bulk fluids at the same temperature, and this has implications for our basic understanding of phenomena such as lubrication, adhesion, surface chemistry, etc [1]. Surface force apparatus (SFA) measurements [2] reveal wall-induced layering of fluid down to a film thickness of few molecular layers. Computer simulation studies [3] have found evidence for layering of the liquid molecules in such liquid films. However, direct structural evidence has not yet been reported. Evidence for layering near a bulk liquid/solid interface has been recently obtained from x-ray reflectivity [4] and more recently the lateral microscopic structure of liquid lead at an interface has been observed by exploiting total internal reflection of evanescent x-rays [5].

EXPERIMENT

As is well known, x-ray reflectivity provides the most direct method for probing the structure of liquid films in the direction normal to the confining surfaces. One may also get insight into the lateral inhomogeneities and roughness correlations in thin films from off-specular diffuse scattering. A typical experimental setup for x-ray reflectivity measurement is shown in figure 1. Such experiments on liquids confined at thicknesses of a few nanometers with x-ray scattering present significant challenges from the technical point of view. Compared to a conventional SFA, a relatively large sampling area is necessary for a reasonable signal to noise ratio in x-ray reflectivity measurements. With the availability of high-energy and high-brilliance beams from the present third generation of x-ray synchrotron sources, very narrow x-ray beams can be used to

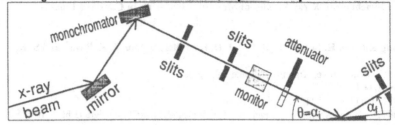

Figure 1. A typical setup of x-ray reflectivity experiment.

penetrate the confining walls and minimize the scattering from the walls. Alignment of the two solid surfaces is another challenging problem since a controlled parallel separation of order nanometers over square milimeters of area is required. Thus the surfaces must be both highly polished and flat over such length scales, as well as dust-free. In addition, the layering is rapidly destroyed by surface roughness, which typically should not exceed ~ 0.3 nm. We have utilized specially designed silicon substrates, shown in figure 2 (a), having diameters of 25.4 mm or 15.0 mm with an rms-roughness of 0.3 nm (determined by x-ray scattering) and a convex curvature with a height variation of less than 10 nm over the whole sample area (determined by interfero-metry). Two grooves were etched in each surface, which left a bridge of the size (2.0_4.5) mm² in the center part. A liquid was spread over the surface of one substrate in a class 1 clean room.

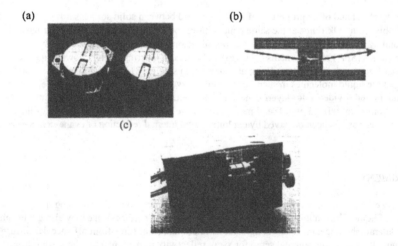

Figure 2. Components of the experimental setup for x-ray scattering on confined liquids. (a) Specially designed silicon substrates (25.4 mm diameter). The area of confinement is the bridge in the center. (b) The schematic diagram of the beam path between the two Si mirrors. (c) The sample cell.

Both substrates were put together so that the grooves formed tunnels that were the paths for the x-rays to the area of confinement (figure 2(b)). The gap distance between both silicon pieces were controlled either by piezodrivers or by applying torque on the screws on top of the cell (figure 2(c)).Octamethylcyclotetrasiloxane (OMCTS) was used due to its nonpolar nature and the large size of its quasispherical molecules [6].

RESULTS

The x-ray reflectivity measurements were performed at 1-ID-C at the Advanced Photon Source (APS) in Argonne National Laboratory with photon energy at 30 keV. Figure 3 shows the

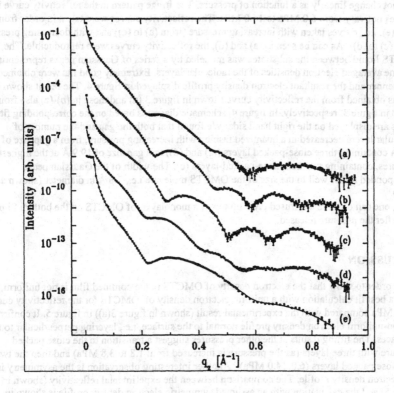

Figure 3. The reflectivities (in symbols) on confined fluids and the fitting results (solid lines on top of the data) at the pressure of (a) 1.8 MPa . (b) 3.8 MPa, (c) 6.0 to 14.0 MPa, and (d) 2.1 MPa. (e) The reflectivity of bottom Si mirror was taken after pressure released. The experiment was performed in the order from (a) to (e). Note that the gap size is decreased in a quantized fashion with increasing pressure.

reflectivity curves (in symbols) on OMCTS confined between the two Si mirrors as a function of pressure: (a) at 1.8 MPa, (b) at 3.8 MPa, (c) from 6.0 to 14.0 MPa, and (d) at 2.1 MPa. In (e), the reflectivity from bottom Si piece is shown. The curves have been separated for clarity by displacing along the vertical axis. One can see the fringe pattern on reflectivity moves out to larger q_z, i.e., the corresponding gap size is smaller with increasing pressure. The electron density profile of the liquid normal to the surfaces maybe obtained by model fitting to those reflectivity curves. The reflectivity data have been analyzed using the following model: the silicon substrates and native oxide layers were represented with the appropriate step functions of electron density normal to the surface suitably smeared to take into account ~ 0.3 nm roughness and obtained by fitting to the reflectivity of the bare substrates. A certain gap between the surfaces was assumed, and the density of the liquid in the gap was modeled as discussed below. Figure 3 shows the fits to the reflectivity curves using this model. Qualitatively, however, we note first that the gap size does not change linearly as a function of pressure. The fringe pattern in the reflectivity curve in (c) does not vary from 6.0 MPa to 14.0 MPa. The reflectivity curves were taken in order from (a) to (e), i.e., curves taken with increasing pressure (from (a) to (c)) and with decreasing pressure (from (c) to (d)). As can be seen in (a) and (d), the reflectivity curves were reproducible. The OMCTS liquid between the substrates was modeled by a series of Gaussian peaks representing in-plane averaged electron densities of the molecular layers. Extremely good fits were obtained in this manner and the resultant electron density profile displayed in figure 4. The result shown in (a) was obtained from the reflectivity curve shown in figure 3 (a) and those in (b)-(e) also from (b)-(e) in figure 3, respectively. In figure 4, schematic diagrams based on the corresponding fitting results are displayed on the right hand side. We found that both the gap and the number of molecular layers decreased in a quantized fashion with increasing pressure, from a gap size of 25.2 Å containing three close-packed layers ((a) and (d)) to a gap size of 19.9 Å at the highest pressures, containing two non-close-packed-layers (c). The width of the Gaussian peaks corresponded rather well to the size of the OMCTS molecule i.e., 7.8 Å in diameter (shown as a circle).

In (e), one can observe a diffused layer on top of a monolayer of OMCTS on the bottom Si piece even after the pressure released.

DISCUSSION

In order to show that the electron density of OMCTS in the confined film is not uniform, we show a best fit calculation with a uniform electron density of OMCTS for the reflectivity curve at 1.8 MPa compared with the experimental result (shown in figure 3(a)) in figure 5. It confirms the non-uniform electron density profile normal to the surface, i.e., layering perpendicular to the interfaces. The fitting results at the other pressures suggest a transition to the close packed structure with three layers (as the pressure is increased from 1.8 to 3.8 MPa) and then the two non-close packed layers (6.0-14.0 MPa). Another interesting observation is the asymmetry in the electron density profile. The comparison between the experimental reflectivity (shown in figure 5) and the calculation with an assumed symmetric electron density profile is shown in figure 6. Up to 0.4 Å^{-1} in q_z, the reflectivity curve is in good agreement with calculation using a symmetric electron density profile. Thus reflectivity values at higher wave vectors are crucial for obtaining detailed information regarding the electron density profile. The origin of asymmetry in the profile is not yet well understood. It might be presumably due to tilting between the mirrors, or differences in their surface properties.

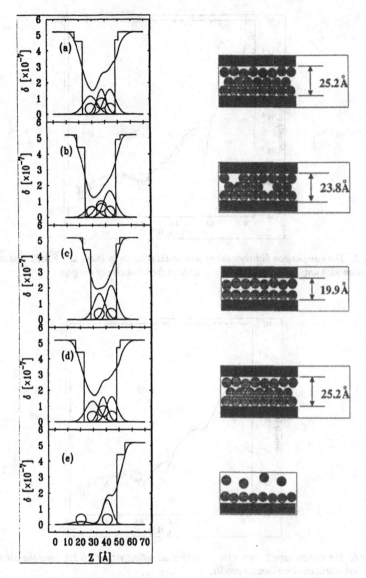

Figure 4. *The electron density profile used in fitting reflectivity curves shown in figure 3 i.e., taken (a) at 1.8 MPa, (b) at 3.8 MPa, (c) at 6.0-14.0 MPa, (d) at 2.1 MPa, and (e) from bottom Si piece after released pressure. The models are schematically presented on the right hand side of each profile. The diameter of the OMCTS molecule is 7.8 Å (shown as a circle).*

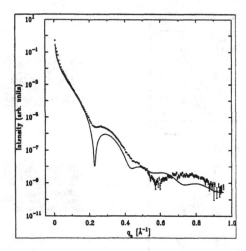

Figure 5. *The comparison between the experimental reflectivity curve at 1.8 MPa and the calculation with uniform electron density profile in the same size of the gap.*

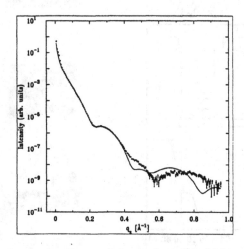

Figure 6. *The comparison between the experimental reflectivity at 1.8 Mpa and the calculation with a symmetric electron density profile.*

In addition to the layering in confined geometry reported in this paper, one may measure in-plane ordering using Grazing Incidence Diffraction (GID). Such measurements are planned in the near future.

ACKNOWLEDGMENTS

We wish to thank Dr. S. Granick for valuable discussions. The work was supported by the U.S. Department of Energy, Office of Science, under Contract No. W-31-109-Eng-38. This work was also supported by the University of Illinois, Champaign under Grant DEFG02-96ER45439.

REFERENCES

1. S. Granick, Physics Today, July 1999, p.26 and references therein.
2. B. Bhushan, J.N. Israelachvili and U. Landman, Nature 374, 607 (1995); L. Demirel and S. Granick, Phys. Rev. Lett. 77, 2261 (1996).
3. P.A. Thompson, G.S. Grest and M.O. Robbins, Phys. Rev. Lett. 68, 3448 (1992; J.P. Gao, W.D. Luidtke and U. Landman, J. Phys. Chem. B 101, 4013 (1997); J.P. Gao et al., Science 270, 605 (1995).
4. C.-J. Yu, A.G. Richter, A. Datta, M.K. Durbin and P. Dutta, Phys. Rev. Lett. 82, 2326 (1999).
5. H. Reichert, O. Klein, H. Dosch, V. Honkimäki, T. Lippmann, and G. Reiter, Nature 408, 839 (2000).
6. J. Klein and E. Kumacheva, J. Chem. Phys. 108, 6996 (1998).

Mat. Res. Soc. Symp. Proc. Vol. 651 © 2001 Materials Research Society

Modeling Gas Separation Membranes

Anthony P. Malanoski[2] and Frank van Swol[1,2,3]

[1]Sandia National Laboratories, Advanced Materials Laboratory,
1001 University Blvd, SE,
Albuquerque, NM 87106, U.S.A.
[2]Chemical and Nuclear Engineering Department, The University of New Mexico
[3]Center for Micro-Engineered Materials, The University of New Mexico

ABSTRACT

Recent advances in the development and application of self-assembly templating techniques
have opened up the possibility of tailoring membranes for specific separation problems. A new
self-assembly processing route to generate inorganic membrane films has made it feasible to
finely control both the three-dimensional (3D) porosity and the chemical nature of the adsorbing
structures. Chemical sites can be added to a porous membrane either after the inorganic
scaffolding has been put in place or, alternatively, chemical sites can be co-assembled in a one-
step process. To provide guidance to the optimized use of these 'designer' membranes we have
developed a substantial modeling program that focuses on permeation through porous materials.
The key issues that need to be modeled concern 1) the equilibrium adsorption behavior in a
variety of 3D porous structures, ranging from straight pore channels to fractal structures, 2) the
transport (i.e. diffusion) behavior in these structures. Enriching the problem is the presence of
reactive groups that may be present on the surface. An important part of the design of actual
membranes is to optimize these reactive sites with respect to their strength as characterized by
the equilibrium constant, and the positioning of these sites on the adsorbing surface. What makes
the technological problem challenging is that the industrial application requires both high flux
and high selectivity. What makes the modeling challenging is the smallness of the length scale
(molecular) that characterizes the surface reaction and the confinement in the pores. This
precludes the use of traditional continuum engineering methods. However, we must also capture
the 3D connectivity of the porous structure which is characterized by a larger than molecular
length scale. We will discuss how we have used lattice models and both Monte Carlo and 3D
density functional theory methods to tackle these modeling challenges.

INTRODUCTION

The success of new self-assembly techniques has indicated the possibility of tailoring
membranes for specific gas separation problems. These new approaches have made it feasible to
finely control the 3D porosity as well as the chemical nature of the adsorbing structures. This
ability to control the resulting structure at such a small scale introduces the possibility of many
different structures each with unique properties for gas separation. In order then to design a
specific membrane with desired properties a means of guiding the choices to be made on the
exact shape of the structure, placement of active sites and the choice of active sites would be
beneficial.

One approach that provides such guidance is the use of theory and modeling to examine potential structures first and provide a means of screening out choices that are not suitable. The theory used can contain a great deal of the specific details of a system so that quantitative comparisons with experimental results are possible. However this comes at a cost, the computation time is large and it may take an unreasonable amount of time to explore many structures. On the other hand, theories can be used that sacrifice some of the details and hence accuracy while still retaining the most important features of the real system. The theory can then explore a great deal more structures in the time that a more detailed model would take to study a single structure. If the theory retains the most important features, then the model would show qualitatively the same trends that would be seen experimentally.

We have chosen to develop a substantial modeling program that focuses on the permeation through porous materials. We chose to use a simplified model and solve it using 3D density functional theory (DFT) and Monte Carlo (MC) simulations so that it is possible to explore many structures and obtain their properties. The program can be used to obtain information on two key issues in permeation, the equilibrium adsorption in a variety of porous structures and the transport (i.e. diffusion) behavior in these structures. In the following sections we describe the model in more detail and the theories used to calculate properties for it. We also present typical results for the case of adsorption and for diffusion.

THE MODEL

The systems studied were modeled on a single occupancy lattice allowing only the possibility of nearest neighbor interactions between molecules that not wall sites. A wall site could be specified as any site in the simulation cell and would not move during an MC simulation and in a DFT calculation would have a site density of unity on the specified site. Wall sites could be placed in a variety of geometries including square pores, random placements of blocks of sites, complete wall along a vector of the simulation box, arrays of cylinders with a specified cross section, and also input from a file generated by a separate program. In this way diffusion limited aggregates generated with another program could be used as a porous structure. The wall sites would be used in generating with any specified external potential an effective external potential used in the model. All the interactions of wall sites with the rest of the system were via this effective external potential. An additional feature incorporated into the model is the possibility of a reaction between certain species [1]. This reaction can be localized to only occur near specific wall sites so that the represented reaction is catalyzed or so that the reaction represents chemisorption if the product only exists on the surface.

When interested in the equilibrium adsorption behavior of a structure, the system is treated as a single region subject to fixed chemical potential, temperature, and volume. When interested in the transport in a structure a means of creating a chemical potential gradient was used. This requires that the system be divided into three regions, two are maintained at fixed chemical potentials while the third legion has its total N determined by the net flux into it from the other two regions.

DENSITY FUNCTIONAL THEORY AND MONTE CARLO

We used a mean-field approximation of the attractive terms in our 3D –DFT so that the grand potential functional is similar to that of Bruno et al. [2,3]. Using this expression for the grand potential we can derive the set of Euler-Lagrange equations [3]. We used a Picard iteration scheme to solve for the site densities for given specified effective external potential, reaction strength and chemical potentials. When a reaction is present[2], one of the species that is a reactant or product does not have its chemical potential specified. Rather, its chemical potential is fully determined by the equilibrium constant of the reaction and the chemical potentials of the other reactants and products. This solution method is applied when we are interested in equilibrium adsorption and in the two fixed chemical potential regions in the case we are interested in transport. In the third region, the density profile is determined using a diffusion plus reaction equation where the flux is written in terms of the site chemical potentials. This site chemical potential was determined by rewriting the DFT site density equations to solve for the chemical potential. We apply a forward time step integration procedure to solve this equation. As an additional case to consider, we also incorporated the ability to solve for the density profile when the total N of a species is fixed rather than its chemical potential. In this case, using undetermined Lagrangian multipliers the new site density is the same expression as for the fixed chemical potential case except that the undetermined multiplier replaces the chemical potential. The undetermined multiplier is the average of the site chemical potentials for a given density profile. This is solved subject to the constraint that the total N of a species is fixed.

MC simulations consisted of four possible moves, a destruction move in which an occupied site is emptied, a creation in which one the components is inserted into a site if empty, a reaction move in which reactants are turned into products, and a translation move in which two lattice sites exchange occupants. When considering the transport in structures the two sites in the translation move must be neighbors and one of the sites must be empty. This is so that the simulation is representative of how a real system would evolve. No creation and destruction moves were performed in the third region.

RESULTS AND DISCUSSION

We will start with the adsorption problem of a single species into two types of porous structure. We choose to generate one porous structure by building a diffusion limited several seeds randomly scattered throughout the box, cutting off any parts of the aggregate that did not fit into the chosen system box of 50 x 50 x 50 lattice sites. Another 'wormlike' structure was generated using several self-avoiding random walks to create vacant sites in an initially perfect solid. The final number of solid sites was nearly identical for both structures. The results shown are for a system at a reduced temperature of $T/T_c = 0.733$ and a wall-molecule interaction strength of unity. The walls interacted with the molecules via a Lennard-Jones potential cut off at a distance of 10 units. The DFT was run to give density profiles with a tolerance of 10^{-6}. In figure 1, results for the adsorption isotherm from DFT calculations. Data points were generated from DFT in increments of 0.01 in the chemical potential so for clarity in the figure the data points were not shown. Also, the results from DFT are in excellent agreement with those from MC simulations (not shown here). Such MC simulations take about a 100 times longer than the equivalent DFT calculation. Another benefit of DFT is that not only are the density profiles solved but the free energy or grand potential of the system is also obtained directly. This is necessary information to

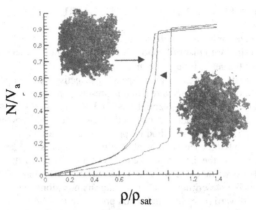

Figure 1 *Adsorption (DFT) curves for two types of large random structures compared to a planar surface. The left most isotherm and structure correspond to a diffusion limited aggregate while the other structure is wormlike. The right most isotherm corresponds to a planar surface. The wall-fluid interaction is the same for all three structures.*

determine the phase transition between different phases for a system. To determine the phase transition using MC simulations requires that additional simulations be done to generate a path suitable for thermodynamic integration.

We now turn to the adsorption/desorption behavior distribution of open-ended pores. Following earlier work by Marini Bettolo Marconi and van Swol [4] we studied a distribution of open-ended square pores all contained in a block (see the inset of figure 2). There were 3 pores of 6 x 6, 2 pores each of 7 x 7 and 5 x 5 and one pore each of 4 x 4 and 8 x 8. We considered two pore lengths: 19 lattice sites and 76 lattice sites. This entire block was contained in a simulation box that was 51 x 29 x 29 or 89 x 29 x 29 depending on the pore length. The DFT runs were set to a tolerance of 10^{-6}. The adsorption branch was determined for a chemical potential of -8 to 0.5 at intervals of 0.01 and a temperature of $T/T_c =0.733$. The desorption branches were generated by starting from certain starting points along the adsorption branch and subsequently decreasing the chemical potential in steps of 0.01. The DFT generates the grand potential at each value of μ on each branch so it is straightforward to determine the thermodynamically stable phase at each value of the chemical potential. The results shown in figure 2 clearly illustrate the hysteresis behavior of a realistic set of narrow pores generated in one single calculation. Note that the adsorption/desorption curves shown reflect the *combined* effects of the pore size distribution, finite pore length, outer surfaces, and pore ends. The curves demonstrate that open-ended pores display very little hysteresis upon desorption, an effect that is more pronounced for longer pores. Clearly, pore connectivity, heterogeneity of the pore surface, and surface roughness are just as easily included in this approach.

As a final illustration we now focus on non-equilibrium behavior, separating two gases, *a* and *b*, with a regular (cubic) porous structure (see figure 3), employing the (three box) chemical potential gradient method described earlier. The diffusion runs where performed in a simulation

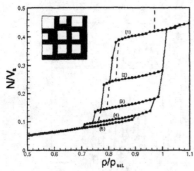 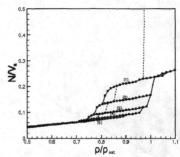

Figure 2 *Adsorption/desorption (DFT) curves for a collection of open-ended square cylinder pores in a block. A plan view of the block is displayed in the inset. The temperatures shown is $T/T_c = 0$. Two pore lengths were studied: 76 lattice sites (left) and 19 lattice sites (right). Several desorption branches are shown (labeled 1-4). They correspond to all pores initially filled (path 1) and combinations of empty larger pores and smaller pores filled (paths 2-4). The adsorption path is labeled (5). The location of the capillary transitions is indicated by the dashed vertical lines.*

cell which was 130 x 10 x 10 in size and at a temperature of $T/T_c = 2$. The pore structure was 100 lattice sites long and in the form of a cubic array of square pores that were 5 x 5 lattice sites in size. The diffusion runs were set to a tolerance of 10^{-5}. The control regions where each 15 x10 x 10 in size. The chemical potential in one box was such as to give a bulk density for component a of 0.0176 and component b of 0.024. The other control box was set to a low value of the chemical potential for both components such that their estimated bulk densities where about 10^{-5}. MC runs were carried out until the system reached steady state. The final averages were then taken from a run that consisted of 1.275×10^9 MC steps. The molecules could not swap positions with each other and where limited to local swaps. Two runs for the averaging where carried out to assure that the system was at steady state. The results are shown in figure 3, where we have plotted the plane (perpendicular to the gradient) averaged densities for both species, as obtained from MC and DFT. Figure 3 demonstrates that the DFT agrees extremely well with the more detailed MC simulations. Both density profiles exhibit a rich detailed periodic structure on top of a global linear profile. This is, of course, an immediate reflection of the 3D connectivity of the periodic cubic array. We note that the steady-state profiles directly determine the fluxes and hence the selectivity, the ratio of the fluxes for species a and b. We currently use these types of calculations to explore how to optimize the performance of a membrane by varying the pore structure, pore size, chemical site placement etc.

CONCLUSIONS

This paper demonstrates that by choosing the appropriate model a program can be developed that will readily obtain the properties of a variety of porous structures. The coarse-grained lattice model that we have focused on is particularly versatile in that both DFT and MC as well as hybrid methods can treat it. Because its computation cost has been kept small, it is possible to

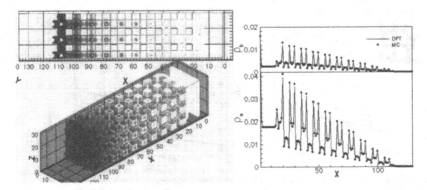

Figure 3 *Transport of an a/b mixture through a membrane block composed of a cubic array of pores. The 3D density profile is displayed on the left. On the right we plot the (yz averaged) steady state profiles of each component, showing both MC and DFT results. The feed is at a typical gas density and contains 88% of species a and 12% of species b.*

use this to use this as an exploratory tool in aiding the design of a tailored structure. Recent enhancements [5] include the possibility of refining the lattice such that one better represent the pore structure and allow for more detailed intermolecular interactions. This feature provides a systematic approach of the limit of continuum space. The study reported here also illustrates that the use of DFT is advantageous, as the loss of accuracy is quite small while the increase in timesavings is substantial. Having two methods available to study systems remains beneficial however because the MC method can probe fluctuations in the system which the density functional theory cannot.

ACKNOWLEDGMENTS

This work was supported, in part, by a grant from the Advanced Technology Program (ATP) and by the DOE. Sandia is a multiprogram laboratory operated by Sandia Corporation, a Lockheed Martin Company, for the U.S. Department of Energy under contract No. DE-AC04-94AL85000.

REFERENCES

1) W.R. Smith and B. Triska, *J.Chem.Phys.*, **100**, 3019-3027 (1993)
2) E. Bruno, U. Marini Bettolo Marconi, and R. Evans, *Physica*, **A141**, 187 (1987)
3) R. Evans, in *Inhomogeneous Fluids*, ed. D. Henderson (Dekker, 1993)
4) U. Marini Bettolo Marconi, and F. van Swol, *Phys. Rev. A*, **39**, 4109 (1989)
5) A.P. Malanoski and F. van Swol, (2000) *to be published.*

Mat. Res. Soc. Symp. Proc. Vol. 651 © 2001 Materials Research Society

Probing dynamics of water molecules in mesoscopic disordered media by NMR dispersion and 3D simulations in reconstructed confined geometries.

P. Levitz[1], J.P. Korb[2], A. Van Quynh[3] and R.G. Bryant[3]

[1]CRMD-CNRS, 1B rue de la ferollerie, 45071 Orléans Cedex 2, France.
[2]LPMC-CNRS , Ecole Polytechnique, 91128 Palaiseau France.
[3]Chemistry Department, University of Virginia, Charlottesville, VA 22901 USA.

ABSTRACT

Disordered mesoporous materials with pore sizes ranging from 2 nm to some 10 nm develop large specific surface areas. These matrices can be easily filled with polar fluids And the interfacial region between the solid matrix and the pore network strongly influences the molecular dynamics of the entrapped fluid. A promising way to probe such a coupling on a large time-scale is to look at the dispersion of the nuclear spin-lattice relaxation rate of the polar liquid using field cycling NMR relaxometry technique. We have performed such an experiment on a fully hydrated porous Vycor glass, free of electron paramagnetic impurities. The proton nuclear magnetic relaxation rate ($1/T_1$) exhibits a logarithmic dependence on Larmor frequency over the range from 0.01 to 30 MHz. A cross-over is observed below 0.1 MHz. In order to understand the relationship between geometric disorder, interfacial confinement, and nuclear magnetic relaxationdispersion (NMRD), we first compute an off-lattice reconstruction of the Vycor glass This model agrees with available experimental data (specific surface, porosity, chord length distributions, small angle scattering and tortuosity). A Brownian dynamics simulation is performed to analyze long time molecular self-diffusion and NMRD data. These later are well reproduced and appear to be connected with the translation diffusion of water near the SiO_2 interface. The logarithmic character of the NMRD is specifically related to the interfacial geometry of the Vycor glass. Several other multiconnected interfacial structures such as periodic minimal surfaces do not exhibit such an evolution. Therefore, NMRD appears to be selectively sensitive to the interfacial geometry of mesoscopic disordered materials (MDM).

INTRODUCTION

Mesoscopic disordered media such as mesoporous materials develop large specific surface area. This interfacial region strongly influences the molecular dynamics of the embedded fluids. The confinement might influence the molecular dynamics on a length scale ranging from 1 nm up to several μm . At short times t and small distances r (t <1 ns, r< 3 nm), neutron spin echoes or quasi elastic scattering [1] can be used to follow local molecular dynamics inside the pore network or nearby the interfacial region. At very long times and large distances (above 1 ms and 1 μm), macroscopic experiments or NMR pulsed field gradient spin echo experiments [2] are currently used. At the mesoscopic scale, few tools provide insight to how homogenization and up scaling are done; however, for example, dielectric relaxation and NMRD are useful[3].
In this paper, we focus on this last experimental approach. During its diffusion, a nuclear spin belonging to a fluid molecule, experiences a fluctuating magnetic interaction directly related to the molecular dynamics. This magnetic "noise" at the Larmor frequency , induces a magnetic relaxation.. On an experimental point of view, NMRD provides a remarkable opportunity to

follow the relaxation rate over a large range of Larmor frequencies, mainly from 10 kHz to tenths MHz [2-5]; this experiment then provides a characterization of the spectral density that characterizes the magnetic noise in the system, which is directly related to the molecular dynamics

EXPERIMENT

Porous Vycor glass (7930, lot 742098,Trademark Corning) was prepared by leaching a phase separated borosilicate glass and appears as a strongly interconnected and almost pure SiO_2 skeleton. This material was previously investigated using transmission electron microscopy (TEM) , small angle X-rays and neutron scattering [6]. The main scattering feature of this porous solid is a strong correlation peak around 0.23 nm^{-1} corresponding to a length of 30 nm [6] . Moreover, the surface appears slightly rough below 1.5 nm and a deviation from the traditional Porod regime is observed. The porosity is close to 0.3. The N_2 BET gives a specific surface of $S_v=157$ m^2/cm^3. The experimental tortuosity of the pore network ranges between 4 and 5. Our sample is free of paramagnetic impurities as shown by ESR spectroscopy performed at 4K.

Proton nuclear magnetic relaxation rates were measured using a field cycling instrument of the Redfield design described elsewhere [5]. The experiment was performed on a fully hydrated Vycor glass. The ^1H NMRD profile of the fully saturated Vycor glass at 298 K is reported in Fig1.

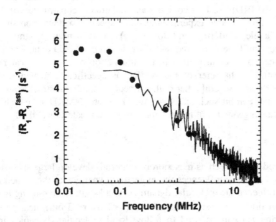

Figure 1: ^1H spin-lattice relaxation dispersion in fully hydrated Vycor glass. Full circles: experimental data at T=298 K. Continuous line: computation using Brownian dynamics inside the smooth Vycor-like pore network reconstruction. $O_2=0.18$, $\tau_{Des}> 10$ μs (see text for details).

At high frequencies, we find a relaxation rate, R_1, slightly higher than the typical value of the bulk liquid. From 10 kHz to 0.1 MHz, R_1 exhibits a logarithmic behavior. Finally, a plateau is observed at very low frequencies (f_{cross}~ 0.1 MHz), that we attribute to a "cross-over" time of 1 µs. We address below, the following questions: How to explain the log evolution ? How is it related to the Vycor interfacial geometry? Why a such low fcross value is found? To answer these questions, we have first tried to obtain a 3D reconstruction of the Vycor glass that agrees with various experimental constraints. Second, we have analyzed the long time dynamics of a water molecule inside the pore network and close to the solid/liquid interface. Finally, we have computed the NMR relaxation rates using a model, the so-called reorientation mediated by translation displacements (RMTD), previously proposed by Kimmich et al [7].

3D RECONSTRUCTION OF M.D.M. FROM EXPERIMENTAL CONSTRAINTS.

We compute an off-lattice reconstruction of the Vycor glass using a correlated Gaussian random field technique [8]. This well documented method used the bulk auto-correlation of the pore network computed from TEM images [6]. This reconstruction is shown in Fig 2A and agrees with available experimental data such as the porosity, the small angle X-ray scattering, and the chord distribution functions of the pore network. The computed tortuosity is on the same order as the experimental one, around 4. As already mentioned, TEM images were recorded with an overall resolution of 1 nm. For these conditions, surface roughness of the Vycor glass is not correctly handled and the specific surface of the reconstructed porous material is too low (105 m^2/cm^3 to be compared to 157 m^2/cm^3).

 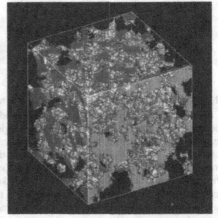

Figure 2: 3D off-lattice reconstruction of a Vycor-like porous glass. The pore network is in white. The edge of the cube is 100 nm. On the left (A): "smooth" interface. On the right (B): "rough" interface.

The reconstruction may be improved in generating a self-affine surface roughness. For this purpose, an algebraic tail is added to the spectral density of the Gaussian field in the high frequency domain [8-9]. We finally obtain a model (a "rough Vycor", see Fig 2B) having the right specific surface (157 m^2/cm^3).

LONG TIME MOLECULAR DYNAMICS AND RELAXATION PROCESSES.

We have simulated the self-diffusion of the embedded fluid molecule in the pore network and close to a SiO_2 surface. Following ref [1], the bulk self-diffusion coefficient D_{bulk} is set to $2\ 10^{-5}\ cm^2/s$ and the surface diffusion coefficient is set to $D_{bulk}/3$. Different average desorption times (τ_{Des}) of the fluid molecule diffusing inside the surface layer were analyzed ranging from 200 ps to something above 10µs . The constraint of a nearly constant fluid density inside the whole pore network permits a definition of an adsorption probability (P_{ads}) inside the surface layer (thickness a=0.3 nm). For a defined interfacial geometry, it is relatively easy to show that Pads evolves as $1/\tau_{Des}$. Brownian dynamics, with a time step of 5 ps, is performed to follow the self diffusion of the fluid molecule inside the pore network and within the surface layer.

We consider that at long time and low frequencies, the mean relaxation process is mainly due to the intra-dipolar proton-proton interaction. In the slow dynamics domain, a formal expression of the spin-lattice relaxation rate is [10]:

$$R_1(\omega) = \frac{1}{T_1(\omega)} \approx \phi_s.C_1.(O_2)^2.[J_1(\omega)+4J_2(2\omega)]$$

with (1)

$$J_{1,2}(\omega) = \frac{24\pi}{5} TF(\langle Y_2^{1,2*}(\Omega_{LD}(t=0)).f(t=0).Y_2^{1,2}(\Omega_{LD}(t)).f(t)\rangle) \quad \begin{array}{ll} f(t)=1 & r(t)\in S \\ f(t)=0 & r(t)\notin S \end{array}$$

$$\phi_S = a.S_v$$

$$C_1 = \frac{9}{8}\left(\frac{\mu_0}{4\pi}\right)^2 \frac{\gamma^4 \hbar^2}{r_{HH}^6} \quad (2)$$

$\Omega_{LD}(t)$ are the Euler angles between the constant magnetic field B_0 and the solid surface director n where the molecule is located at time t. The product, inside the brackets, is equal to 0 when the molecule does not belong to the interfacial region. O_2 is an order parameter that takes into account the fast dynamics of the adsorbed water [10]. Generally, the absolute value of O_2 is lower than 1. At low frequency, R_1 depends on the time correlation of the surface directors as probed by a molecule during its self diffusion. Computed spin-lattice relaxation dispersions are shown in Figure 3 for the "smooth" Vycor-like matrix. The relaxation rate is normalized according to $R_1^N = R_1 / (O_2)^2$ in order to remove the effect of the rapid dynamics. For short τ_{Des} values, molecules start diffusing on the surface and leave for the first time the interfacial region. In the frequency domain, one observes a R_1^N plateau close to $(1/2\pi\tau_{Des})$. The relaxation rate evolves as ln(f) as τ_{Des} increases. Finally, for very long τ_{Des} (above several µs), a plateau at small frequencies is observed around 0.1 MHz and appears to be almost independent of τ_{Des}. Similar observations are also valid for the "rough" Vycor-like matrix. As shown in Figure 6, as expected, R_1^N is systematically lower than the value computed for a smooth interface. Finally it is possible to fit the experimental results as shown in Fig 1. The order parameters, either for the smooth (O_2=0.18) or the rough (O_2=0.28) Vycor-like matrixes are compatible with a "rapid" dynamics of the adsorbed water molecules.

Both, experimental data and computations show a logarithmic evolution of the spin-lattice relaxation rate and a long time cross-over at Larmor frequencies below 0.1 MHz. This cross-over is observed for molecules staying nearby the surface for more than 1 μs before exchange to the pore network. In this time scale, it is interesting to study the molecular self-diffusion propagator and especially the probability to return after a time t to the origin. This is shown in Fig. 4 for a molecule diffusing within the interfacial layer. Note that the probability Gs(0,t) evolves as 1/t for t<3 μs. On this time-scale, a molecule located on the surface has diffused over 40-60 nm, slightly above one "pseudo-period" of the Vycor glass. This 1/t evolution explains the logarithmic law. A cross-over to a 3D exploration is observed around 3 μs . Above this limit, Gs(0,t) starts to evolve as $t^{-3/2}$. This result is compatible with the location of the experimental plateau at low frequencies.

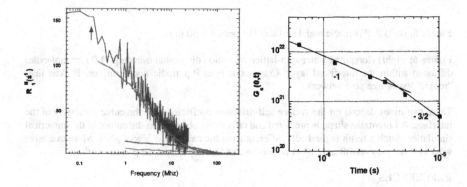

Figure 3: (On the left) Computed water spin lattice relaxation dispersion using Eqs (1-2). From the top to the bottom: $\tau_{Des} > 10$ μs, $\tau_{Des} = 200$ ns, $\tau_{Des} = 2$ ns, $\tau_{Des} = 200$ ps.

Figure 4: (On the right) Probability to return at time t to the initial position for a molecule diffusing inside the interfacial layer of a smooth Vycor-like pore network.

CONCLUSION

NMRD provides a way to understand how a geometric confinement can influence the molecular dynamics at the mesoscopic scale (several 10 nm and up to 1 μs). These results raise the question : Is NMRD selectively sensitive to the confined geometry? An interesting example is shown in Fig 5 for the case of a P periodic nodal surface [11. We show in Figure 6, the spin-lattice relaxation dispersion computed from this structural model. Following Halle et al [12], it is possible to demonstrate in a closed form and to check numerically that the relaxation rate is dominated by two characteristics correlation times τ_E and τ_D.

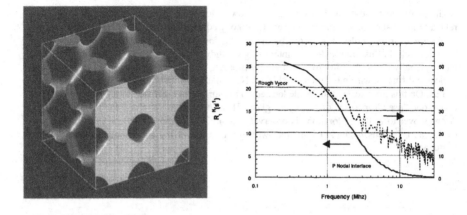

Figure 5: (left) A P periodic nodal surface. The period is 50 nm.

Figure 6: (right) Computed water spin-lattice relaxation dispersion using eqs (1-2) for molecules diffusing within the interfacial layer. Continuous line: P periodic nodal surface. Broken line: "rough" Vycor-like pore network.

These two times depend on the surface self-diffusion coefficient and the cubic symmetry of the interface. A Lorentzian shape is predicted and effectively observed as the output of the numerical simulation. Such a result is markedly different from the case of the Vycor glass. More extensive work is under way on different other pore networks to complement the present work.

REFERENCES

1. M.C. Bellisent-Funel, S.H. Chen, J.M. Zanotti, Phys. Rev. E.,**51**, 4558, (1995).
2. P.T. Callaghan, A.Coy, D. MacGowan, K.J. Packer, F.O. Zelaya. Nature, **351,**467, (1991).
3. S. Noack, Progress in NMR Spectroscopy, **18,**171-276,(1986).
4. S. Staps, R. Kimmich, J. Niess, J. Appl. Phys., **75**, 529, (1994).
5. J.P. Korb, M. Whaley-Hodges, R.G. Bryant, Phys. Rev. E., **56**, 1934,(1997).
6. P. Levitz, G. Ehret, S.K.Sinha, J.M. Drake. J.Chem. Phys., **95**, 6151, (1991).
7. R. Kimmich, H.W. Weber. Phys. Rev. B, **47,**11788,(1993).
8. P. Levitz, Advances in Colloid and Interface Science, **76-77**, 71-106, (1998).
9. N. F. Berk, Phys. Rev. A., **44**, 5069, (1991).
10. B. Halle, H. Wennerstrom, J. Chem. Phys., **75**, 1928 (1981).
11. A.L. MacKey, Proc. R. Soc. London, Ser A, **47**, 442 (1993).
12. B. Halle, S. Ljunggren, S. Lidin, J. Chem. Phys., **97**, 1401 (1992).

Mat. Res. Soc. Symp. Proc. Vol. 651 © 2001 Materials Research Society

Flow, Diffusion, Dispersion, and Thermal Convection in Percolation Clusters: NMR Experiments and Numerical FEM/FVM Simulations

Rainer Kimmich, Andreas Klemm, Markus Weber, and Joseph D. Seymour[1]
Sektion Kernresonanzspektroskopie, Universität Ulm, 89069 Ulm, Germany

ABSTRACT

Based on computer-generated templates, percolation objects were fabricated. Random-site, semi-continuous swiss cheese, and semi-continuous inverse swiss-cheese percolation models above the percolation threshold were considered. The water-filled pore space was investigated by nuclear magnetic resonance (NMR) imaging and in the presence of a pressure gradient, by NMR velocity mapping. The percolation backbones were determined using velocity maps. The fractal dimension of the backbones turned out to be smaller by about 17 % than that of the complete cluster. As a further relation of interest, the volume-averaged velocity was calculated as a function of the probe volume radius. In a certain scaling window, the resulting dependence can be represented by a power law. The experimental results favorably compare to computer simulations with the finite-element method (FEM) or the finite-volume method (FVM). Thermal convection in percolation clusters of different porosities was studied using the NMR velocity mapping technique. The velocity distribution is related to the convection roll size distribution. The maximum velocity as a function of the porosity clearly visualizes a closed-loop percolation transition if the Rayleigh number conditions are appropriate. Percolation theory suggests a relationship between the anomalous diffusion exponent and the fractal dimension of the cluster, *i.e.* between a dynamic and a structural parameter. Interdiffusion between two compartments initially filled with H_2O and D_2O, respectively, was examined by proton imaging. The results confirm the theoretical expectation. Finally, advection driven dispersive transport was investigated in the large Péclet number limit. The superdiffusive transport anomaly was demonstrated and discussed in terms of the non-local advection-diffusion and the fractional diffusion theories.

INTRODUCTION

Transport of gases and liquids through porous media is largely determined by the geometry of the pore space apart from the influence of external fields and the characteristics of the fluid and its adsorption properties at pore walls. In this study, we examine the interrelation of transport and pore space microstructure of percolation clusters as the simplest class of pore space models. The term "transport" refers here to flow, diffusion, dispersion, and thermal convection. Model objects fabricated on the basis of numerically simulated templates have been studied in nuclear magnetic resonance (NMR) experiments.

Three different, well-defined percolation models [1-3] are considered: a) *Random site percolation*: Sites on a square or cubic lattice with the lattice constant a are occupied with a probability (or porosity) p in the vicinity of the percolation threshold given by the critical value $p=p_c$. Neighbouring occupied sites are connected by pores with a cross-section corresponding to the lattice constant a or integer multiples of it. The total subsets of connected lattice sites are called clusters. For $p>p_c$, sample-spanning clusters occur that can be examined with respect to transport properties. b) *"Swiss cheese" percolation* [4]: Circular or spherical obstacles of a certain radius are randomly distributed in a semi-continuous transport medium irrespective of any overlap. The pore space is then formed by the interstitial volume. c) *"Inverse swiss cheese" percolation*: Circular or spherical voids are placed at

---

[1] Permanent address: New Mexico Resonance, Albuquerque NM 87106, USA

random in a semi-continuous matrix irrespective of any overlap. The pore space consists of the entity of the (partially overlapping) voids.

In the latter two cases, "semi-continuous" means that the obstacle or void positions coincide with grid points of a certain square or cubic base lattice. The respective radius of the obstacles and voids was chosen to be $2a$, where a is the base lattice constant [5]. That is, the obstacles or voids are allowed to "continuously" overlap each other. Note however, that the actual "circular" or "spherical" shape of the obstacles and voids can only be approximated by a point-symmetric arrangement of a few pixels or voxels, of course.

In order to put the results concerning coherent flow through the percolation cluster on a reliable and mutually consistent basis we have studied clusters characterized by the *same* parameters both in real NMR microscopy experiments and in FEM/FVM simulations. Three-dimensional percolation clusters were chosen in such a way that any unsuspended "islands" that cannot be fabricated in the model objects were absent. The percolation networks were first defined with the aid of a random-number generator. These data sets were then used as matrix patterns for the FEM/FVM simulations as well as templates for the fabrication of real model objects. The model objects were filled with water or silicon oil. Of course, only the so-called "infinite" cluster can be reached in this way, whereas any finite clusters in the interior remain inaccessible in the NMR experiments. Spin density and velocity maps of the interstitial water were recorded with the aid of four- or six-dimensional space and velocity encoding sequences of radio frequency and field gradient pulses [6] for model objects of two- or three-dimensional percolation clusters, respectively.

EXPERIMENTAL AND NUMERICAL METHODS

For the spin density and velocity mapping NMR experiments, standard radio frequency and field gradient pulse sequences [6] were employed. The spatial resolution of the spin density or velocity maps was better than 300 μm. Velocity fields induced by thermal convection in a percolation cluster were studied in a Rayleigh/Bénard configuration [7]. The temperature difference between the bottom and top heat reservoirs was 45 K, the spatial distance 2 cm.

The percolation model objects were fabricated using a circuit board plotter (for details see Refs [5,8,9]. The "two-dimensional" model objects had a finite width in the third dimension, of course, but were translationally invariant in this direction. In the flow experiments, water was pumped through the objects using a periciclic pump. The percolation backbone, that is the voxel subset of the pore space that contributes to transport across the sample, was determined with the aid of maps of the velocity magnitude for the two- or three-dimensional cases, $v = \sqrt{v_x^2 + v_y^2}$ and $v = \sqrt{v_x^2 + v_y^2 + v_z^2}$, respectively. The velocity noise level in the pore space, v_n, was determined as the rms velocity in stagnant water. The pixel subset that has velocities $v \leq v_n$ was defined as "static". Blackening all such static pore space pixels yields the percolation backbone [9]. Figure 1 shows an example.

FEM and FVM computer simulations of the velocity field in the pore space were carried out using the commercial software packages ANSYS/FLOTRAN 5.5 or FLUENT 5.3, respectively (compare Ref. [10]). The convergence criterium was fixed to a residuum of 10^{-5}. Each matrix point was represented by 3 x 3 (flow) or 7 x 7 (convection) "elements". Obstacles in the pore space are defined by $v = 0$ at the corresponding knots. In the simulations, all pertinent parameters of the NMR experiments were anticipated, that is, the fluid viscosity, the total flow rate, and the object size.

Non-advective diffusion experiments were carried out with the aid of a two-compartment sample initially filled with H_2O and D_2O, respectively. Coherent flow was prevented by

Figure 1. *In perspective rendered three-dimensional spin density maps of an inverse swiss cheese model object filled with water. The object size was 3.2 cm x 3.2 cm x 3.2 cm. The size of the cubic lattice on which the cluster was generated is 64x64x64. The average porosity of the percolation cluster is p=0.087.* **Left:** *Complete percolation cluster as recorded with static water. The fractal dimension is d_f = 2.29 (evaluated from the spin density map) or d_f =2.292 (determined from the computer-simulated template).* **Right:** *Percolation backbone derived on the basis of a velocity map of flowing water in the same object. The detected velocities ranged from 0.85 mm/s (noise level) up to a maximum value of 16.65 mm/s. The overall flow rate was 3.56 ml per minute. The fractal dimension of the backbone was evaluated as d_f^{bb}=1.98.*

stabilizing the water with a gel (Kelcogel). Tests showed that the low concentration gel formation does not affect the water diffusivity. The displacement of the isotopic diffusion front was measured as a function of time using proton imaging. That is, water diffusion was probed on a length scale much larger than the pore dimension, so that geometry induced anomalies could be probed. While the pulsed gradient spin echo (PGSE) technique [6] cannot be utilized to probe molecular diffusion displacements in a liquid on a length scale as large as the pore dimension of our model objects (400 μm), it is well suited for the detection of confinement and dispersion induced anomalies of the dispersion behavior expected if advection is superimposed to molecular diffusion.

With the PGSE method, standard Hahn or stimulated spin echo radio frequency pulse sequences [6] are supplemented by an interdigitated field gradient pulse sequel causing echo attenuation by particle displacements on the time scale of the field gradients, $G=G(t)$. Advective particle transport can be coherent or incoherent depending on whether the velocity changes or not on the time scale of the gradient pulse sequence with respect to magnitude and direction. Echo attenuation by coherent transport components can be compensated for by using gradient sequels $G(t)$ with vanishing zeroth, first and, possibly, higher moments. For example, if z gradient pulses are adjusted in such a way that $\int_0^T G(t')(z+v_z t')\, dt' = 0$, there will be no attenuation effect for particle trajectories starting at an initial position component z with a velocity component v_z stationary on the time scale T of the gradient pulse sequence. This enables one to probe the *incoherent* components of advective transport as a function of time. Fig. 2 shows the typical gradient pulse interval scheme that is essentially employed in this context [6,11,12]. In the present study coherent flow compensated echo attenuation experiments were considered in the limits: A) The experimental parameter is Δ while $\varepsilon' \approx 0$ (i.e. $<< \Delta$); B) The experimental parameter is ε' while $\Delta = const << \varepsilon'$. In either case the field gradient strength was varied in order to probe echo attenuation while all intervals of the pulse sequences were kept constant. The dependence of the echo attenuation on the displacement time intervals Δ or ε' is indicative for precession phase distributions caused by displacement distributions due to dispersion and tortuous flow velocity changes.

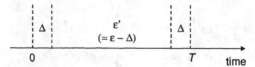

Figure 2. *Scheme of the essential field gradient pulse intervals employed in coherent flow compensating PGSE NMR experiments* [6,11,12]. *Echo attenuation is caused by precession phase shifts due to particle displacements. Identical phase shifts in the two Δ intervals indicate constant particle velocity vectors ("coherent flow") and cancel.*

RESULTS

I. FLOW: The three different percolation models mentioned above are compared in Fig. 3. The computer-generated templates, on the basis of which "two-dimensional" model objects were fabricated, are shown in the first row (a). After filling the objects with water and exerting a stationary pressure gradient, the flow patterns through the clusters were recorded. Blackening the velocity noise contrasts of all pixels that are known from the black-and-white converted spin density maps to belong to the solid matrix, led to the velocity maps shown in the third row (c). The second row (b) shows FEM simulations of these flow patterns. The fourth row (d) finally represents the percolation backbones obtained by blackening all pore space voxels in the NMR spin density maps with velocities below the noise level.

The spin density and velocity maps were evaluated using the so-called sandbox method [8,9]. Quantities of interest are the volume-averaged porosity and the volume-averaged velocity. N_p probe circles (spheres) of varying radius r are first placed randomly at positions r_k within the map in such a way that the probe volumes are completely inside the sample and the center of the probe volume is in the pore space. Then the average values of the observables of interest are formed for the N_V voxels at positions r_j inside the probe volume.

Finally, the arithmetic mean of the data set for the N_p probe volumes with a given radius r is taken. In analytical form, the two quantities are thus defined as

$$\overline{\rho}_V(r) = \frac{1}{N_p}\sum_{k=1}^{N_p}\frac{1}{N_V}\sum_{j=1}^{N_V}\rho(\vec{r}_j) \text{ and } \overline{v}_V(r) = \frac{1}{N_p}\sum_{k=1}^{N_p}\frac{1}{N_V}\sum_{j=1}^{N_V}v(\vec{r}_j), \text{ where } r \geq |\vec{r}_k - \vec{r}_j| \text{ and } \rho \text{ can take}$$

the values 1 (pore) or 0 (matrix). The volume-averaged porosity can be evaluated from black-and-white converted spin density maps [8], whereas the evaluation of the volume-averaged velocity directly refers to maps of the velocity magnitude in two or three dimensions,

$$v = \sqrt{v_x^2 + v_y^2} \text{ and } v = \sqrt{v_x^2 + v_y^2 + v_z^2} \text{ , respectively.}$$

The volume-averaged porosity is characterized by three parameters: The *fractal dimension, d_f,* the *percolation probability, P_∞,* and the *correlation length, ξ.* The fractal dimension is defined in the scaling window, $a < r < \xi$, where the volume-averaged porosity obeys a power law of the form $\overline{\rho}_V(r) \propto r^{d_f - d_E}$ with d_E the Euclidean dimension. This law applies to the scaling range $a < r < \xi$. Above the correlation length, *i.e.* for $r > \xi$, the volume-averaged porosity takes a constant plateau value corresponding to the percolation probability, P_∞. This quantity is defined as the probability that a site belongs to the "infinite" cluster traversing the whole sample [13].

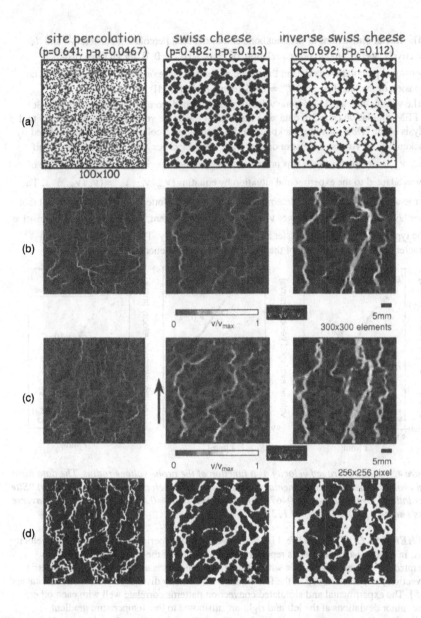

Figure 3. Comparison of two-dimensional percolation networks: **I.** *"Site percolation"*, $p=0.641$, $p-p_c=0.0467$; **II.** *"swiss cheese"*, $p=0.482$, $p-p_c=0.113$; **III.** *"inverse swiss cheese"*, $p=0.692$, $p-p_c=0.112$. **Row a)** *Computer-generated templates (pore space rendered in white).* **Row b)** *FEM simulated velocity maps.* **Row c)** *Experimental velocity maps recorded in water-filled model objects to which a pressure gradient was exerted. The maximum velocity was in the order of 10 mm/s. The arrow indicates the water flow direction.* **Row d)** *Black-and-white converted backbones derived from the experimental velocity maps.*

The fractal dimensions of the backbone and of the total percolation cluster, d_f^{bb} and d_f, respectively, were found to obey a relation $d_f^{bb} = d_f - (0.3 \pm 0.1)$ for all three two-dimensional percolation models in the whole porosity range examined. The average fractal dimension of the backbones is $d_f^{bb} = 1.46$ (compare Ref. [14]).

The volume-averaged backbone velocity was evaluated from the experimental (Fig. 3c) and FEM simulation (Fig. 3b) data as a function of the probe volume radius. In analogy to the analysis of the NMR data, the backbone of the simulated percolation clusters was defined by "blackening" all voxels of the spin density map with velocities $v \leq v_{co}$, where the cut-off value, v_{co}, corresponds to the rms noise value, v_n, in the experiments. The actual value of v_{co} was adapted to the experimental situation by equating $(v_{co} / v_{max})_{sim} = (v_n / v_{max})_{exp}$. The volume-averaged velocity data determined in this way are plotted in Fig. 4. It turned out that a power law is valid in a wide range, $\overline{v}_V(r) \propto r^{-\lambda}$. The exponent, λ, is more or less insensitive to the type of the percolation cluster as well as to the porosity. This suggests a universal character of the power law and of the value range of the exponents, $0.4 < \lambda < 0.75$.

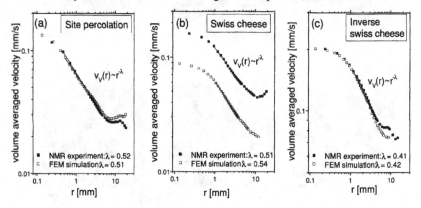

Figure 4. *Volume averaged velocity as a function of the probe volume radius. The data have been evaluated from the experimental and FEM simulated velocity maps (see Fig. 3). a) "Site percolation", p=0.641, p-p$_c$=0.0467; b) "swiss cheese", p=0.482, p-p$_c$=0.113; c) "inverse swiss cheese", p=0.692, p-p$_c$=0.112.*

II. THERMAL CONVECTION: Figure 5 shows typical experimental and simulated velocity maps. In either case the contrasts represent the magnitude of the velocity. The template generated on a square base lattice with 100x50 lattice points is also shown. The symmetric convection patterns observed in the free liquid are obviously distorted by the matrix obstacles [7,15]. The experimental and simulated convection patterns correlate well with each other. Some minor deviations at the left and right are attributed to the temperature gradient distributions that are experimentally ill-defined near the side walls, and to the somewhat different temperature ranges and absolute values.

Convection preferentially occurs on paths that are part of the percolation backbone. With decreasing porosity, sample spanning rolls become less and less likely. Below the percolation threshold all convection rolls are more or less localized according to the finite extension of the percolation clusters. The maximum velocity changes dramatically with the porosity. Fig. 6 shows a plot of the maximum velocity as a function of the porosity. These data demonstrate a

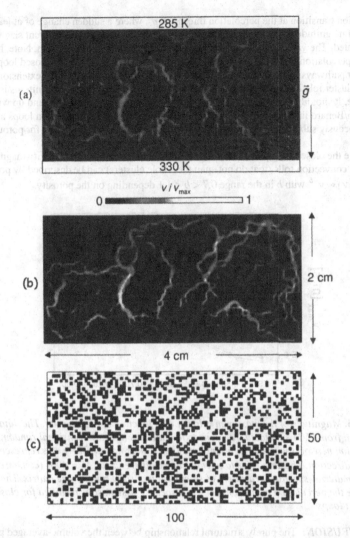

Figure 5. *Typical velocity maps of free convection patterns in a site-percolation network (a) measured and (b) simulated with NMR velocity mapping and FVM simulation techniques, respectively. The maps refer to a Rayleigh/Bénard configuration with the temperature gradient along the y direction. The NMR experiments were performed in a silicon oil filled model object fabricated on the basis of the computer-generated template shown in (c). The base lattice size is 100x50, the porosity p= 0.7 > p_c. The fabrication resolution of the model object was 400 μm, the experimental digital space resolution 190 μm. The experimental velocity resolution was 0.1 mm/s. The velocity range probed was ±0.7 mm/s, the echo time 130 ms, the repetition time 800 ms.*

percolation transition at the percolation threshold, p_c, where a sudden change of at least two orders of magnitude of the maximum velocity occurs. Two systems of different size were investigated. The value of the critical porosity is larger for the smaller system. Note, however, that the percolation threshold defined and determined in this way refers to closed loops of the transport pathways in contrast to the ordinary definition that is based on the extension of the largest cluster relative to the system. A slight deviation between the two definitions is plausible. It should also be kept in mind that the percolation transition is bound up with the Rayleigh/Bénard instability: Percolation clusters permitting larger convection loops are simultaneously subject to larger Rayleigh numbers in their version modified for porous media.

Above the closed-loop percolation transition, the distribution of the velocity magnitude of the local convection rolls (that do not span the whole clusters) can be described by power laws, $n(v) \propto v^{-b}$ with b in the range $0.7 < b < 1.1$ depending on the porosity.

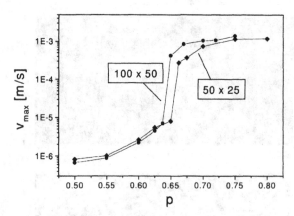

Figure 6. *Magnitude of the maximum velocity as a function of the porosity. The data were evaluated from FVM simulated free-convection patterns in two-dimensional random-site percolation networks with two different base lattice sizes n_x x n_y. The y axis represents the vertical direction. The data refer to a temperature difference of $\Delta T = 40\ K$ (compare Fig. 5) and the material-specific parameters of silicon oil in a solid polystyrene matrix. The lines illustrate the percolation transitions for the respective system sizes modified for closed transport loops.*

III. DIFFUSION: The purely structural relationship between the volume-averaged porosity and the probe volume radius is opposed by the dynamic property for the mean squared displacement of a random walker on the cluster $\langle r_d^2 \rangle \propto \begin{cases} t^{2/d_w} & \text{for } t \ll t_\xi \\ D_{\text{eff}}t & \text{for } t \gg t_\xi \end{cases}$, where $t_\xi \propto \xi^{d_w}$ is the time a random walker needs to explore the correlation length ξ, and d_w is the fractal dimension of the random walk. Strictly speaking, anomalous diffusion is only expected in the scaling range $a < \sqrt{\langle r_d^2 \rangle} < \xi$. The diffusion coefficient becoming effective in the long-time limit, $t \gg t_\xi$, is denoted by D_{eff}. According to the Alexander/Orbach conjecture [16], the quantity d_w is assumed to be related to the fractal dimension by $d_w \approx 1.5 d_f$ for $d_E \geq 2$. That is, the structural parameter d_f characteristic for the volume-averaged porosity is linked to the dynamic parameter d_w specifying anomalous diffusion. Evaluations for both quantities have been carried out, so that a comparison becomes possible. Note, however, that this relation is only an approximate one [1,2,17,18].

Figure 7 shows plots of the mean squared displacement in the unobstructed case and in a random-site percolation object. The data were evaluated from spin density images rendering the isotopic inter-diffusion front between compartments initially filled with H_2O and D_2O, respectively. The displacement of the front was measured as the position at half height of the proton signal projected on the main isotopic gradient direction.

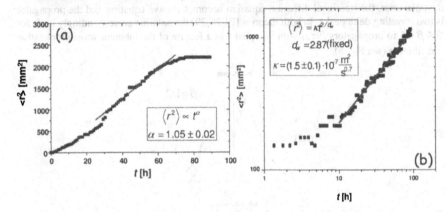

Figure 7. *Mean squared displacement of the propagating proton diffusion front versus time. (a) Unobstructed diffusion. (b) Diffusion in a model object based on a two-dimensional random-site percolation network (matrix size 100x100; p-p_c=0.031; d_f=1.9; correlation length ξ = 7 lattice constants a). The respective time resolutions were 63' and 26'. The plateau reached at long times is due to the finite extension of the sample. The deviations in the short-time limit reflect the initial situation that can only imperfectly be represented by a step function.*

IV. DISPERSION: Transport phenomena due to combined advection and diffusion is known as hydrodynamic dispersion. This so-called Taylor dispersion was experimentally [19] and theoretically [19,20] studied for flow through tubes. In the present context the extension to multiphase systems such as porous media and solid liquid suspensions [21], and the direct connection with stochastic processes and kinetic theory [22] are of interest. For non Fickian, anomalous, or scale dependent transport as occurs in fractal media in the scaling window $a < \sqrt{\langle z^2 \rangle} < \xi$, averaged transport theories based on non-local treatments have been developed from the perspectives of statistical mechanics [23] and ensemble averaged continuum mechanics [24,25].

Dispersion in percolation structure porous media is predicted to be governed by the scaling laws of percolation clusters [25,26]. The regime of interest is the large Péclet number limit where advection is predominant. Considering dispersion only along the backbone, percolation scaling arguments [26] suggest the mean squared displacement in the main direction of flow scales in the two-dimensional case as $< z^2(t) > \propto t^{1.26}$, while incorporating holdup dispersion effects due to dead end pores results in $< z^2(t) > \propto t^{2.3}$. Hence superdiffusive transport is expected for 2D percolation porous media in the large Péclet number (Pe) regime, a result in agreement with computer simulations [25,27].

The theory of Koch and Brady [24] results in a non-local advection-diffusion equation. This can be compared to a recently proposed fractional kinetic equation for sub-ballistic superdiffusion [28,29]. The fractional equation predicts a mean squared displacement $\langle z^2(t) \rangle = \dfrac{2K_{2-\beta}}{\Gamma(3-\beta)} t^{2-\beta}$. The exponent value $\beta = 1$ yields the Fickian diffusion limit characterized by a Gaussian propagator $P(z,t)$, whereas $\beta = 0$ indicates ballistic motions. In the latter case, the fractional diffusion equation becomes a wave equation and the propagator is two traveling delta peaks. In both theories [24,28,29] the solution correspondingly leads for $0 \leq \beta < 1$ to propagators $P(z,t)$ with two peaks as a feature of the inherent wave propagation contribution (see Fig. 8).

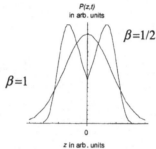

Figure 8: *Propagators P(z,t) for displacements z in a time t. The data are based on the fractional diffusion equation [28,29] for β = 1 (Gaussian limit) and β=1/2 (example for superdiffusive behavior).*

The PGSE method provides direct measurement of the displacement length and time dependent hydrodynamic dispersion for flow through porous media [30]. In the limit of short field gradient pulses (relative to the displacement time intervals) the measured spin echo signal normalized to eliminate magnetic relaxation effects is given as the space Fourier transform of the propagator. Furthermore, in the limit of small values of the reciprocal displacement space wave vector q which is conjugate to the particle displacement component along the field gradient, the mean squared displacement can be evaluated as a function of the displacement time. The wave vector q is adjusted via the gradient strength. Equivalently, an apparent (time dependent) diffusion coefficient can be determined. Fig. 9 shows two examples of such evaluations demonstrating the superdiffusive character of advection in the large Pe limit. The mean squared displacement and apparent diffusion coefficient data plotted in Fig. 9 demonstrate the influence of displacements by incoherent flow which is characterized by particle velocities changing direction and magnitude during the observation time. Note that anomalous scaling is again restricted to a rms displacement range limited by the shortest length of the cluster forming matrix structure on the one hand, and on the effective correlation length, on the other. Larger rms displacements exceeding the scaling window are expected to approach the Einstein relation for ordinary diffusion.

CONCLUSIONS AND DISCUSSION

Flow, diffusion, dispersion and convection through percolation clusters were studied with various NMR techniques. As far as feasible, the results were compared with FEM/FVM simulations. Random site, swiss cheese and inverse swiss cheese percolation clusters were considered as model structures.

Computer-generated clusters were used as templates for the fabrication of model objects for two- and three-dimensional percolation clusters. The pore space of the sample-spanning clusters in these objects was filled with water and rendered in the form of spin density maps.

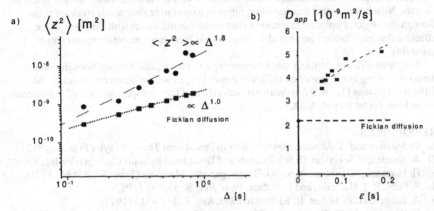

Figure 9. *Advection in the large Pe limit probed with the flow compensated PGSE experiments A) and B) as described above (comp. Fig. 2). a) Anomalous time dependences of superdiffusive mean squared displacements of water flowing through a two-dimensional random site model object with $p\text{-}p_c = 0.03$ and a probe volume extension $L \approx 3\,\xi$. The superdiffusive data correspond to $<z^2> \propto t^{1.8}$ in the anomalous scaling window. The Fickian diffusion limit (measured without flow) is shown for comparison. b) Time dependent apparent diffusion coefficient of water flowing through densely coiled cotton fibers [11]. The average flow velocity was 1 mm/s. Anticipating that the data refer to a scaling window, $a_f < \sqrt{\langle z^2 \rangle} < \xi$, where a_f is the fiber diameter, the superdiffusive data are compatible with a law $<z^2> = \frac{1}{2} D_{app} t \propto t^{1.3}$. The room temperature diffusion coefficient of static water is shown for comparison.*

In the flow experiments, an external pressure gradient was exerted, and velocity maps were recorded within the clusters. On this basis, the percolation backbones were evaluated.

The perfect coincidence of the experimental spin density and velocity maps with the computer-generated templates and FEM/FVM simulated flow patterns, respectively, demonstrate that the fabrication process, the measuring technique, the evaluation procedure, and the simulation technique can be considered to be mutually consistent procedures. One of the conclusions is that parameters such as the fractal dimension, the correlation length, the percolation probability both of the total clusters and of the percolation backbones can safely be determined with computer simulation techniques. The same applies to the volume-averaged velocity which was found to scale with the probe volume radius in a wide range in the form of a power law. The exponent of this law depends only weakly on the porosity and the type of the percolation models. A theoretical linkage to the fractal dimension evaluated from the volume-averaged porosity appears to be plausible, but has not yet been established.

The fractal dimension evaluated from the volume-averaged porosity favorably compares with the value obtained from anomalous-diffusion data on the basis of the Alexander/Orbach conjecture in the frame of the experimental accuracy. That is, values based on the structure and on the dynamics are demonstrated to be consistent even though the rms displacements partly exceeded the correlation length of the cluster significantly. The scaling range exceeds the correlation length significantly in agreement with recent computer simulations [31].

Thermal convection is strongly perturbed by the matrix. A percolation transition for convection loops was demonstrated which correlates with the occurrence of sample spanning clusters. Note however that this transition refers to loops rather than single stranded paths through the object. Furthermore, because the convection roll varies with the porosity the effective Rayleigh/Bénard number also varies so that there is a connection to this sort of instability.

The superdiffusive displacement behavior expected for liquids flowing through porous media according to the advection-diffusion theory [24] as well as based on the fractional diffusion equation [28] was demonstrated with coherent flow compensated PGSE experiments in the large Péclet number limit.

REFERENCES

1. D. Stauffer and A. Aharony, Introduction to Percolation Theory (Taylor Francis, 1992).
2. A. Bunde and S. Havlin, (Eds.), Fractals and Disordered Systems (Springer-Verlag, 1996).
3. H. Hermann, Stochastic Models of Heterogeneous Materials (Trans Tech Publ., 1991).
4. S. Feng, B. I. Halperin, and P. N. Sen, *Phys. Rev. B* **35**, 197 (1987).
5. A. Klemm, H.-P. Müller, R. Kimmich, *Phys. Rev. E* **55**, 4413 (1997).
6. R. Kimmich, NMR Tomography, Diffusometry, Relaxometry (Springer-Verlag, 1997).
7. D. A. Nield and A. Bejan, Convection in Porous Media (Springer-Verlag,1992).
8. H.-P. Müller, J. Weis, and R. Kimmich, *Phys. Rev. E* **52**, 5195 (1995).
9. H.-P. Müller, R. Kimmich, and J. Weis, *Phys. Rev. E* **54**, 5278 (1996).
10. J. S. Andrade Jr., M. P. Almeida, J. Mendes Filho, S. Havlin, B. Suki, and H. E. Stanley, *Phys. Rev. Letters* **79**, 3901 (1997).
11. F. Klammler and R. Kimmich, *Phys. Med. Biol.* **35**, 67 (1990).
12. S. L. Codd, B. Manz, J. D. Seymour, and P. T. Callaghan, *Phys. Rev. E* **60**, R3491 (1999).
13. A. Kapitulnik, A. Aharony, G. Deutscher, and D. Stauffer, *J. Phys. A: Math. Gen.* **16**, L269 (1983).
14. M. Porto, A. Bunde, S. Havlin, and H. E. Roman, *J. Phys. Rev. E* **56**, 1667 (1997).
15. M. D. Shattuck, R. P. Behringer, G. A. Johnson, and J. G. Georgiadis, *Phys. Rev. Letters* **75**, 1934 (1995).
16. S. Alexander and R. Orbach, *J. Physique-Lettres (Paris)* **43**, L625 (1982).
17. D. C. Hong, S. Havlin, H. J. Herrmann, and H. E. Stanley, *Phys. Rev. B* **30**, 4083 (1984).
18. J. G. Zabolitzky, *Phys. Rev. B* **30**, 4077 (1984).
19. G. I. Taylor, Proc. Roy. Soc. Lond. A **219**, 186 (1953).
20. R. Aris, Proc. Roy. Soc. Lond. A **235**, 67 (1956).
21. H. Brenner, J. Stat. Phys. **62**, 1095 (1991).
22. C. van den Broeck, Physica A **168**, 677 (1990).
23. J. H Cushman, B.X. Hu and T.R. Ginn, J. Stat. Phys. **75**, 859 (1994).
24. D.L. Koch and J.F. Brady, J. Fluid Mech. **200**, 173 (1987).
25. J. Koplik, S. Redner and D. Wilkinson, Phys. Rev. A **37**, 2619 (1988).
26. M. Sahimi, Rev. Mod. Phys, **65**, 1393 (1993).
27. H.A. Makse, J.S. Andrade and H.E. Stanley, Phys. Rev. E **61**, 583 (2000).
28. R. Metzler and J. Klafter, Europhys. Lett. **51**, 492 (2000).
29. W. R. Schneider and W. Wyss, J. Math. Phys. **30**, 134 (1989).
30. J. D. Seymour and P. T. Callaghan, *AIChE J.* **43**, 2096 (1997).
31. O. J. Poole and D. W. Salt, *J. Phys. A; Math. Gen.* **29**, 7959 (1996).

Mat. Res. Soc. Symp. Proc. Vol. 651 © 2001 Materials Research Society

Molecular Dynamics Simulation of Confined Glass Forming Liquids

Fathollah Varnik[1], Peter Scheidler[1], Jörg Baschnagel[2], Walter Kob[3], Kurt Binder[1]

[1] Institut für Physik, Johannes Gutenberg-Universität, Staudinger Weg 7, D–55099 Mainz, Germany
[2] Institut Charles Sadron, ULP, 67083 Strasbourg, France
[3] Laboratoire des Verres, Université Montpellier II, 34000 Montpellier, France

ABSTRACT

Two model studies are presented in order to elucidate the effect of confinement on glass forming fluids, attempting to study the effects of the interactions between the confining walls and the fluid particles. In the first model, short bead-spring chains (modelling a melt of flexible polymers) are put in between perfectly flat, structureless walls, on which repulsive potentials act. It is shown that chains near the walls move faster (in the direction parallel to the walls) than chains in the bulk. A significant decrease of the (mode-coupling) critical temperature with decreasing film thickness is found. In the second model, a binary Lennard-Jones liquid is confined in a thin film, whose surface has an amorphous structure similar to the liquid. Although, as expected, the static structural properties of the liquid are not affected by the confinement, relaxation times near the wall are much larger than in the bulk. Consequences for the interpretation of experiments are briefly discussed.

MOTIVATION

Understanding the glassy state of condensed matter and in particular the transition that leads from a supercooled fluid to the amorphous solid is one of the greatest challenges of our time. A particularly controversial concept is the idea of a characteristic "correlation length" ξ that grows as the glass transition is approached. This length is supposed to measure the size of the regions over which cooperative structural rearrangements need to occur to allow the system to relax [1].

The concept of this length scale was a motivation to investigate how the glass transition in thin polymer films depends on their thickness D [2–5]. Here one made use of the key idea of finite size scaling at ordinary static (second order) phase transitions, that "D scales with ξ" [6], i.e., a shifted transition is reached when ξ has grown to about the size $\xi = D/2$. The first measurements (by ellipsometry) [2] could be done only for rather large molecular weights, and found a decrease of the glass transition temperature T_g for $D \leq 50$nm. Although in the meantime there are many more experiments [3–5], many details are still unclear. An intriguing feature clearly is that for long entangled chains the motion near a surface may differ from the one in the bulk [7].

Due to the complexity of the problem it is desirable to separate the problem of the glass transition from another very difficult and incompletely understood problem, namely of whether and how polymers "reptate" [8]. This can be done by studying the glass transition of short polymer chains that are not entangled and which can be studied very well with computer simulations [9]. A further advantage of such simulations is that one can provide boundary conditions at the surface of the thin film at will, and thus understand better the interplay of finite size and surface effects. Thus we can study a model where molecules are confined between absolutely identical, perfectly characterised, ideally flat walls [10], obtaining complete information in both space and time in atomistic detail, as described in the following sections.

A COARSE-GRAINED POLYMER MODEL

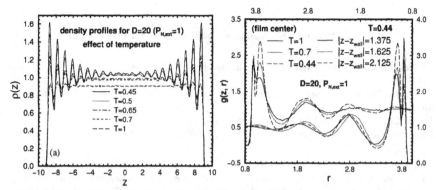

Figure 1: (a) *Density profiles* $\rho(z)$ *vs.* z *for* $D = 20$ *and* $N = 2000$ *monomers in the system, and five temperatures as indicated.* (b) *Radial distribution function* $g(z,r)$ *vs.* r *if the monomers are in the film centre at* $z = 0$ *(lower and left axes labels) and if (for* $T = 0.44$*) z is close to the wall (upper and right axes labels).*

Molecular dynamics has the disadvantage that it is restricted to timescales of (at best) nanoseconds. This problem is somewhat alleviated if chemical detail is abandoned, and one studies a bead-spring model where several monomers are integrated into one effective segment. In the present simulation consecutive monomers along a chain interact with an anharmonic spring,

$$U_F(l) = -0.5kR_0^2 \ln\left[1 - (l/R_0)^2\right],\qquad(1)$$

while all monomers interact with a truncated and shifted Lennard-Jones potential,

$$U_{LJ}(r) = \begin{cases} V(r) - V(r_c) & \text{for } r < r_c \\ 0 & \text{else.} \end{cases}\qquad(2)$$

Here, $r_c = 2 \times 2^{1/6}\sigma$ and $V(r) = 4\epsilon\left[(\sigma/r)^{12} - (\sigma/r)^6\right]$. In order to keep the comparability with results of previous bulk-simulations [11] of the same model, long range corrections have not been employed. The lateral system size was about $L_x = L_y \approx 10\sigma \approx 7R_g$, where R_g is the chain's radius of gyration. At this system length, finite size effects are practically absent for the present polymer model [12].

If the parameters are chosen as $\varepsilon = 1$, $\sigma = 1$ (fixing units of temperature and length), $k = 30$, and $R_0 = 1.5$, the model exhibits two incompatible length scales, the minimum positions of the bond potential and of the Lennard-Jones potential. No simple crystal structure is compatible with these lengths, and hence this frustration suffices to make this model a "good glass former", even if the chains are very short such as $L = 10$ [10]. Previous studies have shown that the static structure factor $S(q)$ (q=wave number) resembles real experiments very closely, that if the system is cooled, the density exhibits a kink at a (cooling rate dependent) $T_g \approx 0.41$, and that an analysis in terms of mode coupling theory [13] yields a critical temperature $T_c \approx 0.45$. The self diffusion constant fitted to a Vogel-Fulcher [1] relation yields a Vogel-Fulcher temperature [11] $T_0 \approx 0.34$.

POLYMER FILMS BETWEEN REPULSIVE WALLS

Choosing the normal pressure $P_{N,\text{ext}} = 1$ in the system exactly at the value chosen in the simulations of bulk behaviour [9, 11], and two walls at $z_{\text{wall}} = \pm D/2$ with a potential $U_{\text{wall}} = (\sigma/z)^9$, $z = |z_{\text{monomer}} - z_{\text{wall}}|$, one observes density profiles with the characteristic layered structure (Fig.1a).

Both amplitude and range of these oscillations increase as the temperature T is lowered. Nevertheless the fluid structure in lateral direction is not much affected, as an analysis of the corresponding radial distribution function $g(r, z)$ shows (Fig.1b). Note that the split nearest neighbour peak of $g(r, z)$ reflects the incompatible length scales mentioned above (the two bonded monomers yield the sharp first peak). Despite the fact that the wall leads to strong local density enhancements at low T, it turns out that there is a speed-up of the dynamics in thin films, Fig.2: Here mean square displacements of all monomers,

$$g_0(t) = \left\langle [\mathbf{r}_i(t) - \mathbf{r}_i(0)]^2 \right\rangle , \tag{3}$$

and the centre of mass,

$$g_3(t) = \left\langle [\mathbf{r}_{\text{cm}}(t) - \mathbf{r}_{\text{cm}}(0)]^2 \right\rangle , \tag{4}$$

of the chains, in directions parallel to the walls, are shown for $T = 0.46$ (curves). In fact, the data superimpose precisely with corresponding bulk data but at shifted temperature, $T = 0.52$. Fig.2 does not show a rare coincidence, but rather we find an approximate superposition for all D if data at the same temperature distance $T - T_c(D)$ are compared. We find $T_c(D = 5) \approx 0.31$, $T_c(D = 10) \approx 0.39$, and $T_c(D = 20) \approx 0.41$, respectively. These estimates stem from an analysis of many relaxation times $\tau_{g,i}$, defined from $3/2\ g_{i,\parallel}(t = \tau_{g,i}) = R_g^2$. Here, the index i can stand for "inner monomer", or end monomer or all monomers or the chains centre of mass (Fig.2). These data are compatible with $\tau_{g,i} \propto (T - T_c(D))^{-\gamma}$, where $\gamma = 2.09$ for all D and all i as found in the bulk [11].

A CONFINED BINARY LENNARD-JONES MIXTURE

For confined small molecular systems it is sometimes found that T_g is larger than in the bulk [14]. To better understand how this can occur, we simulated a binary (80% A, 20% B particles) Lennard-Jones mixture confined into pores [15] and films. The parameters in

$$U_{\alpha\beta}(r) = 4\varepsilon_{\alpha\beta}[(\sigma_{\alpha\beta}/r)^{12} - (\sigma_{\alpha\beta}/r)^6] \tag{5}$$

are $\epsilon_{AA} = 1.0$, $\sigma_{AA} = 1.0$ (choice of units), $\epsilon_{AB} = 1.5$, $\sigma_{AB} = 0.8$, $\epsilon_{BB} = 0.5$, and $\sigma_{BB} = 0.88$ respectively. We use a cut-off radius of $r_{\alpha\beta}^c = 2.5\sigma_{\alpha\beta}$ and the potential is shifted by $U_{\alpha\beta}(r_{\alpha\beta}^c)$. As known from extensive simulations in the bulk [16], with this choice one avoids both crystallisation and phase separation, and hence also this model describes a good glass former.

In order to isolate the effects of the confinement and avoid surface effects as much as possible, we equilibrate a large simulation box with periodic boundary conditions, and introduce at some point an intrinsic boundary in the form of a cylinder [15] or via two parallel planes. All particles outside of these boundaries are then frozen at their liquid-state positions, whereas the particles inside of the boundaries remain mobile. The system is then equilibrated again. The system size in directions where periodic boundary conditions are applied (i.e. along the cylinder axis for the pore and in the film plane) was about $20\sigma_{AA}$ in cylindrical and $12.88\sigma_{AA}$ in film geometry.

Now it is found that there is almost no layering inside the system, unlike Fig.1 for the polymer model and also for corresponding simulations of a Lennard-Jones mixture confined in a pore with smooth walls, and that the lateral correlations are indistinguishable from their bulk counterparts.

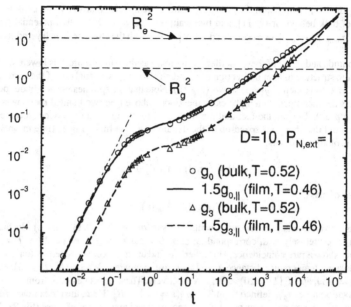

Figure 2: *Log-log plot of the mean square displacement for $D = 10$ at $R_{N,ext} = 1$ versus time at $T=0.46$, comparing to corresponding bulk data at $T = 0.52$. Also the bulk mean square end-to-end distance ($R_e^2 \approx 12.3$) and the gyration radius ($R_g^2 \approx 2.09$) are indicated.*

However, studying the local intermediate scattering function

$$F_S(q, z, t) = \left\langle \exp[i\mathbf{q} \cdot (\mathbf{r}_{j,\|}(t) - \mathbf{r}_{j,\|}(0)] \, \delta\left(z_j(0) - z\right) \right\rangle, \tag{6}$$

with the wave vector $\mathbf{q}$ parallel to the wall, one finds a dramatic slowing down if z is close to the boundary. Fig.3 shows the "profiles" of the corresponding relaxation time $\tau(z,T)$, for $q=7.2$ (where the static structure factor has its peak), and using the definition $F_S(q, z, t = \tau) = F_S(q, z, t = 0)/e$. The cause for this enhancement of $\tau(z,T)$ (and hence also of T_g) is the microscopic roughness of the rigid walls: particles sitting in cages at a rough wall are more effectively confined than in the bulk. Hence a rough wall has an opposite effect than a smooth wall.

CONCLUSION

While the simulations thus elucidate that in a confined fluid T_c (and T_g) can be either depressed or enhanced, depending on the precise "character" of the confining surface (e.g. smooth or rough), the consequences for the behaviour of the length ξ are not so clear yet. There is a static length that increases when T decreases - e.g. describing the range of the layering in Fig.1 - but it is not clear whether it is relevant for the glass transition. For fluids in pores and also in thin films the behaviour analogous to Fig.3 was interpreted as

$$\tau(\rho, T) = f_q(T) \exp\left[\Delta_q/(\rho_p - \rho)\right], \tag{7}$$

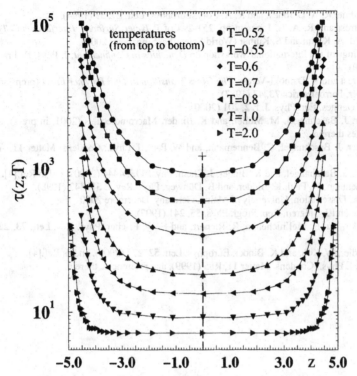

Figure 3: $\tau(z, T)$ plotted vs. the distance z from the centre of the film, for a film thickness $D = 10$ and various temperatures.

ρ being the the distance from the pore centre, and Δ_q, ρ_p being T-independent parameters [15]. However, neither of these results can exclude that an analysis of different quantities could yield evidence for a growing length, of course. More theoretical guidance on this problem is clearly desirable.

ACKNOWLEDGEMENTS

We are grateful to the Deutsche Forschungsgemeinschaft (SFB262/D1,D2) for financial support, as well as to the European Science Foundation (ESF) under the SUPERNET program, and the Bundesministerium für Forschung und Technologie (BMBF). Generous grants of computing time from the NIC Jülich and the ZDV Mainz are acknowledged.

REFERENCES

[1] E. Donth, *Relaxation and Thermodynamics of Polymers: Glass Transition* (Akademie-Verlag, Berlin, 1992).

[2] J.-L. Keddie, R. A. L. Jones, and R. A. Cory, Europhys. Lett. **27**, 59 (1994).

[3] J. A Forrest and R. A. L. Jones in pp. 251-294 of *Polymer Surfaces, Interfaces and Thin Films*, Ed. A. Karim and S. Kumar (World Scientific, Singapore, 2000).

[4] Proceedings of the International Workshop on *Dynamics in Confinement*, J. Phys. IV France **10** (2000).

[5] J. A Forrest and K. Dalnoki-Veress, *The Glass Transition in Thin Polymer Films* (preprint).

[6] K. Binder, Ferroelectrics **73**, 43 (1987).

[7] P. G. de Gennes, Eur. Phys. J. E **2**, 201 (2000).

[8] T. Kreer, J. Baschnagel, M. Müller, and K. Binder, Macromolecules (2001, in press), and references therein.

[9] K. Binder, J. Baschnagel, C. Bennemann, and W. Paul, J. Phys.: Condens. Matter **11**, A47 (1999).

[10] F. Varnik, J. Baschnagel, and K. Binder, J. Chem. Phys. **113**, 4444 (2000), and in Ref.[4].

[11] C. Bennemann, W. Paul, K. Binder, and B. Dünweg, Phys. Rev. E **57**, 843 (1998).

[12] F. Varnik, Dissertation, University of Mainz, Germany, Decembre 2000

[13] W. Götze and J. Sjögren, Rep. Progr. Phys. **55**, 241 (1992).

[14] J. Schüller, Yu. B. Mel'nichenko, R. Richert, and E. W. Fischer, Phys. Rev. Lett. **73**, 2224 (1994).

[15] P. Scheidler, W. Kob, and K. Binder, Europhys. Lett. **52**, 277 (2000), and in Ref.[4].

[16] W. Kob, J. Phys.: Condens. Matter **11**, R85 (1999), and references. therein.

Mat. Res. Soc. Symp. Proc. Vol. 651 © 2001 Materials Research Society

Sliding friction between polymer-brush-bearing surfaces: crossover from brush-brush interfacial shear to polymer-substrate slip.

Jacob Klein*
Weizmann Institute of Science, Rehovot 76100, Israel

ABSTRACT

A model is presented for the shear (or frictional) forces F_s between two surfaces a distance D apart as they slide past each other while bearing mutually compressed polymer brushes, on the assumption that sliding takes place at the brush-brush interface. The predictions of the model for the rapid increase in F_s at increasing compressions are in reasonable agreement with experiments on polystyrene brushes immersed in toluene over two decades in F_s. At higher compressions (smaller D) the experimental shear forces increase only slowly, and diverge from the calculated ones which continue to increase rapidly; at the same time the form of the shear force response at these higher compressions reverts from a viscous-like one to a stick-slip behaviour. These observations strongly indicate that at sufficiently high compressions the plane of slip crosses over from the brush-brush interface to the polymer-solid surface.

INTRODUCTION AND EXPERIMENTAL BACKGROUND

The friction between two solid surfaces coated with polymer brushes interacting across a good solvent may be dramatically reduced relative to the bare (polymer-free) surfaces at the same pressures [1-3]. The reason for this strong lubrication effect is because – over a substantial regime of compressive loads – the brushes are able to support a large normal stress while maintaining a rather fluid interfacial region between them when sliding. This results in very little frictional drag opposing the sliding, hence the low friction coefficients. The origin of the fluid interface is the weak mutual interpenetration of two compressed polymer brush layers: Due to the strong excluded volume repulsion that an outside chain encounters when penetrating into an existing brush [4,5], the interpenetration of two brush layers is strongly resisted [6,7]. It can be shown [2,8,9] that the extent of interpenetration d between two surfaces a distance D apart bearing brush layers each of unperturbed thickness L depends on the extent of compression only weakly, as

$$d \cong s\beta^{1/3} \qquad (D < 2L) \qquad (1)$$

Here $\beta = (2L/D)$ is the compression ratio, D = 2L being the the surface separation when the brushes just touch each other, and s is the mean interanchor spacing of the brush chains. Thus, a fivefold-compression of the brushes relative to the separation at which they just touch ($\beta = 5$), can support a large normal load, but will only increase the width of the interpenetration zone by some 1.7-fold. Within this relatively narrow zone the interpenetrated moieties may be short and barely entangled, so that the viscous dissipation taking place within it is low.

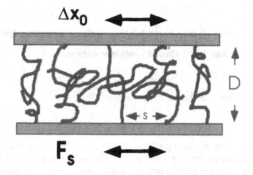

Figure 1. *The shear force F_s is transmitted to the lower surface via the mutually compressed brushes as the top surface is moved back-and-forth by Δx_0.*

Previous studies [1] on polystyrene (PS) brushes using a surface-force balance (SFB) showed that up to compression ratios of around 7 or 8 the frictional forces between the brush-bearing surfaces were below the resolution of the experiment. At higher loads (and compression ratios), as one surface was made to slide sinusoidally back-and-forth past the other, the resulting shear forces showed two distinct forms of behaviour [10]. The experimental configuration is illustrated in fig. 1, while characteristic shear-force vs. applied-lateral-motion traces are reproduced in figs. 2 and 3.

Figure 2. *Shear force response F_s (lower traces) to applied back and forth lateral motion Δx_0 (upper traces) for surfaces bearing polystyrene brushes (M = 141,000) immersed in toluene, at surface separation D = 146 – 139Å [10]*

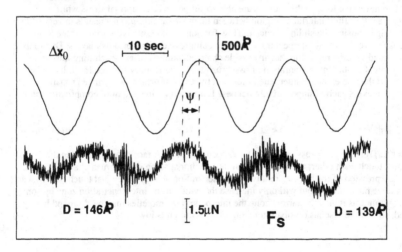

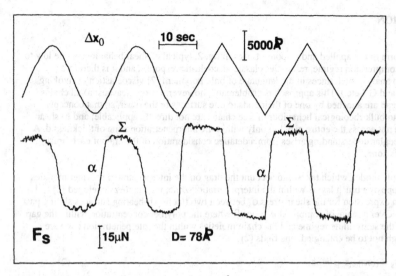

Figure 3. *As fig. 2, but for surface separation D = 78Å. Regions marked α denote that the surfaces are rigidly adhered and move in tandem, which those marked S denote the top surface is sliding freely past the lower one [10].*

Fig. 2, which is typical of shear behaviour in the PS-brush system [10] up to moderate compressions, β < ca. 10, shows a characteristic dissipative response, where the shear force lags the applied lateral motion by a phase angle ψ close to 90°. This is due to the polymer brushes dragging past each other: all chains are attached to one surface or the other, and the shear force is transmitted through the zone of interpenetration (thickness d) where moieties from opposing brushes are in overlap. The largest values of F_s occur close to the point of highest lateral velocity, while F_s goes through zero close to where the lateral applied motion changes direction (top of the sinusoidal curve, upper trace in fig. 2), indicating an almost purely viscous response.

The frictional response at much higher compressions (typical of β > ca. 10 for the PS-brush system), shown in fig. 3, is very different. Here on initial lateral motion of the top layer the frictional resistance to sliding exceeds F_s, so that the surfaces move in tandem, as the shear force F_s rises (regime α); at some point F_s exceeds the frictional force between the surfaces and they slide freely past each other (regime Σ). Such stick-slip behaviour is characteristic of solid-solid friction [11]: it suggests that the interacting and highly interpenetrated brush layers within the gap are now sliding not past each other at the midplane of the interpenetration zone (interfacial shear), but rather that one or both of the surface-attached polymer layers may be slipping across the mica substrate surface. In the following section we analyse this behaviour in more detail

DISCUSSION

The form of the applied and response traces in fig. 2, typical of shear behaviour in the low to moderate compression regime, resembles classic viscoelastic response, and it is therefore tempting to analyse such traces in the language of bulk polymer [12] viscoelasticity (invoking moduli G' and G'' etc.), This approach is problematic, however, for two reasons: a) all chains within the gap are attached by one of their ends to one surfaces or the other, so that concepts pertaining to bulk rheological behaviour of free chains are not directly applicable; and b) shear takes place not across the entire gap, but only within the interpenetration zone of thickness d. A better molecular understanding arises from a detailed consideration of the drag of each brush through this zone.

A simple model, which takes into account the drag on the interpenetrated polymer moieties from the opposing brush layers within the interpenetration zone, was earlier developed [2] and provides an expression for the shear stress σ_s between two flat brush-bearing surfaces sliding past each other. Over a range of compression ratios β where the polymer concentration within the gap remains in the semi-dilute regime and the chain moieties within the interpenetration zone are short enough not to be entangled, one finds [2]:

$$\sigma_s = (6\pi\eta_{eff}v_s\beta^{7/4})/s \tag{2}$$

Here η_{eff} is the effective viscosity responsible for the drag on the polymer moieties sliding through the interpenetrated region when the surfaces are sliding past each other at velocity v_s. Over a substantial range of compressions (typically $\beta <$ ca. 4 or 5), d remains small, the interpenetrated polymer moieties remain unentangled, and the magnitude of η_{eff} is not far above that of the pure solvent itself. Hence the low shear stress required for sliding, and the very low friction coefficients. At sufficiently high loads and thus compressions, the mean segmental concentration in the gap will exceed the semi-dilute regime and d becomes larger so that overlapping chain moieties from opposing brushes become entangled; indeed, beyond a sufficiently high compression, β^* say, the entire gap will be mutually interpenetrated by the chains, and d = D will decrease rather than increase at $\beta > \beta^*$. For these reasons one then expects that eq. (2) no longer holds at the highest compression ratios ($\beta > \beta^*$). If we make the reasonable assumption that on initial overlap d • s (the assumption underlying eq. (1)), then $\beta = \beta^*$ when d = D, yielding at once from eq. (1) that $\beta^* \cong (2L/s)^{3/4}$. For the system illustrated in figs. 2 and 3, a polystyrene brush of M = 141,000 in toluene, 2L = 1250Å, s = 85Å, giving β^* • 7 – 8. That is, for compressions of this brush system by a factor of order 7-fold or greater from overlap, the gap will be fully interpenetrated by the two opposing brushes. It is of interest that this is the compression ratio beyond which, for this polymer brush system, the shear force begins to grow measurably (within the resolution of our experiments). For this regime, which is of interest here, a different model to the one leading to eq. (2) is required.

Detailed theoretical models have been proposed for the shear of interacting polymer brushes in the melt [9,13]. Our system is rather different in that we start, at low compressions, with a highly solvated brush (ca. 97% solvent in the unperturbed brush), and both the extent of interpenetration and – particularly – the mean concentration and therefore the friction in the interpenetration zone increase with compression. We extend the model of ref. [2] to the

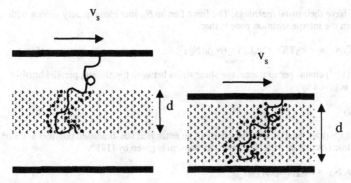

Figure 4. *Illustrating the entanglement of the brushes within the mutual interpenetration zone of width d (shaded), which may be partial (left) or may extend across the gap D (right) at sufficiently high compressions ($\beta > \beta^*$). For clarity only a single brush is shown; the brush can relax its configuration within this zone by retracting back through its entanglement 'tube', indicated by the locus of bold dots.*

case where both the mean polymer density in the gap and the extent of brush interpenetration are large enough that significant entanglement effects are present as the brushes slide through each other.

In fig. 4 we illustrate the entanglement of the brushes within the mutual interpenetration zone d. We need to evaluate the frictional drag on the chains from one surface as they slide through the network formed within the interpenetration zone. Since the chains are tethered at one end to one of the surfaces, then if they are entangled the only way in which they may relax is by arm retraction through a tube formed by topological constraints due to the surrounding chains. This mechanism is essentially identical to the relaxation mechanism of the arms of an entangled star-branched polymer. We may then proceed as follows to estimate the friction on the chain moieties in the interpenetration zone.

Let the time for relaxation by retraction of the chain moiety within the interpenetration zone d be $\tau_d(N_d)$, where N_d is the number of monomers on the moiety within this zone (clearly for compression ratios $\beta > \beta^*$, $d = D$, the surface separation, and $N_d = N$, the total number of monomers on a brush chain). At the relatively high concentrations in the gap we expect quasi-ideal chain configuration so that $d = N_d^{1/2}a$, where a is the monomer size. Over a time $\tau_d(N_d)$ the mean square displacement of the center-of-mass of the relaxing moiety will be $R^2(N_d) = N_d a^2$. This corresponds to a diffusion coefficient given by

$$D \bullet R^2(N_d)/\tau_d(N_d) \tag{3}$$

From the fluctuation-dissipation theorem [14] we then expect the friction coefficient $\zeta(N_d)$ (i.e. force per unit velocity) on the chain moiety N_d within the interpenetration zone to be

$$\zeta(N_d) = k_B T/D \tag{4}$$

where k_B and T have their usual meanings. The force f on an N_d-mer chain moiety sliding with velocity v_s within the interpenetration zone is then

$$f = v_s \zeta(N_d) = v_s k_B T/D \bullet v_s k_B T \tau_d(N_d)/(N_d a^2) \tag{5}$$

Since there are $(1/s^2)$ chains per unit area, the shear stress between the sliding, parallel brush-bearing surface is given by

$$\sigma_s = f/s^2 \tag{6}$$

To obtain $\tau_d(N_d)$ we recall that the viscosity of an entangled star-branched polymer solution, of polymer volume fraction ϕ, with N_d monomers per arm is given by [12]

$$\eta_{star}(\phi, N_d) = E_{star}(\phi)\tau_d(\phi, N_d) \tag{7}$$

where $E_{star}(\phi)$ and $\tau_d(\phi, N_d)$ are the corresponding plateau modulus of the solution and the relaxation time of an individual arm respectively. To map the star-polymer solution viscosity onto the frictional properties of the interpenetrated brushes, we may safely assume that at the higher surface compressions the mean segmental volume fraction ϕ within the gap is uniform, and given by $\phi = 2\Gamma/D\rho$ (where Γ is the brush adsorbance on each surface and ρ is the polymer density). We expect the modulus E_{star} in a concentrated polymer solution to vary as $E_{star} \bullet k_B T/N_e \xi^3$, where N_e and ξ are the corresponding entanglement degree of polymerisation and correlation length respectively [15]; in the melt $\xi \to a$, while in concentrated solutions of volume fraction ϕ around $0.2 - 0.5$, the regime of interest here, we may estimate $\xi \bullet a/\phi$. Substituting from from equations (5) and (7), we obtain an expression for the shear stress in terms of parameters that are known from the experiment or may be extracted from the literature on polymer rheology:

$$\sigma_s = f/s^2 \bullet (v_s a/s^2)(N_d/N_e)^{-1}(\phi)^{-3}\eta_{star}(\phi, N_d) \tag{8}$$

To convert this shear stress to predict the measured forces F_s in the SFB experiments, where the crossed cylinder configuration of the opposing surfaces is equivalent to a sphere, radius R, on a flat, we need to integrate over the area of interaction, as illustrated in fig. 5. From fig. 5

Figure 5. *Illustrating the geometry of contact for a sphere radius R on a flat used for evaluating the shear force (eq. (9)).*

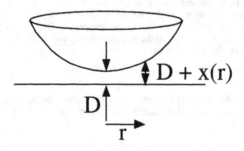

we have $r^2 = 2Rx(r)$ (for $R \gg D$, easily valid for the SFB experiments). Thus the circular area element at r, $2\pi r dr = 2\pi R dx$. When $(D + x)$ exceeds $2L$ the brushes no longer interact; integrating the stress over the interaction area gives the overall shear force

$$F_s = \int_0^{2L-D} 2\pi R \sigma_s dx \qquad (9)$$

where σ_s is a function of the surface separation $(D + x)$ across the interaction zone through eq. (8) via N_d and ϕ. Note that the upper limit of integration implies that the expression for σ_s from eq. (8) is being used also for low values of the compression ratio, for which the expression from eq (2) should be more applicable. This simplification is likely to make little practical difference however, since the magnitudes of the shear stress at the high compression ratios will dominate the integral. The viscosities of star-branched polystyrene solutions of concentrations ϕ corresponding to the different compressions β used in the experiments, and to the the different values of N_d, have been extracted from the literature [16,17], and are used to calculate values of F_s from eq. (9).

In fig. 6 we compare values of F_s from experimental traces such as shown in figs. 2 and 3 with those calculated for each set of experimental conditions using eq. (9). While there are several simplifying assumptions used in deriving the value of σ_s appearing in eqs. (8) and (9),

Figure 6. *Experimental values (open circles and broken curve to guide eye) of the maximum shear force F_s taken from traces such as figs. 2 and 3, for different surface separations and values of sliding velocity v_s. The solid points (one for each experimental point, joined by lines to guide eye) are calculated from eq. (9) using the corresponding values of D (yielding d) and v_s.*

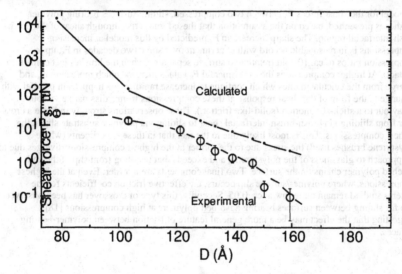

the dominant factor as D changes is the viscosity $\eta_{star}(\phi, N_d)$. This is because at the lowest D values (strongest compressions) the mean polymer segment volume fractions are of order $\phi \bullet$ 0.3 - 0.5, and the viscosity of polystyrene, which diverges as it approaches the glassy concentration at $\phi \bullet 0.8$, increases exponentially with $\phi \propto (1/D)$ [16,17]. Thus changes in $\eta_{star}(\phi, N_d)$ at changing D are huge compared with changes in the other parameters. We note that the experimental and calculated values track each other over some 2 decades variation in F_s, down to $D \bullet 120 \text{Å}$, which gives us some confidence that our essential idea – that the viscous drag dominating the shear force varies as the corresponding star-branched viscosity - is correct. At higher compressions, however, there is a clear and increasing divergence between the calculated and observed values, the latter increasing much more slowly than the former. As hinted earlier when remarking on the change in the form of the shear force traces at the highest compressions, this divergence is likely to be due to the transition from sliding via shear at the midplane between the interacting brushes, to sliding via slip of the polymer at the solid substrate. Elsewhere [10] we have considered the shear stresses associated with the frictional forces at the point (D = ca. 120Å for the system of fig. 6) where this crossover occurs, and have shown that they are a substantial fraction of the critical shear stress required to detach the zwitterionic end-groups anchoring the polystyrene chains to the mica surface. The reason therefore for the crossover from sliding via interfacial brush shear to sliding via polymer-substrate slip is that the latter becomes the mechanism of least resistance at sufficiently high compressions.

CONCLUSIONS

A model for the shear forces F_s between two compressed, sliding surfaces bearing polymer brushes is presentted, based on the assumption that these forces arise through shear of the brush-brush interfacial region. The sharp increase in F_s predicted by this model at increasing compressions is in reasonable accord with experiment over some two decades in F_s, up to compression ratios of ca. 10-fold (relative to surface separation when the brushes just come into contact). At higher compressions the experimental F_s values increase much more slowly, and diverge from the calculated ones which continue to increase rapidly: this happens in parallel with a change in the form of the shear response at these compressions from a dissipative viscous behaviour to a stick-slip motion (solid-like friction). These observations strongly indicate a cross-over from sliiding via brush-brush interfacial shear to sliding via polymer-substrate slip at the highest compressions. This is most likely due to the fact that in these experiments (where polystyrene brushes used) the stress due to the former at the highest compressions diverges due to to approach to glassiness of the polystyrene, and exceeds that resulting from slip of the end-attached polymer chains on the surface. Two final comments are in order: Even at the highest compressions, where polymer/surface slip occurs, the effective friction coefficients for the systems studied remain quite low at ca. 0.05. Secondly, this type of crossover has been observed also for sliding between surfaces bearing adsorbed polymers at high compressions [18], suggesting that the effect may be a more general feature of friction between polymer-bearing surfaces.

ACKNOWLEDGEMENTS

Support of this work by the Deutsches-Israel Program (DIP), the US-Israel Binational Science Foundation and the Minerva Foundation is acknowledged with thanks.

FOOTNOTES AND REFERENCES

* Correspondence to current address at Physical and Theoretical Chemistry Laboratory, Oxford University, Oxford OX1 3QZ , UK

1. J. Klein, *et al.*, *Nature* **370**, 634-636 (1994).
2. J. Klein, *Ann. Rev. Mater. Sci.* **26**, 581-612 (1996).
3. P. Schorr, S. M. Kilbey and M. Tirrell, *Polymer Preprints (Abstr. Amer. Chem. Soc)* **218: 278-POLY , Part 2 AUG**, U477 (1999).
4. S. Milner, T. A. Witten and M. Cates, *Macromolecules* **21**, 2610 (1988).
5. E. B. Zhulina and A. Semenov, *Poly. Sci USSR* **31**, 196 (1989).
6. S. Alexander, *J.Phys (Paris)* **38**, 983-987 (1977).
7. P. G. de Gennes, *Adv. Colloid Interface Sci.* **27**, 189-209 (1987).
8. C. M. Wijmans, E. B. Zhulina and G. J. Fleer, *Macromolecules* **27**, 3238-3248 (1994).
9. T. Witten, L. Leibler and P. Pincus, *Macromolecules* **23**, 824 - 829 (1990).
10. J. Klein, E. Kumacheva, D. Perahia and L. J. Fetters, *Acta Polymerica* **49**, 617-625 (1998).
11. D. Tabor, *Friction* (Doubleday, New York, 1973).
12. J. D. Ferry, *Viscoelastic properties of polymers, 3rd Edn.* (Wiley, New York, 1985).
13. J.-F. Joanny, *Langmuir* **8**, 989 - 995 (1992).
14. M. Doi and S. F. Edwards, *The Theory of Polymer Dynamics* (Oxford University Press, New York, 1986).
15. P. G. de Gennes, *Scaling Concepts in Polymer Physics* (Cornell Univ. Press, Ithaca, N.Y., 1979).
16. W. W. Graessley, R. L. Hazleton and L. R. Lindeman, *Trans. Soc. Rheology* **11**, 267 - 285 (1967).
17. L. A. Utracki and J. E. L. Roovers, *Macromolecules* **6**, 366 - 372 (1973).
18. U. Raviv, R. Tadmor and J. Klein, *J. Phys. Chem.* **Submitted**, (2000).

Mat. Res. Soc. Symp. Proc. Vol. 651 © 2001 Materials Research Society

Probing Confining Geometries with Molecular Diffusion: A Revisited Analysis of NMR-PGSE Experiments

Stéphane RODTS[1] and Pierre LEVITZ[2]

[1]Laboratoire Central des Ponts et Chaussées, Ministère de l'Equipement, Paris, FRANCE
[2]Centre de Recherches sur la Matière Divisée, CNRS, Orléans, FRANCE

ABSTRACT

We present an interpretation of the self diffusion propagator for a molecular fluid confined in a porous medium. Breaking with conventional routes of interpretation, we focus in reciprocal space on the time dependence of the propagator at a fixed wave vector q. New theoretical results are reported, as well as NMR measurements on a water-saturated packing of glass beads and on a stack of plastic platelets with rough surfaces. It is shown that at least three time regimes may be distinguished, characteristic of new ways in which the propagator is affected by the geometry of the system: a short-time exponential regime of almost unrestricted diffusion, a pseudo-exponential regime probing the transport process across the material at the length scale $\lambda = 2\pi/q$, and an algebraic regime at long times with exponent $-d/2$, where d is the dimensionality of connected parts of the pore space.

INTRODUCTION

The understanding of the relationship between the self diffusion of a fluid embedded in a porous material and the geometry of the pore network is an active field of research for many scientific communities. Indeed, this question is of fundamental interest in various practical areas such as molecular exchange in biological tissues, chromatography, heterogeneous catalysis, durability of building materials, waste storage... During the last two decades, magnetic nuclear relaxation has proved to be a powerful tool for studying diffusion in such systems thanks to the pulsed gradient spin echo technique [1] (NMR-PGSE). This technique gives direct measurements of the 3D-Fourier transform $G(\vec{q},t)$ of the self diffusion propagator $P(\vec{r},t)$, the probability density function that a molecule undergoes a displacement $\vec{r}$ from its origin in a time t.

Mitra & al. [2] suggested that in a fluid-saturated macroporous material, the propagator may be regarded as a probe of the geometry of the porous network. Conventional approaches consider features of the q-dependence of the propagator at a given diffusion time t [3]. By focusing on the curvature of the propagator at $q=0$, one usually defines a time-dependent diffusion coefficient $D(t)$ whose behavior at short and long times gives estimates of the specific surface area and the tortuosity of the material, respectively. In the high-q limit, one may find a q^{-4} asymptote characteristic of a smooth solid-pore interface [4]. Eventually, in the long time limit, the propagator tends to behave as a diffraction pattern for connected parts of the pore network, and exhibits correlation peaks in some model systems [2,5]. Unfortunately, these approaches consider only restricted parts of the available data, and the full interpretation of the propagator remains a challenging problem.

It was shown recently that the time dependence of the propagator at a fixed wave vector q may provide an alternative route of interpretation. A length-scale dependent diffusion coefficient $D(q)$ was introduced to characterize the propagator in the time interval where $G(\vec{q},t) \geq 0.1$. It proved [6] to give a relevant insight into the micro-macro transition for diffusion in the material, and especially [6-7] to be sensitive to the so-called dead-end pore effect [8].

In this paper, we address the more general question of the complete time dependence of the propagator at a fixed q. Theoretical developments are reported, as well as experiments on both water-saturated packing of glass beads and on a system of parallel plastic platelets with rough surfaces. It is shown that, for each value of q, at least three time regimes should be distinguished, each probing different aspects of diffusion in a confined geometry: a) a short time regime, independent of the geometry of the system, only characteristic of the fluid; b) a pseudo-exponential regime at intermediate times, probing the transport process across the material at the length scale $\lambda = 2\pi/q$, and relevant for a proper definition and interpretation of $D(q)$; c) an algebraic asymptotic regime at long times, sensitive to the dimensionality of the system.

THEORY AND EXPERIMENT

Self-diffusion propagator in the short time-limit

In the following, diffusion is assumed to occur in the isotropic molecular regime inside the pore space, and without adsorption. For any saturating fluid in a finite-size porous sample, the propagator is given by an exact expression in the Laplacian eigenstate formalism:

$$G(\vec{q},t) = \sum_{n=0}^{+\infty} \left|\tilde{\varphi}_n(\vec{q})\right|^2 e^{-D_b \lambda_n t} \tag{1}$$

The $\tilde{\varphi}_n(\vec{q})$ are the Fourier transforms of a basis set of real-valued eigenstates $\varphi_n(\vec{r})$ with eigenvalues λ_n. They are defined by the following definitions and orthonormalisation conditions:

$$\Delta \varphi_n(\vec{r}) = -\lambda_n \varphi_n(\vec{r}) \quad \vec{r} \in \Omega, \quad \vec{u}.\vec{\nabla}\varphi_n(\vec{r}) = 0, \quad \vec{r} \in \partial\Omega \tag{2}$$

$$\int_{\vec{r}\in\Omega} \varphi_n(\vec{r})\varphi_m(\vec{r})d\Omega = \delta_{n,m}, \quad \tilde{\varphi}_n(\vec{q}) = \frac{1}{\sqrt{\Omega}} \int_{\vec{r}\in\Omega} \varphi_n(\vec{r})e^{i\vec{q}.\vec{r}}d\Omega \tag{3}$$

Ω is the pore space, $\partial\Omega$ the interface, $\vec{u}$ a vector perpendicular to the interface, $\delta_{n,m}$ the Kronecker symbol, Δ and $\vec{\nabla}$ the Laplacian and the gradient operators, and λ_n are real-positive numbers. In this context, the following Parseval relationships relative to the decomposition of the function $\exp(i\vec{q}.\vec{r})$ on the eigenstates of the Von-Neumann-Laplace operator hold:

$$\sum_{n=0}^{+\infty} \left|\tilde{\varphi}_n(\vec{q})\right|^2 = \frac{1}{\Omega} \int_{\vec{r}\in\Omega}\left|e^{i\vec{q}.\vec{r}}\right|^2 d\Omega = 1 \quad \text{and} \quad \sum_{n=0}^{+\infty} \left|\tilde{\varphi}_n(\vec{q})\right|^2 \lambda_n = \frac{1}{\Omega} \int_{\vec{r}\in\Omega}\left\|\vec{\nabla}(e^{i\vec{q}.\vec{r}})\right\|^2 d\Omega = q^2 \tag{4}$$

As a consequence, one is able to take the time derivative of the infinite sum in (1). At fixed q, one thus *proves* the following short time expansion characteristic of unrestricted diffusion:

$$G(\vec{q},t) = 1 - D_b\left(\sum_{n=0}^{+\infty} \left|\tilde{\varphi}_n(\vec{q})\right|^2 \lambda_n\right)t + o(t) = e^{-D_b q^2 t} + o(t) \tag{5}$$

We emphasize that no information is obtained about the length of the time interval where this expansion is valid, and that this time interval may also depend on q.

The q-dependent diffusion coefficient

The q-dependent diffusion coefficient $D(q)$ was introduced previously [6,7] as an approximated estimate of the mean rate of decrease of the propagator at fixed q. It is numerically derived using a least-squares fit to $G(\vec{q},t)$ (Eq. 6), where $D(q)$ and $A(q)$ are fitted parameters:

$$G(\vec{q},t) \approx A(\vec{q})exp(-D(\vec{q})q^2 t) \quad \text{(fixed } q, G \text{ considered as a function of } t) \tag{6}$$

It was observed experimentally in fluid saturated packings of glass beads as well as for water diffusing in oil-in-water emulsions that the data could indeed be fitted by such a single exponential function. It was shown [6] that $D(q)$ characterizes diffusion at the tunable length scale $\lambda=2\pi/q$, and gives some relevant information about the transition from microscale to macroscale for diffusive transport. However, in those works, all the measurements used corresponded to the time interval where the propagator was greater than approximately 0.1.

Self diffusion propagator in the long time limit

In the case of diffusion in a homogeneous isotropic connected material ("3D" system), Mitra & al. [2] proposed an ansatz for the propagator at any time in direct space. We notice that their approach simplifies in the long time limit to the following expression:

$$P(\vec{r},t) \approx \phi^{-2}(4\pi D_b t / \tau)^{-3/2} exp\left[-|\vec{r}|^2 \tau / (4D_b t)\right]\phi_2(\vec{r}) \tag{7}$$

τ is the tortuosity of the material and ϕ its porosity and $\phi_2(\vec{r}) =<\chi(\vec{r}')\chi(\vec{r}+\vec{r}')>_{\vec{r}'}$ is the isotropic two point correlation function of the pore space. In the average brackets, $\vec{r}'$ runs over the whole space, $\chi(\vec{r})$ is 1 in the pore space and 0 elsewhere. In reciprocal space (7) becomes:

$$G(\vec{q},t) \approx \exp(-D_b|\vec{q}|^2 t/\tau) + C \int_{\vec{q}'} \exp(-D_b|\vec{q}-\vec{q}'|^2 t/\tau) S(\vec{q}')d^3q' \tag{8}$$

The C constant depends only on how the Fourier transforms are normalized. $S(\vec{q})$ is the Fourier transform of $\phi_2(\vec{r}) - \phi^2$. In disordered systems, $S(\vec{q})$ is a continuous function, so that the gaussian in the convolution product of the rhs of Eq. (8) may be approximated at long times by a 3D-Dirac function weighted by $[\tau\pi/(D_b t)]^{3/2}$. One thus obtains at long times:

$$G(\vec{q},t) \approx C\,S(\vec{q})\,(\tau\pi/D_b)^{3/2}\,t^{-3/2} \quad \text{(fixed } \vec{q}, \text{ long times)} \tag{9}$$

When the system is isotropic on the macroscale in two directions x and y, and disconnected in the z direction ("2D" system), we generalize the idea of Mitra & al. and propose the following ansatz as an approximate form of the propagator:

$$P(\vec{r},t) \approx C_1(4\pi D_b t / \tau)^{-1} exp\left[-(r_x^2+r_y^2)\tau / (4D_b t)\right]\phi'_2(\vec{r}) \tag{10}$$

C_1 is a normalization constant independent of time, and $\phi'_2(\vec{r})$ stands now for the correlation function of the connected parts of the material, that is $\phi'_2(\vec{r}) =< \omega(\vec{r}',\vec{r}+\vec{r}')>_{\vec{r}'}$; $\omega(\vec{r}_1,\vec{r}_2)=1$ when $\vec{r}_1$ and $\vec{r}_2$ lie in the same connected part of the porous network and is zero otherwise; τ is the tortuosity of the system in the x and y directions. After Fourier transformation of (10), we calculate in a similar way to the 3D case that for any wave vector out of the z axis in reciprocal space one must have in the long time limit the algebraic behavior:

$$G(\vec{q},t) \approx C\,S'(\vec{q})\,t^{-1} \quad \text{(fixed } \vec{q}, \vec{q} \text{ out of z axis, long times)} \quad (11)$$

$S'(\vec{q})$ is the regular part of the Fourier transform of $\phi'_2(\vec{r})$, and C is a new $\vec{q}$-independent constant only depending on C_1 and on the normalization convention for the Fourier transform.

We demonstrate in a similar way that in (a) a disordered anisotropic system disconnected in two directions y and z, and with a finite tortuosity in the third direction x ("1D" system), and (b) in a totally disconnected system ("0D" system), one should have respectively:

$$G(\vec{q},t) \approx C\,S'(\vec{q})\,t^{-1/2} \quad \text{(fixed } \vec{q}, \vec{q} \text{ out of yz plane, long times)} \quad (12)$$
$$G(\vec{q},t) \approx C\,S'(\vec{q}) \quad \text{(any } \vec{q}, \text{ long times)} \quad (13)$$

$S'(\vec{q})$ still represents the Fourier transform of the correlation function of connected parts of the pore space. To our knowledge, we are the first to point out the systematic algebraic time dependence of the propagator with a power $-d/2$, where d is the 'connected' dimensionality of the system. We emphasize that, for this asymptotic regime to occur, some "disorder" is required in the system in order not to get trivial zero values for $S'(\vec{q})$.

NMR-PGSE experiments

Proton NMR-PGSE measurements were performed in water saturated systems at 20°C on a Bruker DSX100 spectrometer operating at 100MHz in a Bruker Micro-05 probe head generating field gradients up to 180 G/cm. Two different systems were used in the experiments. The first was a random close packing of glass beads (Ø40µm±10µm), which will be considered here as a prototype of isotropic 3D-connected material. Its tortuosity was deduced from $D(q)$ values at low q values [6] and was found to be close to 1.55.

The second case was a packing of plastic platelets with rough surfaces, considered hereafter as 2D-connected. A 0.1mm thick plastic sheet was first hand-scratched on both sides with a 150µm grained sandpaper. Atomic force microscopy of the surface estimated the depth of the scratches to about 10µm. 6x10mm platelets were then cut, 60 of which were stacked together and imbibed with water. The average void space between neighboring platelets was estimated to be about 15µm by comparing the global thickness of the sample in dry and wet states. The stack was oriented in the NMR probe so as to have q-vectors almost parallel to the platelets. The scratching was necessary to create the required disorder to make the algebraic asymptotic behavior appear. Tortuosity in the direction parallel to the platelets was measured to be 1.25.

PGSE measurements for both systems are displayed in fig 1. We ensured that the size of each sample in the direction of wave vectors (~10mm) was big enough to prevent any finite size artifact to occur at very low propagator intensities. We also emphasize that with the PGSE technique, the time t is only known with an unknown constant offset of the order of 1ms. This offset explains why some curves in figures 1b and 1d do not exactly converge towards 1 at $t=0$.

DISCUSSION

The long-time data in figure 1 are first compared with the theoretical predictions (9) and (11). In the packing of glass beads (fig. 1a), one clearly observes that the continuous curves

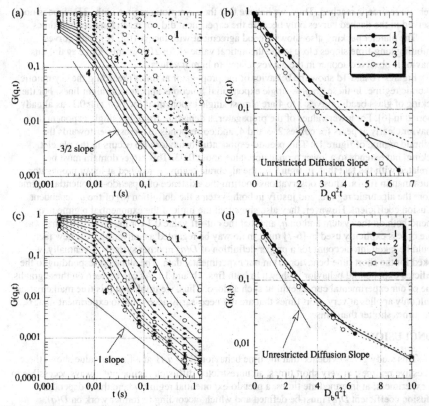

Figure 1. *NMR-PGSE measurements of water self diffusion propagator versus time at fixed wave vector. (a) and (b): saturated random close packing of Ø40μm glass beads. (c) and (d): saturated stack of platelets with rough surfaces -wave vector parallel to the platelets-. In figures (a) and (c), each of the 16 curves is associated to a wave vector. 16 equally spaced wave vector values are involved, ranging between 0 and 375000 rad/m. Curves with opened circles are numbered 1, 2, 3, 4 and correspond to values of 50000, 125000, 250000 and 375000 rad/m respectively. Figure (b) and (d) show for each system the curves 1 to 4 versus the reduced time $D_b q^2 t$. This reduced time was chosen so as to visualize the behavior of every curves at 'intermediate times' on the same graph.*

–which correspond here to high q values- behave at long times almost as a set of parallel straight lines with slopes close to the theoretical value -3/2. Dashed curves however -which correspond to lower q values- don't exhibit so clearly such behavior. This may be explained on the basis of Eq. (8). Indeed, at fixed q, the propagator at long times may roughly be written as:

$$G(\vec{q},t) \approx \exp(-D_b|\vec{q}|^2 t/\tau) + C (\tau\pi/D_b)^{3/2} S(\vec{q}) t^{-3/2} \qquad (14)$$

For a given order of magnitude for $S(\vec{q})$, the smaller the wave vector, the slower the decreasing rate of the exponential term and, as a consequence, the longer the time at which the algebraic regime is reached. Thus, we believe that the asymptotic regime with -3/2 slope is not observed in the dashed curves only because the experiment did not reach long enough times.

The platelet packing also shows a good agreement with theory (fig. 1c), with curves exhibiting long-time slopes close to the theoretical value -1 for 2D systems. Once again, this behavior is observed sooner in the curves related to high q values.

Figures 1b and 1d show the behavior of the propagator at times preceding the asymptotic algebraic regime. In these graphs, single exponential functions appear as straight lines. For the packing of glass beads, curves 1 to 4 are almost single exponential for $G(\vec{q},t) \geq 0.1$, as already reported in [6]. For lower values of the propagator, a departure from this single exponential behavior clearly appears for curves 2, 3 and 4, and corresponds to a crossover towards the algebraic regime of figure 1a. This pseudo-exponential behavior also occurs in the platelet packing, but extends to a lower propagator value about 0.03. This larger domain must be correlated with the fact that, in figure 1c, the algebraic regime is reached at lower propagator values than in fig 1a. Those observations confirm the existence of a pseudo-exponential regime before the algebraic regime, and justify in both systems the definition (6) of the q-dependent diffusion coefficient. However, they also show that the limit of the exponential regime may depend both on the system and on q, and that the cut-off value 0.1 for the propagator that was more or less explicitly used in [6-7] must in no way be regarded as universal. This new fact should be taken into account for a proper definition of $D(q)$ in other systems. Eventually, we looked for the short time behavior (5) in our experimental data. The slope corresponding to the predicted exponential behavior is shown in both figs. 1b and 1d. As can be seen on those graphs, none of our experimental curves exhibit such a slope. Thus, we think that this true mathematical result only applies at very short times that are not necessarily accessible to experiment -that is, here, times shorter than 1ms.

CONCLUSION

This study gives evidence that the time behavior of $G(\vec{q},t)$ at a fixed q value shows three successive regimes: at very short times, an unrestricted diffusion regime not always accessible to the experiment; at intermediate times, a pseudo-exponential regime where the q-dependent diffusion coefficient $D(q)$ must be defined and which, according to former work on $D(q)$, contains fundamental information on the transport properties of the material at different length scales; and a long-time algebraic regime, originating from diffraction effects from the pore structure, and probing the dimensionality of connected parts of the system. The limits between these regimes depend both on the system and on q, and should be carefully taken into account for a proper definition of $D(q)$. We hope this work will help further development of theoretical tools aiming at interpreting NMR-PGSE experiments.

REFERENCES

1. Stejskal E.O., Tanner J.E., *J. Chem. Phys.*, **42**, 288-292 (1965).
2. Mitra P.P., Sen P.N., Schwartz L.M., Le Doussal P., *Phys. Rev. Lett.*, **68**, n°24, 3555, (1992).
3. Mitra P.P., *Physica A*, **241**, 122-127 (1997).
4. Sen P.N., Hurlimann M.D., De Swiet T.M., *Phys. Rev. B*, **51**, n°1, 601-604 (1995).
5. Callaghan P.T., Coy A., Mac Gowan D., Packer K., Zelaya F., *Nature*, **351**, 467-469 (1991).
6. Rodts S., Levitz P., accepted in *Magnetic Resonance Imaging*
7. Hills B.P., Manoj P., Destruel C., *Magn. Res. Imag.*, **18**, 319-333 (2000)
8. Lever D.A., Bradbury M.H., Hemingway S.J., *Journal of Hydrology*, **80**, 45-76 (1985).

Mat. Res. Soc. Symp. Proc. Vol. 651 © 2001 Materials Research Society

Proton Spin-Relaxation Induced by Localized Spin-Dynamical Coupling in Proteins And in Other Imperfectly Packed Solids

J.-P. Korb[1], A. Van-Quynh[2], R. G. Bryant[2]

[1]Laboratoire de Physique de la Matière Condensée, CNRS UMR 7643, Ecole Polytechnique

91128 Palaiseau, France

[2]Chemistry Department, University of Virginia, Charlottesville, VA 22901 U.S.A.

ABSTRACT

The magnetic field dependence of ^1H spin lattice relaxation rates in noncrystalline macromolecular solids including engineering polymers, proteins, and biological tissues is described by a power law, $1/T_1 = A\omega_0^{-b}$, where ω_0 is the Larmor frequency, A and b are constants. We show that the magnetic field dependence of the proton $1/T_1$ may be quantitatively related to structural fluctuations along the backbone that modulate proton-proton dipolar couplings. The parameters A and b are related to the dipolar coupling strength, the energy for the highest vibrational frequency in the polymer backbone, and the fractal dimensionality of the proton spatial distribution.

INTRODUCTION

The proton nuclear spin-lattice relaxation rate $1/T_1$ carries information about both the structure and dynamics of solids. For a number of noncrystalline macromolecular solids, including proteins and engineering polymers, the magnetic field dependence of the proton spin-lattice relaxation rate $1/T_1$ is describable by a power law,

$$\frac{1}{T_1(\omega_0)} = A\omega_0^{-b}, \tag{1}$$

where ω_0 is the Larmor frequency, and A and b are constants [1-4]. The value of b is often near 0.5 which suggests that the spin-lattice relaxation is dominated either by a spin-diffusion process [5, 6a] or by a quasi mono-dimensional process [7]. However, in polymers like polycarbonate, b is about 0.78 [8] and in a rotationally immobilized protein, b is reported between 0.65 and 0.85 [1, 3, 9].

Kimmich and coworkers have suggested that backbone fluctuations dominate the proton spin relaxation process [2]. Here we describe a quantitative model for the proton nuclear spin-lattice relaxation based on the coupling between a partially connected spin network and localized structural fluctuations. The model shows that the magnetic field dependence of the spin-lattice relaxation rate $1/T_1$ in the low field range is determined by low frequency structural fluctuations along the backbone, the strength of the dipolar couplings among the protons, and the fractal dimensionality of the proton spatial distribution. The correspondence between $1/T_1$ and the dynamical characteristics of the protein demonstrates that the field dependence of $1/T_1$ provides a scaled report of the low frequency structural motions of the macromolecule. The approach is general and the extension to solid polymers is presented as an example.

EXPERIMENT

Figure 1 shows the magnetic field dependence of protein proton $1/T_1$ for dry solid lysozyme measured using a field cycling instrument of the Redfield design [10]. This instrument permits measurements of $1/T_1$ from 0.01 to 30 MHz with nearly constant sensitivity by monitoring the approach to equilibrium following a rapid field switch. For the instrument employed, 10 ms is required for the magnetic field to switch and settle; within this constraint, the magnetization decay is exponential within experimental error.

Measurements at high fields have demonstrated that spin diffusion is efficient among protons in solid proteins and the spin-lattice relaxation is dominated by coupling to the rapidly rotating methyl group protons [11,12]. However, the methyl groups relax the protein protons so slowly that no spin temperature gradient is established. This situation is opposite to that of slow spin diffusion to paramagnetic relaxation sinks in ionic solids [5]. Although the relaxation by methyl motions is not disputed, the correlation time for methyl group rotation is too short (~10^{-10}s) to contribute to the low-field relaxation power-law dispersion shown in Figure 1; another relaxation process is required.

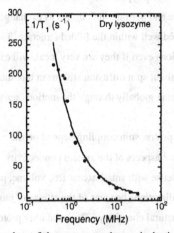

Fig. 1 Semilogarithmic plots of the proton nuclear spin-lattice relaxation rates as a function of the magnetic field strength expressed as the proton Larmor frequency for protons in dry lysozyme (●). The quadrupole peaks have been suppressed. The solid line is obtained from an *ab initio* calculation from Eq. (6 and 7) as discussed in the text.

The folded tertiary structure of the protein is maintained by a large number of London dispersion forces supplemented by hydrogen bonds and salt bridges (Fig. 2). Because there are so many atoms (N), the number of vibrational modes (3N) is very large and must extend to low frequencies. The slow chemical exchange of amide hydrogen, which occurs in the range from ms to several weeks, is striking evidence for low frequency fluctuations in a folded protein [13]. The detailed nature of the slowest motions is not known but must include contributions from backbone reorientation and side-chain motions. Structural fluctuations in the protein are generally thought to be relatively rare. That is, the protein has a structure that remains more or less fixed. However, the energies that hold the folded protein in this structure are weak. The total free energy of protein folding is the sum of a great many small contributions. The free energy minimum for the folded structure is shallow and may have a considerable number of local minima of nearly the same energy. In consequence, it is easy to imagine that the protein structure fluctuates among these nearly isoenergetic states and also samples higher energy states less frequently. To approach a model for the nuclear spin relaxation, we assume that relatively rare defect structures occur where some portion of the structure becomes transiently more mobile [2, 9, 14]. We suppose that the localized backbone defects transiently modulate the dipole-dipole interactions among the neighboring solid protein protons, thus causing global nuclear spin relaxation. This leads to the idea of a mobile defect hypothesis where such a structural defect diffuses around within the protein structure and permits such things as exchange of amide hydrogen atoms even though the site is buried well within the folded structure. This structural defect with its associated molecular motions even if they are very local, will cause nuclear spin relaxation. The above arguments about spin diffusion still mean that there is only one proton spin temperature, and spins will relax globally through the motions sensed in the defect region of the protein.

The defect movement and the proton-spin coupling depend on the contacts between atoms, or in some sense on different aspects of the system connectivity. Although the protein system may be relatively dense with small excess free volume, proton-proton contacts cannot be uniform in all directions because of the effects of molecular packing and the geometrical constraints of structural chemistry. Thus, the inter proton dipole-dipole coupling may form a network, the connective details of which depend on the conformation of the macromolecule as well as the way molecules pack in the solid matrix. Based on the X-ray structure [15] of the protein lysozyme, we have verified that the protons form a tridimensional percolation network. To determine the fractal dimensionality of the proton network in lysozyme, we computed the number of protons, N_H, within the mean radius of the protein and located at a given distance, R, from a reference proton in the protein as shown in Fig. 3. We find $N_H \propto R^{d_f}$ with a fractal dimension of mass $d_f=2.50\pm0.05$. This value is characteristic of a 3-dimensional percolation network at the percolation threshold [16]. The same fractal dimension is found for ribonuclease A and alpha chymotrypsin. To confirm the percolation character of the system studied, we used the same method to find that the fractal dimension of mass (Fig. 4), for α-carbon of the protein backbone is $d_f=1.76\pm0.1$ for lysozyme (Figs. 4 and 5). This value of d_f is very close to 1.74 ± 0.04, which is characteristic of a backbone skeleton close to the percolation threshold [17]. The uncertainty in the determination of d_f is probably due to the tertiary structure of the protein which creates some oscillations in the data in Fig. 5.

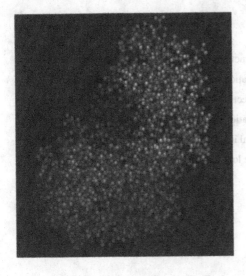

Fig.2 Structure of the lysozyme based on the X-ray structure of the crystalline hydrated protein. Hydrogen atoms are shown in white (15).

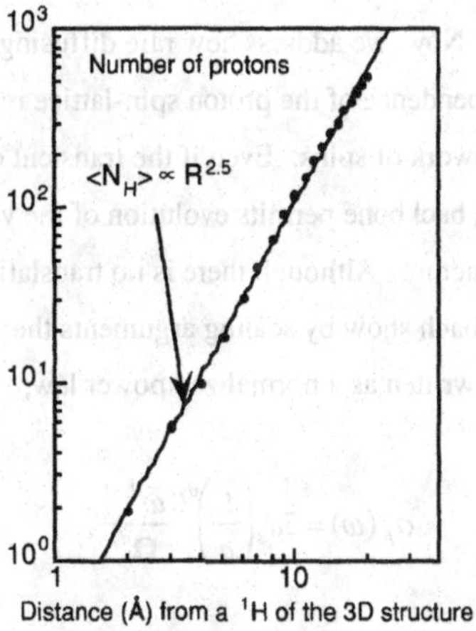

Fig. 3 Logarithmic plot of the fractal proton distribution N_H in the lysozyme based on the X-ray structure of the crystalline hydrated protein given in Fig. 2.

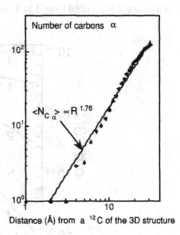

Fig.4 Structure of the backbone of the lysozyme based on the X-ray structure of the crystalline hydrated protein (15).

Fig. 5 Logarithmic plot of the fractal alpha carbon distribution $N_{C\alpha}$ in the lysozyme based on the X-ray structure of the backbone of Fig. 4.

THEORY

Now, we address how rare diffusing structural defects can influence the magnetic field
dependence of the proton spin-lattice relaxation rate in a semi-solid disordered fractal
network of spins. Even if the transient defects are localized, the structural connectivity of
the backbone permits evolution of the vibrational disturbances throughout the whole protein
structure. Although there is no translational invariance in the fractal regime, Alexander and
Orbach show by scaling arguments that the low-frequency density of vibrational states can
be written as a normalized power law,

$$\sigma_{fr}(\omega) = 3d_s \left(\frac{L}{a}\right)^{d_f} \frac{\omega^{d_s-1}}{\Omega_{fr}^{d_s}}, \tag{2}$$

where the spectral dimension, $d_s \sim 4/3$, replaces the Euclidean dimensionality, d [18], a and L
are the interproton distance and the maximal size of the system, respectively. Integration of
Eq. (2), from 0 to Ω_{fr}, the upper limit of the frequency of the fractal regime, gives the total
number of modes $3N=3(L/a)^{df}$ available in the system. Figure 6 demonstrates that the
normalized density of states at 10 kHz changes by 18 orders of magnitude when the
dimensionality is reduced from 3 to 1.

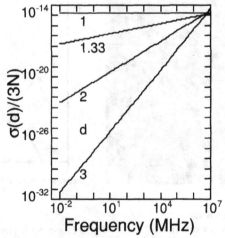

Fig. 6 Logarithmic plot of the normalized vibrational density of states as a function of the
frequency for Euclidean dimensions of 3, 2, d_s =1.33 and 1.

There are two extremes, extended and localized, that describe the spatial extent of such vibrational states. The former is the condition for homogeneous structure, where a plane wave is a useful approximation leading to phonons. The latter is relevant for strong disorder, where the wave function decays exponentially (or more rapidly) with distance thus leading to elementary localized excitations termed "fractons" [18, 19]. Here we follow partly the localized description and consider the exponential form of the localized normalized wave function:

$$\phi_\alpha(\omega_\alpha, r) = \Gamma(d_f)^{-1/2} \ell_\alpha(\omega_\alpha)^{-d_f/2} \exp\left(-\frac{r}{2\ell_\alpha(\omega_\alpha)}\right) \tag{3}$$

where Γ is the gamma function, $\ell_\alpha(\omega_\alpha)$ represents the spatial extent (or localization length) of the vibrational state of frequency ω_α, and r is the radial distance from the center of the excited domain. One can deduce from Eq. (2) and from scaling arguments about boundary conditions that $\ell_\alpha(\omega_\alpha)$ depends on the frequency ω_α through the following anomalous dispersion relation [20]:

$$\ell_\alpha(\omega_\alpha) \propto a\left(\frac{\omega_\alpha}{\Omega_{fr}}\right)^{-d_s/d_f}. \tag{4}$$

Substitution of Eq. (4) into Eq. (3) shows the strong frequency dependence of the wave function. A strong frequency dependence is thus expected in the relaxation of nuclear spins by localized vibrations. A previous theoretical treatment of direct and Raman processes has been proposed to calculate the relaxation rate $1/T_1$ of localized electronic states by fractons and localized phonons [21]. In particular, the Raman relaxation process was useful for the interpretation of the strong temperature dependence of the electron spin relaxation rate $1/T_1$ of low-spin Fe^{3+} hemoproteins [22]. However, for nuclear spin systems, both direct and Raman relaxation processes seem to be quite inefficient because of the very low Debye-type density of vibrational states, $\sigma(\omega) \propto \omega^{d-1}$ present at low frequency for systems having an Euclidean dimensionality d=3 [6b]. Effectively, we find a negligible value, $1/T_1 \sim 10^{-17} s^{-1}$, for the proton spin-lattice relaxation rate caused by a direct process at 10 MHz and room temperature for a dipolar interaction $H_{dip} = \hbar\omega_{dip} = \hbar^2\gamma^2/r_m^3$ between two protons at the van der Waals contact distance, $r_m = 2.2A$. However, Figure 6 demonstrates that the normalized density of states at 10 kHz changes by 18 orders of magnitude when the dimensionality is reduced from 3 to 1.

We consider the direct process for the calculation of the nuclear spin-lattice relaxation rate of a proton due to a localized structural defect centered a distance r away and described by Eqs. (2-4). A Raman process supposes the presence of two localized excitations coupled to a nuclear spin and the integration of the whole spectrum eliminates the frequency dependence [6b]. We use the well-known "golden rule" to calculate the proton spin-lattice relaxation rate per unit of volume $W(\omega_0, r)$ induced by a fluctuating proton dipolar interaction for a direct process. We take the localized backbone vibration to be proportional to the gradient of the wave function given in Eqs. (3 and 4) and use the usual Bose factors for quantized vibrations. For the direct process, only the vibrational states matching the Larmor frequency induce relaxation; we take $\omega_\alpha = \omega_0$ and find, at high temperature ($\hbar\omega_0 \ll kT$), the rate per unit of volume $W(\omega_0, r)$

$$W(\omega_0, r) = \pi\omega_{dip}^2 \left(\frac{\omega_0}{\Omega_{fr}}\right)^{2d_s/d_f} |\phi_0(\omega_0, r)|^2 \frac{k_B T}{Ma^2\omega_0^2} \sigma_{fr}(\omega_0), \tag{5}$$

where $\omega_{dip} = \gamma^2\hbar/r_m^3$ and M is the total mass of the N atoms. Because of fast spin diffusion, the protons relax globally and exponentially in time, we thus average Eq. (5) over all r, using the normalization condition: $\int_0^\infty dr \; r^{d_f-1}\phi_0^2(r) = 1$. Substitution of Eq. (2) into Eq. (5) yields,

$$\frac{1}{T_1} = \frac{3}{2}\pi d_s \frac{E_v^0}{\hbar}\left(\frac{\hbar\omega_{dip}}{E_v^0}\right)^2 \left(\frac{k_B T}{E_v^0}\right)\left(\frac{\hbar\omega_0}{2E_v^0}\right)^{-\left(3-\frac{2d_s}{d_f}-d_s\right)}, \tag{6}$$

where $E_v^0 = \hbar\Omega_{fr}/2$ is the maximum vibrational energy. By comparison with Eq. (1) the exponent b is

$$b = 3 - \frac{2d_s}{d_f} - d_s. \tag{7}$$

One notes that Equations (6) and (7) achieve the usual form of $1/T_1$ found for a one spin-phonon relaxation process in a system of Euclidean dimensionality d [6b], provided that we

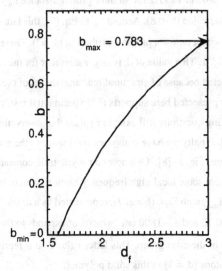

Fig.7 Variation of the exponent b (Eq.7) of the power law (Eq. 6) as a function of the fractal dimension d_f.

replace d_s and d_f by d and Ω_{fr} by Ω, where Ω is the phonon frequency with $a\Omega=v$, and v is the sound velocity in the material. The fractal model leading to Eq. 6 preserves the temperature dependence of the Euclidean result, but the frequency dependence is drastically changed.

DISCUSSION

Equation (7) connects the experimental value of b with the structural and dynamical characteristics of the material studied. For instance, b achieves a maximal value of 0.783 for $d_f = 3$ which corresponds to a uniform distribution of protons (Fig. 7). This maximal value is similar to those reported for dry proteins [9]. It makes sense that b achieves a minimal value of 0 in the extreme limit of self-avoiding chains for which $d_f = 5/3$. In that case there are no couplings between chains and the process at the origin of the dispersion of $1/T_1$ disappears. In order to explain the data presented in Fig. 1 for the dry lysozyme, we perform an *ab initio* calculation of proton $1/T_1$ with Eqs. (6 and 7). We approximate E_v^0 in Eq. (6) as the frequency of the amide(II) vibrational mode at 1560 cm^{-1} [23] then take $\omega_{dip} = \hbar\gamma^2 / r_m^3$ giving a dipolar frequency around 11 kHz for an inter-proton distance $r_m = 2.2$Å, we fix $d_s = 1.33$ and b to its maximal value (b~0.8). According to Eq. (7), this latter value of b is expected for a dry protein with a uniform proton distribution ($d_f = 3$). These substitutions produce the solid line in Fig. 1. This value of d_f is larger than that for the hydrated crystalline protein as expected because of structural rearrangements of the protein caused by hydration [24]. The theory presented here supports the experimental results obtained at room temperature. Interesting questions still exist to explain the observation that b decreases below 200 K [9]. Finally we have compared our theory to the proton dispersion curve of solid polycarbonate (Fig. 8) [8]. One notes an asymptotic constant value at high frequency, which comes from some local high frequency motion, which we subtracted from the data. The best fit of $1/T_{1,corr}$ with Eqs. (6 and 7), represented as a dashed line in Fig. 8, has been obtained with b=0.78 and $E_v^0 \sim 1800$ cm^{-1} which corresponds to the highest frequency elongated mode of the ester group. This latter value of b is representative of a uniform distribution of protons ($d_f = 3$) in this solid polymer.

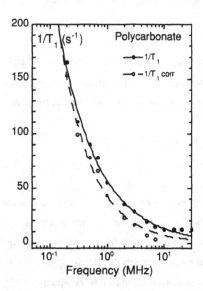

Fig. 8 Semilogarithmic plots of the proton nuclear spin-lattice relaxation rates as a function of the magnetic field strength expressed as the proton Larmor frequency for protons in solid

polycarbonate (●) at 291 K (8). The experimental data (○) have been corrected by subtracting the asymptotic constant value at high fields.

Because the fractal system is more dense at small length scale, the elastic force constant k_α should also be scale dependent. For instance, for a subsystem of size $\ell_\alpha(\omega_\alpha)$, the force constant is defined as $k_\alpha = m_\alpha \omega_\alpha^2$ where m_α is the mass of the subsystem and according to Eq. (4) $k_\alpha(\omega_\alpha) \propto \ell^{d_f} \omega_\alpha^2 \propto \omega_\alpha^{2-d_s}$. Because we detect only the vibration occurring at the Larmor frequency ω_0, one has $k_0(\omega_0) \propto \omega_0^{2-d_s}$. A direct comparison with Eq. (6) leads to the general scaling relation $\dfrac{1}{T_1} \propto \dfrac{k_B T}{k_0(\omega_0)^{b/(2-d_s)}}$ which simplifies to $1/T_1 \propto k_B T/k_0(\omega_0)$ for a percolation network at the percolation threshold where $d_s/d_f \sim 0.5$. The correspondence between $1/T_1$ and the dynamical characteristics of the protein thus demonstrates that the field dependence of $1/T_1$ provides a scaled report of the low frequency structural motions of the macromolecule.

For systems hydrated with H_2O, proton spin-lattice relaxation is complicated by cross-relaxation between the liquid and solid protons. For proteins and tissues over most of the frequency range studied, the relaxation coupling causes the magnetic field dependence of the solid spins to dominate that of the liquid protons [3]. Thus, these results provide a quantitative foundation for interpretation of field effects in magnetic imaging of tissues.

CONCLUSION

Incorporation of fractal dimensionality and the concept of localized dynamical disturbances into the theory for relaxation by spin-"phonon" direct processes accounts quantitatively for the field dependence of the proton spin-lattice relaxation rates $1/T_1$ for proteins and other chain molecules. These concepts provide a general framework for interpretation of spin relaxation dispersion measurements on macromolecular solids. The correspondence between $1/T_1$ and the dynamical characteristics of the protein demonstrates that the field dependence of $1/T_1$ provides a scaled report of the low frequency structural motions of the macromolecule.

ACKNOWLEDGMENTS : We thank, J.-F. Gouyet, M. Guéron, P. Levitz, D. Petit, B. Sapoval (École Polytechnique), and A. Redfield (Brandeis University) for stimulating discussions. This work was supported in part by the University of Virginia and the National Institutes of Health, under Grants GM34541 and GM39309.

REFERENCES

[1]   R. Kimmich, F. Winter, W. Nusser, K.H. Spohn, J. Magn. Reson. **68**, 263 (1986).
[2]   R. Kimmich, F. Winter, Progr. Colloid Polym. Sci. **71**, 66 (1985).
[3]   C. C. Lester and R.G. Bryant, Magn. Reson. Med. **21**, 117 (1991); *ibid*. **22**, 143 (1991).
[4]   D. Zhou and R.G. Bryant, Magn. Reson. Med. **32**, 725 (1994).
[5]   W.E. Blumberg, Phys. Rev. **119**, 79 (1960)
[6]   A. Abragam, *The Principles of Magnetic Resonance*, Oxford, The Clarendon Press, 1961, (a)   Ch IX, p 386; (b) Ch IX, Section IV.C.
[7]   J.-P. Korb and J.-F. Gouyet, Phys. Rev B **38**, 493 (1988); J.-P. Korb, M. Whaley, Hodges, Th. Gobron, R.G. Bryant, Phys. Rev. E **60**, 3097 (1999).
[8]   R.G. Bryant, J. Schaeffer, unpublished.
[9]   W. Nusser and R. Kimmich, J. Phys. Chem. **94**, 5637 (1990).
[10] A. G. Redfield, W. Fite, H. E. Bleich, Rev. Sci. Instrum. **39**, 710-15 (1968).
[11] R.G. Bryant, Annu. Rev. Biophys. Biomol. Struct. **25**, 29 (1996).
[12] (a) E.R. Andrew, D.N. Bone, D.J. Bryant, E.M. Cashell, R. Gaspar, Q.A. Ming, Pure Appl. Chem. **54**, 584 (1982) ; (b) R. Gaspar, E.R. Andrew, D.J. Bryant and E.M. Cashell, Chem. Phys. Lett. **86**, 327 (1982).
[13] S. W. Englander, N.R. Kallenbach, Q. Rev. Biophys. **16**, 521-655 (1984).
[14] R. B. Gregory, R. Lumry, Biopolymers **24**, 301-326 (1985).
[15] Protein Data Bank (Web site of the Rutgers University).
[16] D. Stauffer, *Introduction to percolation theory* (Taylor and Francis, London, 1985).
[17] S. Havlin, and D. Ben Avraham; Adv. Phys. **36**, 695 (1987).
[18] S. Alexander and R. Orbach, J. Phys. Lett **43**, L625 (1982).
[19] E. Courtens, ,J. Pelous, J. Phalippou, R. Vacher, Th. Woignier, Phys. Rev. Lett. **58**, 128 (1987); E. Courtens, R. Vacher, J. Pelous, Th. Woignier, Europhys. Lett. **6**, 245 (1988).
[20] S. Alexander, Phys. Rev. B **40**, 7953 (1989).
[21] S. Alexander, O. Entin-Wholman, R. Orbach, J. Phys. Lett. **46**, L549 (1985); ibid., L555   (1985).
[22] H. J. Stapleton, J.-P. Allen, C.P. Flynn, D.G. Stinson, and S. R. Kurtz, Phys. Rev. Lett. **45**, 1456 (1980); J.-P. Allen, J.T. Colvin, D.G. Stinson, C.P. Flynn, H.J. Stapleton, Biophys. J. **38**, 299 (1982).
[23] T. Mizyazawa, T. Shimanouchi, S. Mizushima, J. Chem. Phys. **29**, 611, (1958).
[24] G. S. Kachalova, V. N. Morozov, T. Y. Morozova, E. T. Myachin, A. A. Vagin, B. V. Strokopytov, Y. V. Nekrasov, FEBS Lett. **284**, 91(1991).

Mat. Res. Soc. Symp. Proc. Vol. 651 © 2001 Materials Research Society

Reaction Kinetics Effects on Reactive Wetting

Marta González [1] and Mariela Araujo [2,1]
[1] Escuela de Física, Universidad Central de Venezuela
[2] Reservoir Department, PDVSA Intevep, Caracas, Venezuela

ABSTRACT

Chemical reactions at the interface between two liquids and a solid can lead to very complex situations. For example, after the reaction the substrate may become less wettable allowing the formation of running droplets. We propose a model where droplets of reactive fluids of a given radius, are injected into a porous system. In their advance, the droplets react chemically with the solid surface, making it less wettable. The velocity of the droplet after the reaction is calculated for different reaction kinetics including adsorption and solute solubility, for situations where gravity effects are negligible. In all cases, a transition from a Brownian motion regime to a ballistic displacement is observed.

INTRODUCTION

The spreading of liquid droplets on solid surfaces is fundamental to many technological processes. There are many factors that may change the wetting condition of a physical system. An interesting situation occurs when the liquid changes the properties of the solid as it advances, this may happen in various ways such as: a) by the effect of a surfactant present in the liquid, which is able to modify the wetting of the surface, b) by a reactive solute in the liquid, which again alters wettability, or c) if the surface is porous, and it is progressively invaded by the liquid. Here, we consider case b) which has been reported as specially interesting [1,2] since the reaction speed can be monitored by changes in the concentration of reactive species.

When a drop of liquid is free in space, it is drawn into a spherical shape by tensile forces on its surface. However, when it is brought in contact with a flat solid surface, the final shape taken by the drop depends on the relative magnitudes of the molecular forces that exist within the liquid and between the liquid and the solid. A measure of this effect is reflected on the value of the contact angle, which the liquid subtends with the solid. The equilibrium condition among the forces is expressed by the Young equation which establish

$$\gamma_{SG} = \gamma_{SL} + \gamma_{GL} \cos\theta_e \qquad (1)$$

where γ_{LG} is the liquid/air interfacial tension, γ_{SL} (γ_{GL}) are the interfacial energies of the solid/liquid (air/liquid) phases. The equilibrium angle θ_e is the result of the force balance of the interfacial tensions. If the surface is smooth and chemically homogeneous we have one value of θ_e along the entire surface (negligible hysteresis), and if we impose an angle θ different from θ_e we have a "non-compensated Young force" [3].

$$F = \gamma_{SG} - \gamma_{SL} - \gamma_{GL} \cos\theta \qquad (2)$$

Most models of reactive wetting consider a simple first order chemical reaction for the deposition of the solute on the solid surface. In this work, we extend the study to more general reaction kinetics associated to adsorption of the reactive liquid on the solid surface, and the case of equilibrium due to solubility of the solute.

MODEL

The basic model is De Gennes's model [1] where a droplet of a reactive fluid is trapped inside a capillary or a Hele-Shaw cell. Essentially the system is a column of reactive fluid, with length L much larger than its diameter d, moving with certain velocity V. As shown in figure 1, the equilibrium contact angle at the advancing edge B, has the unreacted angle θ_{Be}, smaller than the equilibrium angle θ_{Ae} at the receding edge A, because the solid surface has become less wettable.

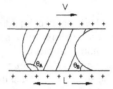

Figure 1. Schematic diagram of a reactive droplet inside a capillary.

Previous works used a first order chemical reaction, where a reactive solute anchors on the surface and generates a surface concentration to model the reaction kinetics. Assuming a small effect on the wettability and the Young equation in combination with the hydrodynamic equations for the porous system, a relation between the drop velocity and the parameters of the system may be found. In this work, we combine the reaction kinetics of different chemical reactions, specifically adsorption and solute solubility, with the hydrodynamic equations to derive expressions for the velocity of the fluid as a function of the system parameters.

REACTION KINETICS

For the kinetics, we assume that initially the system has a given concentration of a reactive solute c, which anchors on the surface and generates a normalized surface concentration ϕ. As a function of time, this solute is consumed. For a first order kinetics we can write

$$\frac{d\phi}{dt} = k_{IA}(c - \phi) \tag{3}$$

where k_{IA} is the reaction constant. The solution to this equation as a function of time is,

$$\phi = \left(1 - e^{-tk_{IA}}\right). \tag{4}$$

It can also be written in terms of the relaxation time given by $\tau = (k_{IA})^{-1}$. We consider the effect of two different reaction mechanisms: a) adsorption and b) solute solubility. In general we may anticipate an inverse relationship between the extent of adsorption of a solute and its solubility – the stronger the solute-solvent interaction the smaller the extent of adsorption. The situation is not simple since the competition for the surface between solute and solvent coupled with solute/solvent interactions are major factors in the determination of the direction the adsorption will take.

Adsorption

The simplest situation occurs when the solute occupies single sites at the surface. Simple calculations allow determining the surface coverage as a function of time

$$\theta = bc_2^L / \left(1 + bc_2^L\right) \tag{5}$$

here, c_2^L is the solute concentration in the bulk, and parameter b is related to the ratio of the reaction constant and the solvent activity.

Some often, adsorption is accompanied by dissociation of solute particles after they have reached the solid surface. In this case, at a given time, the surface coverage is smaller than for the case of single adsorption

$$\theta = \sqrt{bc_2^L} / \left(1 + \sqrt{bc_2^L}\right) \tag{6}$$

There are also more complex situations when the surface is heterogeneous, the b parameter is not constant but varies with the coverage θ leading to $\theta = \alpha\left(c_2^L\right)^{1/n}$, with α and n constants for the system [4]. For these adsorption processes the pre-treatment of the surface before exposure to solution is very important.

Solute Solubility

Consider the situation in which c_2^L changes in time are due to the solubility reaction which may be of first, second or even of nth-order depending on the reaction involved in the solubility process. We have calculated how the evolution of the surface coverage changes in time for several cases by using the integration method. The case of first order reaction was presented in equation (4) and may correspond to various stoichiometries such as $A \rightarrow Z$, $A \rightarrow 2Z$, $A + B \rightarrow Z$, and $2A + B \rightarrow Z$ [5].

a) *Second-Order Reactions*

There are two possibilities: $2A \rightarrow Z$ and $A + B \rightarrow 2Z$. If the initial concentration of A and B are the same, the rate of reaction may be expressed as

$$\frac{d\phi}{dt} = k_{IIA}(c - \phi)^2 \text{ which integrates to } \phi = \frac{k_{IIA}tc}{(1 + k_{IIA}tc)} \tag{7}$$

the variation of ϕ with t is no longer exponential. However, if the initial concentrations are c_A and c_B and these are not the same, the rate after an amount ϕ of A has reacted is

$$\frac{d\phi}{dt} = k_{IIA}(c_A - \phi)(c_B - \phi) \tag{8}$$

which integrates to

$$\phi = \frac{c_B c_A \left(1 - e^{(c_A - c_B)k_{IIA}t}\right)}{c_B - c_A e^{(c_A - c_B)k_{IIA}t}} \tag{9}$$

b) *nth-Order Reactions*

For an nth-order reaction which involves a single reactant of concentration c, or reactants of equal concentration and with stoichiometry $A + B + \ldots \rightarrow Z$, the integration gives

$$\phi = c\left(1 - \exp\left(\frac{1}{1-n}\ln\left(1 + \left(\frac{n-1}{c^{1-n}}\right)k_{nA}t\right)\right)\right) \tag{10}$$

A comparison of the adsorption curves for the two cases is shown in figure 2.

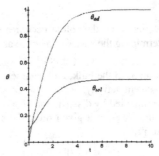

Figure 2. Comparison of surface coverage $\theta(t)$ when solubility equilibrium governs the kinetics of the process ($\theta_{Sol}(t)$) and when adsorption governs the process ($\theta_{ad}(t)$).

VELOCITY RESULTS

We couple the reaction mechanism with the hydrodynamic behavior to determine the droplet velocity as a function of the different parameters of the system. This is done combining the fraction of sites occupied by solute as a function of time ($\theta(t)$) with the time variation of the solute concentration, a value that depends on the rate constants of the two reactions: the solubility of solute and the kinetics of adsorption. Depending on the type of reactive present (solute, solvent and surface), one mechanism may prevalence over the other. The first case is likely to occur in the presence of strong acids in solution, whereas the second case with weak acids or with slightly soluble solutes.

For the case of first order reaction, we can get the droplet velocity using the following procedure:
- Assume that the effect of ϕ on wettability is weak, and that it can be linearized, thus

$$\gamma_{SL}\mid_A - \gamma_{SL}\mid_B = \gamma_1 \phi \tag{11}$$

where γ_1 is a positive coefficient determined experimentally, which measures the increase in γ_{SL} achieved for a complete reaction.
- Substitute into Young's equation

$$\gamma_{GL}(\cos\theta_{Be} - \cos\theta_{Ae}) = \gamma_1 \phi \tag{12}$$

- The Laplace pressure difference between both ends is [3]

$$P_A - P_B = \frac{4\gamma_{GL}}{d}(\cos\theta_{Be} - \cos\theta_{Ae}) \tag{13}$$

- From the solution of the hydrodynamic equations for the capillary, and using the fact that the velocity profile inside the capillary is parabolic, it can be shown that

$$\frac{P_A - P_B}{L} = \frac{32\mu V}{d^2}. \tag{14}$$

The combination of equations (13) and (14) gives a self-consistent condition on V

$$\frac{32V}{V^*}\frac{L}{d} = \left(1 - e^{-k_{IA}\frac{L}{V}}\right) \tag{15}$$

where $V^* = \gamma_1 / \mu$ is a characteristic velocity in the range of *10 m/s* for simple fluids. For De Gennes model two regimes are found:

(i) if $\dfrac{L}{dV^*} \gg \dfrac{1}{k_{IA}L}$, the exponential in (15) is negligible, and we expect

$$V = V^* \frac{d}{32L} \tag{16}$$

(ii) if $\dfrac{L}{dV^*} \ll \dfrac{1}{k_{IA}L}$, we can linearized the left-hand size of Eq. (15), and reach

$$V = \sqrt{\left(\frac{dV^* k_{IA}}{32}\right)} \tag{17}$$

independent of L. Typically $d = 1mm$, $\tau = 100\ s$, $V^* = 10\ m/s$, therefore $V \sim 0.2\ cm/s$ and $(dV^*\tau)^{1/2} \sim 1m \rightarrow L^2 / dV^*\tau \sim 10^{-4}$ for $L = 1\ cm$.

Case I: Equilibrium due to Kinetics of Adsorption

If the adsorption mechanism governs the variation of solute the velocity of reaction is given by $v = k_{ad}\theta$ where $v \equiv -\dfrac{1}{\xi}\dfrac{dc_2^L}{dt}$ with ξ as the stoichiometric coefficient of the solute and c_2^L the fraction of moles of solute in the bulk solution. For single adsorption the resultant equation for $c_2^L(t)$ with the initial condition $c_2^L(0) = c_{20}$ is given in terms of the special *LambertW* function as

$$c_2^L(t) = \frac{LambertW\left(bc_{20}\exp(bc_{20} - k_{ad}t)\right)}{b} \tag{18}$$

A series expansion to first order of the *LambertW* $(x(t))$ function gives

$$\frac{32V}{V^*}\frac{L}{d} = \frac{a_1^L c_{20}}{n^S}\left(1 - \exp(bc_{20} - k_{ad}\frac{L}{V})\right) \tag{19}$$

Thus, the corresponding limit values for: (a) $\dfrac{L}{V^*d} \gg \dfrac{1}{k_{ad}L}$ when the exponential term is negligible and leads to

$$V = \frac{a_1^L c_{20}}{n^S}\frac{V^* d}{32L} \tag{20}$$

Whereas for $\dfrac{L}{V^*d} \ll \dfrac{1}{k_{ad}L}$ we have that $V = \sqrt{\dfrac{a_1^L c_{20}}{n^S}\dfrac{k_{ad}V^* d}{32}}$ (21)

Case II: Equilibrium due to Solubility of solute

We proceed similarly to the adsorption case to get as limit velocity expressions:

(a) if $\dfrac{L}{V^*d} \gg \dfrac{1}{k_{IIS}L}$ $\rightarrow$ $\dfrac{32V}{V^*}\dfrac{L}{d} = \left(\dfrac{bc_{oR}c_{oS}}{c_{oR} + bc_{oR}c_{oS}}\right)$ so that,

$$V = V^* \left(\frac{bc_{oR}c_{oS}}{c_{oR} + bc_{oR}c_{oS}} \right) \frac{d}{L} \tag{22}$$

(b) if $\dfrac{L}{V^*d} \ll \dfrac{1}{k_{IIS}L}$ we expand in series the exponential and we have

$$V = \sqrt{\frac{2bc_{oR}c_{oS}(c_{oR} + c_{oS})}{bc_{oS}(c_{oR} + c_{oS}) + 3c_{oR} - c_{oS}} \frac{k_{IIS}dV^*}{32}} \tag{23}$$

independent of L. In equations (22) and (23) c_{oR} and c_{oS} are the initial reactive solute and solvent concentration, and k_{IIS} is the reaction solubility constant.

It is interesting to note that in both cases we find that V ~ d for low velocities whereas V ~ $d^{1/2}$ for high velocities, thus there is a transition from a ballistic to a brownian type of behavior. We performed experiments with two different types of silanes (diathoxydimethylsilane and 3-aminopropyltriethoxysilane), one with a lower solute solubility at room temperature, and observed that the droplet velocity is smaller for the less soluble solute (3-aminopropyltriethoxysilane). Other experimental observations are reported in references 6 and 7.

CONCLUSIONS

- Reaction kinetics (for adsorption and solute solubility) can modify the asymptotic values of the velocities of a running droplet in a porous structure.
- The velocity dependence with the characteristic size of the porous system (here the diameter of the capillary or the width of the Hele-Shaw cell) is not modified by the particular type of reaction kinetics. The transition from ballistic to brownian type of behavior is observed.
- In the running droplet regime, the velocity is independent of the droplet size.

Acknowledgment

We thank PDVSA Intevep for permission to publish this paper.

References

1. P.G. de Gennes, Physica A **249**, 196 (1998).
2. Davies J. and Rideal in: *Interfacial Physics*, 2nd ed., (Academic Press, 1963). Chapter 6.
3. J. Bear, The dynamics of fluids in porous media, Dover Publications (1972).
4. Jaycock and Parfitt, *Chemistry of Interfaces*, John Wiley and Sons (1981).
5. Keith J et. al, *Chemical Kinetics*, ed. Third Edition. Harper NY pp 18-32 (1989)
6. Bain C., Burnett-Hall G., Montgomerie R., Nature **372**, 414(1994) C. Redon, F. Brochard and F. Rondelez, Phys. Rev. Lett. **66**, 715 (1991).
7. Domingues F., Ondarcuhu T., Phys. Rev. Lett. **75**, 972 (1995).

Mat. Res. Soc. Symp. Proc. Vol. 651 © 2001 Materials Research Society

Friction and the Continuum Limit – Where is the Boundary?

Yingxi Zhu and Steve Granick
Department of Materials Science and Engineering
University of Illinois, Urbana, IL 61801 USA

ABSTRACT

The no-slip boundary condition, believed to describe macroscopic flow of low-viscosity fluids, overestimates hydrodynamic forces starting at lengths corresponding to hundreds of molecular dimensions when water or tetradecane is placed between smooth nonwetting surfaces whose spacing varies dynamically. When hydrodynamic pressures exceed 0.1-1 atmospheres (this occurs at spacings that depend on the rate of spacing change), flow becomes easier than expected. Therefore solid-liquid surface interactions influence not just molecularly-thin confined liquids but also flow at larger length scales. This points the way to strategies for energy-saving during fluid transport and may be relevant to filtration, colloidal dynamics, and microfluidic devices, and shows a hitherto-unappreciated dependence of slip on velocity.

INTRODUCTION

The assumption that flowing single-component viscous liquids come to rest within 1-2 molecular dimensions of a solid boundary, so that fluid molecules at the boundary move on average with the same tangential velocity as that boundary, lies at the heart of our understanding of the flow of simple low-viscosity fluids and comprises a bedrock for much sophisticated calculation [1,2]. Its validity has been debated for over 250 years [3] but justified on practical grounds by the fact that its phenomenological predictions appear to agree with experiments, even for flow through thin capillaries [4]. This "stick" contrasts with "slip" characteristic of highly viscous polymers when the shear stress is high [5], gas flowing past solids [6], superfluid helium [7], and moving contact lines of liquid droplets on solids [8]. Recently much interest has been given to sheared films of molecularly-thin simple liquids, where slip is also believed to occur, but these interesting anomalies, which are relevant primarily to friction, disappear when films are thicker than 5-10 molecular dimensions [9]. Any breakdown of the stick boundary condition at larger length scales would have potential implications in areas such as groundwater transport, filtration, colloidal dynamics, and microfluidic devices, and might form the basis of a strategy for saving energy in the transport of fluids such as oil and gasoline. For these practical reasons as well on scientific grounds, failure of the no-slip flow boundary condition for low-viscosity fluids would be of considerable interest.

Here we put to direct test recent theoretical predictions [10,11] and molecular dynamics simulations [14] that predict its breakdown when low-viscosity fluids fail to wet a smooth solid boundary. We confirm the generality of a study that found slip of hexadecane based on fluorescence studies [12], of another recent study that found toluene to slip on mica that had been modified with adsorbed C_{60} [13], as well as of controversial earlier literature [14]. The physics appears to be somewhat different than has been supposed, however. There is a long tradition of believing that slip velocity would be proportional to shear stress [10,11]. We find instead that these effects emerge only when a critical wall pressure is exceeded and that they grow nonlinearly when the wall pressure exceeds this level.

EXPERIMENTAL DETAILS

The modified surface forces apparatus used for these dynamic experiments has been described in detail elsewhere [16]. The geometry is cylindrically-shaped surfaces, with radius of curvature $R \approx 2$ cm, that are oriented at right angles to one another. To form surfaces whose interaction with fluids are neutral or weakly repulsive, atomically-smooth and step-free mica was coated with a methyl-terminated monolayer of condensed octadecyltriethoxysilane (OTE) [17]. The amplitude and frequency of oscillatory modulation of film spacing were controlled independently, thereby making it possible to vary the mean velocity of translation over a wide range without concomitant large change of the film thickness.

RESULTS

Solids of mean radius of curvature R, at spacing D, encounter a hydrodynamic force F_H when they approach one another (or retreat from one another) dynamically in a liquid medium. This force F_H is proportional to rate at which spacing changes, dD/dt (t denotes time), and inversely proportional to D. High-order solutions of the Navier-Stokes equations essentially confirm this expression, known as the Reynolds Equation [1,2]:

$$F_H \propto 1/D(dD/dt) \qquad . \qquad (1)$$

The constant of proportionality for the case of the stick boundary condition contains geometrical factors specified in the figure caption. A sinusoidal oscillatory drive generates an oscillatory hydrodynamic force whose peak we will denote as $F_{H,peak}$.

We observed quantitative agreement with Eq. (1) at the largest spacings, followed by strong deviations. This is illustrated first in Fig. 1 for the case of deionized water between OTE surfaces. The hydrodynamic force ($F_{H,peak}$) needed to cause drainage of water under conditions specified in the figure caption is plotted against film thickness. The hydrodynamic forces at spacing $D < 600$ Å are systematically less than predicted. To show this more clearly, the Inset shows a linearization of Eq. (1). A quantity proportional to $1/F_{H,peak}$ is plotted against D: whereas the data obtained at the largest spacings extrapolate linearly to the origin in classical fashion, they deviate subsequently and show decided curvature. Thus the same system both obeyed and failed to obey Eq. (1), depending on the surface spacing. When one considers that the characteristic dimension of a water molecule is ≈ 2-3Å, it is apparent that the breakdown of Eq. (1) appeared at extraordinarily large separations. More detailed study, in which the rate of spacing change was varied, showed that the amount of discrepancy depended more directly on the stress at the surface, as will be quantified below.

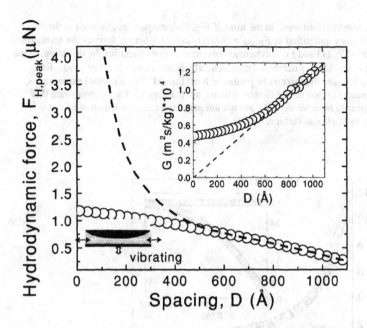

Figure 1. Hydrodynamic force $F_{H,peak}$ (circles) is plotted as a function of film thickness D for deionized water between two crossed cylinders coated with self-assembled methyl-terminated monolayers of condensed octadecyltriethoxysilane (OTE). A schematic diagram of the experiment is shown in the bottom left. The data are compared to the hydrodynamic force expected from the equation (dashed line), $F_{H,peak} = \dfrac{6\pi R^2 \eta}{D} \cdot \dfrac{dD}{dt}$, where R is the mean radius of curvature of the two cylindrical surfaces, D is the closest *spacing, and η is the viscosity of the liquid in between. The peak velocity of vibration was $v_{peak} = d \cdot \omega$ where d is vibration amplitude and ω the radian frequency of vibration, 6.3 rad-s^{-1}. In the Inset panel, the damping function $G = \dfrac{6\pi R^2 v_{peak}}{F_{H,peak}} = \dfrac{D}{\eta}$ is plotted against D. The reciprocal of the slope at large film thickness gives the known viscosity of water, $\eta = 0.83\pm0.06$ cP at 25º C. Deviations from Eq. (1) were characterized by a parameter f, ratio of measured F_H to the value expected from Eq. (1). The f parameters are 1.0, 0.99, 0.877, 0.628, and 0.435 at D= 500, 300, 200, 100, and 50 Å, respectively.

The generality of these findings, confirmed by using a nonpolar fluid, tetradecane ($C_{14}H_{30}$), is shown in Fig. 2. For tetradecane between wetting surfaces of freshly-cleaved mica, the data obeyed Eq. (1) at every spacing. This behavior was known from prior experiments by others [18-20] and its observation here lends credibility to the present experiments. The new results concern nonwetting surfaces. In the Inset of Fig. 2, discrepancies observed in the nonwetting situation are quantified as $F_{H,peak} = fK(1/D)(dD/dt)$, where K denotes the geometrical factors referred to in Eq. (1) and f is a dimensionless number that quantifies the deviation. One sees that, starting at D≈800 Å, f decreased monotonically below unity. In other words, flow became increasingly easier than would be predicted from Eq. (1). These control experiments, showing similar results in two very different systems, are reassuring. They demonstrate that these findings appear to be general: they are not unique to the complex situation of water interacting with a hydrophobic surface.

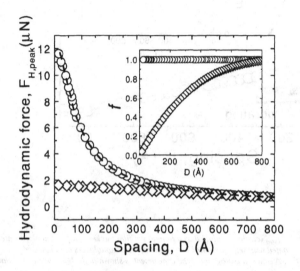

Figure 2. *Control experiments concerning tetradecane, $C_{14}H_{30}$, between mica surfaces (circles) and between methyl-terminated OTE surfaces (diamonds). The vibration frequency was 250 rad-s^{-1}. The result f=1 concerning mica surfaces signifies quantitative agreement with the classical expectation, Eq. (1). In the Inset panel, the f parameter is plotted against film thickness.*

Further control experiments showed that discrepancies from Eq. (1) were larger, the more rapid the rate at which surface spacing was varied.

CONCLUSIONS

Whereas these data clearly point to perturbation of the velocity field near the point of contact with a nonwetting surface, several scenarios of microscopic interpretation are possible. It has been proposed that a thin gas gap separates water from a hydrophobic surface [15,21]; we are not aware of direct evidence supporting this but a low-viscosity boundary layer would be consistent with the data. Alternatively, there is a tradition in fluid dynamics to infer the "slip length", the fictive distance inside the solid at which the stick flow boundary condition would hold. Straightforward mathematical manipulation shows that deviations from Eq. (1) can be

quantified from the equation $f = 2 \cdot \dfrac{D}{6b} \cdot \left[\left(1 + \dfrac{D}{6b} \right) \ln \left(1 + \dfrac{6b}{D} \right) - 1 \right]$, where b is the slip length [10].

The implied slip length was not constant, however. For example, at D=100 Å, b = 2, 62, and 83 Å at peak velocities v_{peak}=50, 160, and 650 Å-sec^{-1}, respectively. These inferences of variable slip length contrast with the common theoretical assumption that the slip length of low-viscosity fluids is a constant number. The phenomenology is reminiscent of the flow of polymers, whose flow also obeys Eq. (1) up to a critical stress but deviates when the applied stress exceeds a critical value [5].

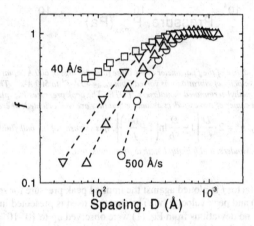

Figure. 3. Influence of peak velocity on deviations from Eq. (1), considered for the case of deionized water between methyl-terminated OTE surfaces. The f parameters are plotted on log-log scales against film thickness. The peak velocity of vibration was 40 Å-s^{-1} (squares), 125 Å-s^{-1} (down triangles), 315 Å-s^{-1} (up triangles), and 500 Å-s^{-1} (circles). The vibration frequency was 6.3 rad-s^{-1}.

The hydrodynamic force between two surfaces may also be interpreted as the integrated stress on the surfaces [20]. Having already fit this data with an empirical slip length (it is worth emphasizing that the slip length is a fitting parameter to which we do not assign physical meaning), known models permit one to calculate the implied peak pressure on the coincident center of the two cylinders [10].

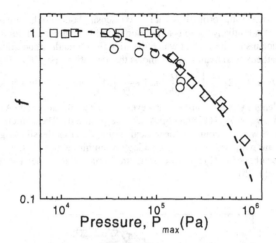

Figure 4. *Pressure dependence of the f parameter at the three film spacings of 600 Å (squares), 300 Å (circles) and 100 Å (diamonds) as the velocity of vibration was varied from v_{peak}=40 Å-s^{-1} to 500 Å-s^{-1}. The data concern deionized water between methyl-terminated monolayers of OTE. The peak pressures P_{max} plotted here refer to the position of the coincident center of two cylinders assuming that they are well oriented, where*

$$P_{max} = \frac{3\eta R v_{peak}}{D^2} \cdot p^*, \quad p^* = 2 \cdot \frac{D}{6b}\left[1 - \frac{D}{6b}\ln\left(1 + \frac{6b}{D}\right)\right], \quad \eta \text{ is viscosity of the bulk fluid, } R \text{ is mean radius of}$$

curvature of the crossed cylinders, and b is slip length defined in the text.

In Fig. 4, the factor f is plotted against the implied peak pressure for several different values of spacing (D) and peak velocity (v). The equation used is presented in the caption of Fig. 4. One observes that no deviations from Eq. (1) were observed up to 10^4-10^5 Pa, which amounts to 0.1-1 atmosphere, but that f decreased thereafter in nonlinear fashion.

It is tantalizing that these data, obtained at different rates of surface spacing change, superpose within experimental uncertainty when considered this way. It has often been assumed that stress-induced slip is a consequence of peculiarities associated with the molecular makeup of polymeric fluids [5] and many system-specific models have been proposed. The present experiments support the view that this effect is more general. Further experiments will be desirable to determine quantitatively how the critical pressure for onset of slip depends on molecular details of wall-fluid interactions.

OUTLOOK

Looking to the future, other experiments (not shown) suggest the following four noteworthy observations:

- We find that the perfection of the OTE monolayer has a large influence on the magnitudes of the f-factor used here to quantify amount of slip. Smoother monolayers have shown magnitudes of f as small as 10^{-4}.
- As has been considered theoretically [23,24], if patchiness (chemical as well as topograpical) were introduced deliberately, the presence of surface patchiness would introduce another length scale into this problem.
- The system of water against OTE is complicated by the presence of hydrophobic attraction. But we find similar behavior when hydrocarbons such as tetradecane are placed between OTE surfaces – cases where the surface forces are zero except at molecularly-thin spacings.
- It is still unclear whether the controlling parameter is hydrodynamic pressure (as supposed by the analysis in Figure 4), or alternatively shear rate, which may be quantified by the ratio of peak velocity to film thickness. In preliminary experiments we find that the data obtained over a wide range of peak velocity and film thickness scale well according to this latter variable.

ACKNOWLEDGEMENTS.

We thank Olga Vinogradova for helpful comments. This work was supported in part by the Ford Motor Co., the National Science Foundation (Tribology Program) and the U.S. Department of Energy, Division of Materials Science through the Frederick Seitz Materials Research Laboratory at the University of Illinois at Urbana-Champaign.

REFERENCES

1. J. Happel and H. Brenner, *Low Reynolds Number Hydrodynamics*, Klumer, Netherlands (1983).
2. S. Kim and S. Karrila, *Microhydrodynamics*, Butterworth-Heinemann, Newton, MA (1991).
3. For a historical review, see S. Goldstein, *Modern Developments in Fluid Dynamics*, Oxford, Clarendon Press (1938), Vol. II , p. 677-680.
4. T. G. Knudstrup, I. A. Bitsanis, G. B. Westermann-Clark, Langmuir 11, 893 (1995).
5. L. Léger and E. Raphaël, Adv. Polym. Sci. 138, 185 (1999).
6. D. A. Noever, J. Colloid Interface Sci. 147, 186 (1991).
7. S. M. Tholen and J. M. Parpia, Phys. Rev. Lett. 67, 334 (1991).
8. C. Huh and L. E. Scriven, J. Colloid Interface Sci. 35, 85 (1971).
9. B. Bhushan, J. N. Israelachvili, and U. Landman, Nature 374, 607 (1995).
10. O. I. Vinogradova, Langmuir 11, 2213 (1995).
11. O. I. Vinogradova, Int. J. Miner. Process. 56, 31 (1999).
12. R. Pit, H. Hervet, and L. Léger, Phys. Rev. Lett. 85, 980 (2000).

13. S. E. Campbell, G. Luengo, V. I. Srdanov, F. Wudl, and J. N. Israelachvili, Nature **382**, 520 (1996).
14. J.-L. Barrat and L. Bocquet, Phys. Rev. Lett. **82**, 4671 (1999).
15. E. Ruckenstein and P. Rajora, J. Colloid Interface Sci. **96**, 488 (1983).
16. A. Dhinojwala and S. Granick, Macromolecules **30**, 1079 (1997).
17. J. S. Peanasky, H. M. Schneider, S. Granick, and C. R. Kessel, Langmuir **11**, 953 (1995).
18. J. N. Israelachvili, J. Colloid Interface Sci. **110**, 263 (1986).
19. D. Y. C. Chan and R. G. Horn, J. Chem. Phys. **83**, 5311 (1985).
20. J. M. Georges, S. Millot, J. L. Loubet, and A. Tonck, J. Chem. Phys. **98**, 7345 (1993).
21. K. Lum, D. Chandler, and J. D. Weeks, J. Phys. Chem. B **103**, 4570 (1999).
22. S. Richardson, J. Fluid Mech. **59**, 707 (1973).
23. D. Einzel, P. Panzer, and M. Liu, Phys. Rev. Lett. **64**, 2269 (1990).

Mat. Res. Soc. Symp. Proc. Vol. 651 © 2001 Materials Research Society

Local Environment of Surface-Polyelectrolyte-Bound DNA Oligomers

Sangmin Jeon, Sung Chul Bae, Jiang John Zhao, Steve Granick
Department of Materials Science and Engineering
University of Illinois, Urbana, IL 61801

ABSTRACT

Two-photon time-resolved fluorescence anisotropy methods were used to study the dynamical environment when fluorescent-labelled DNA oligomers (labelled with FAM, 6-fluorescein-6-carboxamido hexanoate) formed surface complexes with quaternized polyvinylpyridine (QPVP) cationic layers on a glass surface. We compared the anisotropy decay of DNA in bulk aqueous solution, DNA adsorbed onto QPVP, and QPVP-DNA-QPVP sandwich structures. When DNA was adsorbed onto QPVP, its anisotropy decay was dramatically retarded compared to the bulk, which means it had very slow rotational motion on the surface. Motions slowed down with increasing salt concentration up to a level of 0.1 M NaCl, but mobility began to increase at still higher salt concentration owing to detachment from the surface-immobilizing QPVP layers.

INTRODUCTION

The dynamics of substances that are confined in one or more directions to molecular dimensions is receiving considerable attention for several reasons. First, this confined situation serves as an interesting extension of classical questions of the bulk liquid state. Secondly, it falls naturally into the growing interest in the dynamics of interfaces. Third, it presents clear connections to a problem of obvious economic importance, the friction and lubrication of surfaces in moving contact.

The desirability of spectroscopic measurements of confined liquids, within the geometry of a surface forces apparatus or another device in which to produce confinement and controlled deformations of the confined state, has long been clear, but faces the central difficulty of limited signal-to-noise. The meager abundance of sample, buried between two bulk condensed phases with thickness <1-2 nm, sets harsh requirements on the experimental methods that would be sensitive to this small sample size. The mass present is no more than approx. 1-2 ng-cm^{-2}. Methods of fluorescence spectroscopy satisfy this requirement of sensitivity[1].

Here we present a progress report concerning ongoing studies that take this approach.

- Fluorescence correlation spectroscopy. In one implementation, we use fluorescence correlation spectroscopy (FCS). FCS is an experimental method to extract information on dynamical processes from the fluctuation of fluorescence intensity, has enjoyed widespread application in chemical biology. For example, it has been used to study binding of drugs to DNA molecules[2], to determine the density of macromolecules in biological systems[3], to measure the rate constants of chemical reactions, and so forth[4]. The fluctuation of fluorescence is due to dynamic processes such as diffusion, aggregation, and chemical reactions, which occur in very small volumes (~10^{-15} L) in very dilute systems (from pM to nM). By calculating the autocorrelation function $(G(\tau))$ of this fluctuation (F),

$G(\tau) = \langle \delta F(t) \delta F(t+\tau) \rangle / \langle F(t) \rangle^2$, and by choosing a suitable model to analyze the autocorrelation function, the rate of the dynamic process is obtained. Compared to other techniques for studying diffusion problems, such as quasi-elastic light scattering, fluorescence recovery after photo-bleaching, laser-induced transient grating spectroscopy and so forth, FCS presents unique advantages in measuring very dilute systems with high spatial resolution (down to the optical diffraction limit) and also in sensitivity.

- Time-resolved fluorescence anisotropy. This communication focuses on this, another method that is capable of quantitative and systematic direct measurements of the motions and relaxations of individual molecules. Applied to the motions of DNA when it is localized at the solid-liquid interface, preliminary findings are presented. The method is expected to also find wider application to other systems by rational extension[5-10].

DNA SURFACE IMMOBILIZATION

DNA surface immobilization is an integral aspect of many routine DNA assays. The most common way of DNA immobilization is by direct chemisorption of thiol-labeled DNA onto a gold surface[11]. However, this method of chemical attachment does not carry over to nonmetallic surfaces. It is difficult to adopt fluorescence measurement to this system because nonradiative resonance energy transfer between the probe and the metal reduces the fluorescence signal[12].

Electrostatics provides a different method of surface immobilization. In previous work, Higashi et al[13,14] immobilized double-stranded DNA using the electrostatic interaction between DNA and a cationic self-assembled monolayer. Electrostatic attraction to a supported bilayer of cationic lipid has also been employed by Schouten et al.[15]

Here, we have used quaternized polyvinylpyridine(QPVP) to immobilize DNA onto glass surfaces. Since QPVP carries positive surface charges and DNA has a negatively-charged phosphate backbone, strong electrostatic attraction produces uniform DNA layers on a layer of previously-adsorbed QPVP. This approach has several advantages. First, it is so simple that any solid surface can be used. Secondly, the conformations of adsorbed DNA molecules can be easily modified by varying the ionic strength of the surrounding aqueous medium. Third, it is easy to control DNA adsorption by changing the fractional quaternization of the QPVP; this is effective because the amount of DNA adsorbed depends on the surface charge of the substrate.

This method has not been employed previously, probably because of concerns that complexation of single-stranded DNA with a polyelectrolyte would inhibit hybridization to form double-stranded DNA. Indeed, complexation of a polyelectrolyte with DNA has been employed to protect DNA from splitting by cell nuclease[16]. However, our preliminary experiment using a surface forces apparatus show that hybridization can occur when the ionic strength is properly controlled.

The study that follows concerns probes of the local environment of adsorbed DNA, measured by time-resolved two-photon fluorescence anisotropy. Anisotropy, r, is defined as

$$r = (I_{VV} - I_{VH}) / (I_{VV} + 2I_{VH}) \qquad (1)$$

where I is intensity. The first subscript refers to the laser polarization and the second to fluorescence polarization. For example, I_{VH} is the intensity of horizontally-polarized fluorescence excited by a vertically-polarized laser beam. We employed time-resolved fluorescence anisotropy measurement (TRFAM) to investigate the local dynamical environment of surface-immobilized DNA oligomers.

EXPERIMENTAL DETAILS

Materials. Gel-purified DNA oligomer with FAM was purchased from Research Generics Inc. (Huntsville, AL). FAM is 6-(Fluorescein-6-carboxamido) hexanoate and its excitation and emission wavelengths are 492 nm and 520 nm respectively. DNA oligomer has 15 base pairs and its sequence is 5'-FAM-GATGATGAGAAGAAC-3'. The molecular weight is 5,734g/mol and the melting temperature is 42°C.

Quaternized polyvinylpyridine (QPVP) was prepared from poly(1,4vinyl)pyridines (PVP) and quaternized with ethyl bromide in the same way described previously[16]. Infrared spectroscopy confirmed that it was completely quaternized (>98%). The final weight-average molecular weight was 35,100g/mol. To reduce photobleaching during the laser experiments from ambient oxygen, 1wt.% 2-mercaptoethanol(Aldrich) was added to the 0.1M Na_2HPO_4 buffer solutions.
Instruments. Two photon excitation was achieved using a femtosecond Ti:Sapphire laser(Mai Tai, Spectra-Physics) of which the FWHM (full width at half maximum) per pulse was measured to be ca. 100 femtoseconds. The repetition rate was 80 MHz and the wavelength was 800nm.

The homemade microscope setup shown in Fig 1 was designed to combine this measurement with the surface forces apparatus (SFA). However, a cylindrical glass cell was used in the present experiments in order to simplify the surface geometry. The vertically polarized laser beam was split into two beams and one of them was introduced to the objective lens (Zeiss, N.A. = 0.7) and focused onto the sample. The other beam was used as a trigger signal for the single photon counting system(Becker & Hickl GmbH). The fluorescence collected by the objective lens was focused again by a tube lens to increase the response of the final photomultiplier tube detectors.

Since the adsorbed molecules were vulnerable to photobleaching, time drift of the fluorescence intensity was a serious problem in our initial experiments when a single photodetector was used. Instead, a polarizing beam splitter was adapted to measure vertical and horizontal components of the fluorescence at the same time. The PMT (Hamamatsu, R5600) and a photodiode were used to detect the fluorescence and the trigger signal respectively. The PMT signal was input to the time-to-amplitude converter as a start signal followed by a constant fraction discriminator (Becker & Hickl GmbH, TCSPC730). The total instrument response function was around 180 ps.
Sample Preparation. QPVP (20μM) was introduced into the 1cm diameter cylindrical cell and 30 min were allowed for adsorption to the glass surface. After rinsing out the QPVP solution with deionized water (NanoPure, 18MΩ), DNA solution (1μM) was poured into the cell and then another QPVP layer was allowed to adsorb. Each adsorption step took 30 min. The cell was filled with pure water or buffer solution and measurements were repeated in the bulk solution and in the surface-immobilized state to

confirm that no DNA molecules were present free in bulk solution to contribute to the signal from the surface-immobilized DNA.

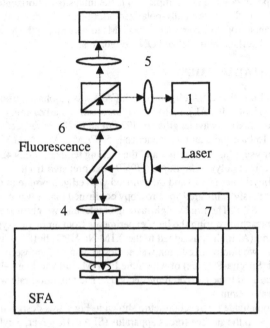

Figure 1. *Two-photon time resolved fluorescence anisotropy setup within a surface forces apparatus (SFA). The optical arrangement is similar to a confocal microscope setup because it was designed for combination with a surface force apparatus. The numbers denote the following components: 1, PMT; 2, polarizing beamsplitter; 3, beamsplitter; 4, objective lens; 5, filter; 6, tube lens; 7, inchworm; 8, leafspring; 9, cylindrical lens; 10, polarizer.*

RESULTS AND DISCUSSION

Fig 2A shows the decays of the vertical and horizontal components of fluorescence from DNA in bulk aqueous solution. Fluorescence intensity is plotted against time on the nanosecond time scale. Since the bulk DNA oligomers rotated freely, the two polarized components decayed quickly and soon overlapped.

However, when the DNA was confined by complexation with QPVP layers, its rotational motion was dramatically hindered and the two polarized components failed to overlap for a considerably longer time, as shown in Fig 2B.

Since the surface charge of the probe molecule was different from that of the DNA oligomer and it was loosely connected with DNA via a spacer five carbons long, it is worth mentioning that its measured anisotropy was not exactly the same as for DNA. However, the local environment of probe was mainly determined by DNA rather than by the probe because the size of DNA far exceeded that of the probe molecules and the DNA was more strongly charged. Therefore, this anisotropy experiment did not give direct information about the dynamics of DNA, but instead information about the interaction

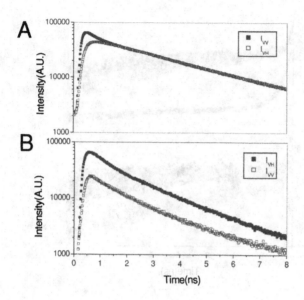

between DNA and QPVP.

Figure 2. *Fluorescence intensity decay of labeled DNA in bulk aqueous solution (panel A) and confined DNA between QPVP layer(panel B). Polarized intensity is plotted against time on the nanosecond time scale. Fluorescence decays overlapped rapidly in bulk solution, indicating rapid rotational diffusion, but the anisotropy decay on the surface was slower. ■: vertically polarized fluorescence, □: horizontally polarized fluorescence. All measurements were performed at room temperature and pure water.*

Fig 3A compares the anisotropy decays of labeled DNA in deionized water. Anisotropy is plotted against elapsed time on the nanosecond time scale. For DNA adsorbed to the QPVP surface, its anisotropy decay was significantly retarded. It was further retarded when DNA is covered with a second QPVP layer. We note that time-zero anisotropy, r(0), was larger than 0.4 because two photon excitation was used.

Fig 3B shows the effect of the ionic strength on the local environment of DNA. NaCl salt with 0.1M Na_2HPO_4 buffer solution was added to the cell. As the salt concentration increased to 0.1M, the anisotropy decay was increasingly retarded. This indicated that the probe molecules were increasingly slow to rotate. It may seem surprising to see slowing down when one considers that electrostatic interactions are weakened as ionic strength increases.

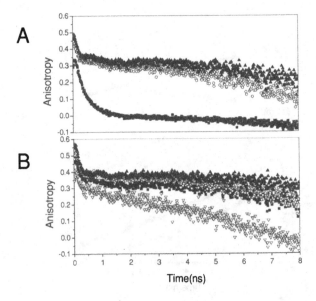

Figure 3. Panel A: Anisotropy decay of DNA in deionized water. ■ : Bulk, ○ : DNA adsorbed on QPVP, ▲ : DNA confined between two QPVP layers. Panel B: Anisotropy decay of labeled DNA at various salt concentrations (NaCl buffered to pH = 8) : ■ : No salt, ○ : 0.01M, ▲ : 0.1M, ∇ : 0.2M

The seeming dilemma can be explained as sketched schematically in Fig 4. In deionized water (Fig 4A), the QPVP layer has few loops and dangling chains so that DNA molecules are confined relatively uniformly. As salt concentration increases to 0.1M, the QPVP conformation changes to have more loops and a looser structure. But

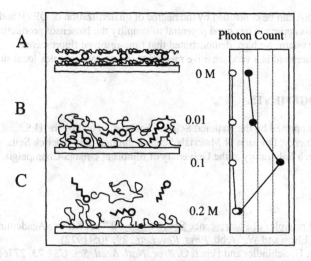

Figure 4. *Hypothetical surface conformations of QPVP and DNA at various ionic strength. Probes are more confined by QPVP up to 0.1M salt concentration and DNA detaches from the surface at higher salt concentration. Right inset: fluorescence count rate measured at various salt concentrations as indicated. ○ : Bulk solution; ● : Surface. At higher than 0.2M concentration, surface fluorescence count rate was same as bulk.*

because the electrostatic interaction between QPVP and DNA is still large enough to produce complexation, the probe has progressively less space to move around. It was confirmed that no fluorescence was detected in the bulk solution and that the surface photon count increased, which signifies that the quantum efficiency of probes increased as commonly occurs when dye molecules bind to a solid surface. Compared to this, significant fluorescence was detected in bulk solution when the ionic strength was raised still more, to 0.2M salt concentration. This signifies that DNA molecules detached from QPVP at this concentration. That is, the QPVP layer was less effective in confining the DNA molecules. When the salt concentration was increased still more, there was no detectable difference in photon count between surface and the bulk solution, but a prohibitively long time was needed to acquire low-noise data because the concentration of DNA on the surface was now so dilute.

CONCLUSION

It is a critical issue to surface-immobilize DNA for DNA recognition and biosensor applications. Here we have employed a cationic polyelectrolyte to surface-immobilize DNA for the first time as far as we are aware and have shown how its conformation changes with alterations of the ionic strength. Since the coverage and orientation of DNA can be controlled by the degree of quaternization of QPVP and salt concentration, this approach has good potential to simplify the biosensor production process. In this system, we have demonstrated that time-resolved fluorescence anisotropy measurement is a very sensitive method to investigate complex local surface environments.

ACKNOWLEDGEMENTS.

This work was supported by the National Science Foundation and by the U.S. Department of Energy, Division of Materials Science through the Frederick Seitz Materials Research Laboratory at the University of Illinois at Urbana-Champaign.

REFERENCES

1. J. Lakowicz, Principles of Fluorescence Spectroscopy, 2nd ed. Kluwer Academic,**1999**
2. D. Magde, E. Elson and W. Webb, *Phys. Rev. Lett.*, **29**, 705(1972).
3. M. Weissman, H. Schindler and Feher, G. *Proc. Natl. Acad. Sci. USA* **73**, 2776(1976)
4. R. Icenogle and E. Elson, *Biopolymers,* **22**, 1919(1983).
5. T. Sasaki, M. Yamamoto and Y. Nishijima, *Macromol.* **21**, 610-616 (1988).
6. K. Ono et al., *Macromol.* **27**, 6482-6486 (1994).
7. T. Smith et al., *Coll. Polym. Sci.* **276**, 1032-1037 (1998).
8. A. Volkmer et al., *Biophys. J.* **78**, 1589-1598 (2000).
9. J. Horinaka, S. Ito and M. Yamamoto, *Macromol.* **32**, 2270-2274 (1999).
10. S. Pauls, J. Hedstrom and C.Johnson, *Chem. Phys.* **237**, 205-222 (1998).
11. R. Nuzzo, D. Allara, *J. Am. Chem. Soc.* **105**, 4481-4484 (1983).
12. L. Meuse et al., *Langmuir* **16**, 4672-4677 (2000).
13. N. Higashi, T. Inoue and M. Niwa, *Chem. Commun.* 1507-1508 (1997).
14. N. Higashi, M. Takahashi and M. Niwa, *Langmuir* 111-115 (1999).
15. S. Schouten, P. Stroeve and M. Longo, *Langmuir* **15**, 8133-8139 (1999).
16. S. Sukhishvili, S. Granick, *J. Chem. Phys.* **109**, 6861 (1998).

Mat. Res. Soc. Symp. Proc. Vol. 651 © 2001 Materials Research Society

Towards a Microscopic Description of Friction

Markus Porto,[1] Veaceslav Zaloj,[2] Michael Urbakh,[3] and Joseph Klafter[3]

[1] *Max-Planck-Institut für Physik komplexer Systeme, Nöthnitzer Str. 38, 01187 Dresden, Germany*

[2] *Cornell University, Department of Computer Science, 4130 Upson Hall, Ithaca, NY 14853-4130, USA*

[3] *School of Chemistry, Tel Aviv University, 69978 Tel Aviv, Israel*

(November 23, 2000)

Abstract

We investigate the response of an embedded system subject to an external drive using a microscopic model. The shear is shown to excite "shearons", which are collective modes of the embedded system with well defined spatial and temporal patterns that dominate the frictional properties of the driven system. We demonstrate that the slip relaxation in stick-slip motion and memory effects are well described in terms of the creation and/or annihilation of shearons.

I. INTRODUCTION

The field of nanotribology evolves around the attempts to understand the relationship between macroscopical frictional response and microscopic properties of sheared systems [1]. Experimental tools such as the surface forces apparatus (SFA) are used to explore shear forces between atomically smooth solid surfaces separated by a nanoscale molecular film [2]. These experiments have unraveled a broad range of phenomena and new behaviors which help shed light on some traditional concepts which have been considered already textbook materials, such as static and kinetic friction forces, transition to sliding, thinning, and memory effects. What one wishes to deduce from the experimental observations [3] are new insights that will help establish the basics of nanotribology, differentiate among different embedded systems, and enable to control friction and maintain the desired type of motion. Hence, these observations have motivated theoretical efforts [4], both numerical and analytical, but many issues have remained unresolved. In particular the relation between the macroscopic observables and the microscopic properties of the embedded system under shear is not well understood, for example in the stick-slip regime. Different theoretical models have been proposed to account for this type of motion including spring-block models [5], chains adsorbed on a substrate [6], models with melting-freezing transitions [7] and an embedded particle model [8]. However, only a few studies have focused on the time-dependent relaxation patterns of the 'slip' part between two successive 'stick' events in the stick-slip regime. The latter may help provide a new insight in the dynamics of the embedded system, which are otherwise hidden due to the macroscopic nature of the probing method. Experimentally it has been observed that, under overdamped conditions, 'slip' relaxation manifests more than one characteristic time scale [9]. Typically a sharp decrease at early times is followed by a tail of a slow decay. In order to approach this problem, we have introduced a model of a sheared system that concentrates on the microscopic properties of the embedded part [8,10–12] and leads to the experimentally observed behaviors [9]. We have defined within the model the concept of "shearons", which are shown to dominate the frictional properties of the sheared system.

II. MODEL

Let us start from a model which includes both lateral and normal motions of the top plate and the particles, see Fig. 1. The model consists of two rigid plates and of a monolayer of N particles, each with a mass m and lateral and normal coordinates $x_{\parallel,\perp}^{(i)}$, embedded between them. The top plate of mass M and lateral and normal center of mass coordinates $X_{\parallel,\perp}$ is connected to a stage at coordinates $X_{\parallel,\perp}^{(0)}$, so that it can be moved laterally via a linear spring of spring constant $K_{\parallel}$ and feels a normal load by a linear spring of spring constant $K_{\perp}$. In the cases considered here, the stage remains at a constant normal position $X_{\perp}^{(0)} = \text{const}$ and is laterally pulled with constant velocity $V \equiv \dot{X}_{\parallel}^{(0)}$. This system is described by $2N + 2$ equations of motion ($2N + 1$ equations for lateral and normal coordinates, respectively)

$$M\ddot{X}_{\parallel,\perp} + \sum_{i=1}^{N} \eta_{\parallel,\perp} \left[\dot{X}_{\parallel,\perp} - \dot{x}_{\parallel,\perp}^{(i)}\right] + K_{\parallel,\perp} \left[X_{\parallel,\perp} - X_{\parallel,\perp}^{(0)}\right] +$$

$$\sum_{i=1}^{N} \frac{\partial \Phi(x_{\parallel}^{(i)} - X_{\parallel}, x_{\perp}^{(i)} - X_{\perp})}{\partial X_{\parallel,\perp}} = 0 \tag{1}$$

$$m\ddot{x}_{\parallel,\perp}^{(i)} + \eta_{\parallel,\perp} \left[2\dot{x}_{\parallel,\perp}^{(i)} - \dot{X}_{\parallel,\perp}\right] + \sum_{\substack{j=1 \\ j \neq i}}^{N} \frac{\partial \Psi(\sqrt{[x_{\parallel}^{(i)} - x_{\parallel}^{(j)}]^2 + [x_{\perp}^{(i)} - x_{\perp}^{(j)}]^2})}{\partial x_{\parallel,\perp}^{(j)}} +$$

$$\frac{\partial \Phi(x_{\parallel}^{(i)}, x_{\perp}^{(i)})}{\partial x_{\parallel,\perp}^{(i)}} + \frac{\partial \Phi(x_{\parallel}^{(i)} - X_{\parallel}, x_{\perp}^{(i)} - X_{\perp})}{\partial x_{\parallel,\perp}^{(i)}} = 0 \quad i = 1, \ldots, N. \tag{2}$$

The second term in Eqs. (1) and (2) describes the dissipative forces between the particles and the substrate/slider and is proportional to their relative velocities with proportionality constants $\eta_{\parallel,\perp}$, accounting for dissipation that arises from interactions with phonons and/or other excitations. The interaction between the particles and the surfaces is represented by the laterally periodic potential $\Phi(x_{\parallel}, x_{\perp}) = -\Phi_0 \left[\cos(2\pi x_{\parallel}/b) \exp(-x_{\perp}/\Lambda) - (\Lambda'/\Lambda) \exp(-x_{\perp}/\Lambda')\right]$, with b being the lateral periodicity and Λ and Λ' the decay lengths perpendicular to the surface. Concerning the interparticle interaction, we take a nearest-neighbor harmonic interaction $\Psi(r) = (k/2)[r - a]^2$, where

k the force constant and a the free equilibrium distance. The direct slider/substrate interaction is assumed to be negligible compared to the slider/chain and chain/substrate interactions. The model contains several 'natural' units given by the surface potential $\Phi(x_{\parallel}, x_{\perp})$: a unit length b, a unit frequency $\Omega_0 \equiv (2\pi/b)\sqrt{N\Phi_0/M}$, a unit force $F_0 \equiv 2\pi N\Phi_0/b$, and two unit force constants $K_0 \equiv (2\pi/b)^2 N\Phi_0$ and $k_0 \equiv (2\pi/b)^2\Phi_0$.

The basic frequency in the problem is chosen as the frequency of the top plate oscillation in the periodic potential $\Omega \equiv (2\pi/b)\sqrt{N\Phi_0/M}$. The other frequencies in the model are the frequency of the particle oscillation in the potential $\omega \equiv (2\pi/b)\sqrt{\Phi_0/m}$, the characteristic frequency of the inter-particle interaction $\hat{\omega} \equiv \sqrt{k/m}$, and the frequencies of the free lateral and normal oscillations of the top plate $\hat{\Omega}_{\parallel,\perp} \equiv \sqrt{K_{\parallel,\perp}/M}$. Our calculations demonstrate that under experimental conditions where $Nm/M \ll 1$, the motion of the top plate is not sensitive to the microscopic frequencies ω and $\hat{\omega}$, except for the limit $a/b \to 1$, which corresponds to monolayer-potential commensurability, and/or when $k \to 0$. In these limits the system exhibits the independent particle behavior discussed in [8].

To simplify the discussion, we introduce the unitless coordinates $Y \equiv X/b$, $Y' \equiv X'/b$ ($Y_0' \equiv X_0'/b$), and $y_i \equiv x/b$ of the top plate and the particles, respectively, as well as the unitless time $\tau \equiv \Omega t$. We define the following quantities: The misfit between periods of the substrate and the inter-particle potentials $\Delta \equiv 1 - a/b$, the ratio of masses of the particles and the top plate $\varepsilon \equiv Nm/M$, the unitless dissipation coefficients $\gamma_{\parallel,\perp} \equiv N\eta_{\parallel,\perp}^0/(M\Omega)$, the unitless damping lengths $\lambda \equiv \Lambda/b$ and $\lambda' \equiv \Lambda'/b$, the ratios of frequencies of lateral and normal free oscillations of the top plate and the oscillation of the top plate in the potential $\alpha_{\parallel,\perp} \equiv \hat{\Omega}_{\parallel,\perp}/\Omega$, the ratio of the frequencies related to the inter-particle and particle/plate interactions $\beta \equiv \hat{\omega}/\omega$, and the dimensionless velocity $v \equiv V/(b\Omega)$. In addition, the friction force f_k and the normal force f_n, both per particle, are defined as $f_k \equiv F_k/(NF_0)$ and $f_n \equiv F_n/(NF_0)$, respectively, where F_k and F_n are the total forces measured with the external lateral ($K_{\parallel}$) and normal ($K_{\perp}$) springs. The forces are displayed relative to $F_0 \equiv 2\pi\Phi_0/b$, the force due to the plate potential.

III. RESULTS

In order to characterize the response of the embedded system in different regimes of motion, we define the particles density ρ by representing each particle by a Gaussian of width $\sigma = 1$

$$\rho(y - y_{cms}, \tau) \equiv \sum_{i=1}^{N} \exp\left\{-\left[\frac{y - y_{cms} - \delta y_i}{\sigma}\right]^2\right\}. \tag{3}$$

Here, we separate the center of mass $y_{cms} \equiv 1/N \sum_{i=1}^{N} y_i$ of the particles in the monolayer from the fluctuations $\delta y_i \equiv y_i - y_{cms}$. Through Eq. (3) we introduce the concept of shearons, which are shear-induced collective modes of the embedded system [12]. As we see later, shearons display well defined spatial and temporal patterns in the particles density ρ due to correlations in the motion of the neighboring and next-neighboring particles on lengths of the order 1. These patterns, which dominate the frictional behavior of the system [12], become visible by introducing the finite width $\sigma = 1$. We find that the observed mean friction force per particle $\langle f_k \rangle$ ($\langle \cdot \rangle$ denotes time average), which is directly related to the spatial/temporal fluctuations by [13]

$$\langle f_k \rangle = \pi \gamma v + \frac{2\pi \gamma}{v} \left\langle \sum_{i=1}^{N} \delta \dot{y}_i^2 \right\rangle, \tag{4}$$

decreases with increasing shearon wave vector q for fixed stage velocity v, i.e. $(\partial \langle f_k \rangle / \partial q)_v < 0$ [12]. It has been already observed in similar models [14] that different modes of motion of the embedded system can coexist for a given set of parameters, and can lead to different frictional response.

We concentrate below on the stick-slip regime typical to low driving stage velocities, and analyze a time window which corresponds to 'slip' motion between two 'stick' events. Namely, just as the top plate overcomes the static friction and starts moving, dissipation sets in and the plate relaxes toward another stick event. This slip relaxation pattern depends on the conditions under which the stick-slip motion is being studied [10]. In the present work we consider the *overdamped* case, where the characteristic slip time is much longer than the response time of the mechanical system $\propto 2\pi \sqrt{M/K_{\parallel}}$. The overdamped regime is realized experimentally if the spring constant $K_{\parallel}$ is sufficiently weak, and/or if the friction constants $\eta_{\parallel, \perp}$ are large. Fig. 2 display the time evolution of the lateral and normal spring forces f_k and f_n during the slip motion. One clearly notices more than one time scale in the relaxation process. The behavior observed for the spring force has its origin in the dynamics of the chain. In Fig. 3 shown are the density ρ and the lateral and normal spring forces f_k and f_n enlarged for three time windows: the fast relaxation part (see Fig. 3(a)), a part of the slow relaxation part, and the part where the slip event ends. During the fast relaxation, the density is almost uniform, being a fingerprint of a chain trapped by one of the plates, in this case the upper one. At the beginning of the slower relaxation, a shearon with a certain wave vector q (see Fig. 3(b)) is created, persisting until the end of the second relaxation. Hence, in the present case, the two relaxations are the manifestation of two different dynamical behavior of the embedded system, namely a first mode of motion where the chain sticks to the top plate, and a second mode, where the chain follows both the potential corrugation of the bottom and top plate and a shearon is created. This shearon persits, with the same wave vector, until the end of the slip event, where the top plate comes to rest again (see Fig. 3(c)).

Using stop/start experiments with a smoothly sliding upper plate, it has been demonstrated [15] that if the external drive is stopped for a certain time, smaller than some upper limit, and then reinitiated with the same velocity, the additional static force $\Delta F_s \equiv F_s - F_k$ (F_s and F_k denote the static and kinetic friction force, respectively) to be overcome might vanish, giving rise to an interesting memory effect in the embedded system closely related to stick-slip motion. In Fig. 4 we present qualitatively equivalent results for the case that the external drive applied on the top plate, which slides smoothly with a given velocity v, is stopped for a certain waiting time $\Delta \tau$ and is then restarted with same velocity v. For relatively long waiting times, $\Delta \tau > \delta \tau$, a finite additional static friction has to be overcome, $\Delta f_s \equiv \Delta F_s/(NF_0) > 0$. For shorter waiting times, $\Delta \tau < \delta \tau$, the motion continues smoothly without any additional static friction after reinitiating the drive, i.e. $\Delta f_s = 0$. This behavior can be interpreted in terms of a shearon that exists during the smooth sliding. Shearons have a lifetime for decay after the drive ceases. For waiting times shorter than this lifetime, $\Delta \tau < \delta \tau$, the initial shearon still exists in the embedded system when the drive is restarted. However, since the shearon is annihilated at $\delta \tau$, it has to be recreated for long waiting times $\Delta \tau > \delta \tau$. The maximum waiting time allowed for a vanishing static friction is therefore determined by the lifetime $\delta\tau$ of the shearon, which is in our example $\delta\tau \cong 347$. We note that the lifetime $\delta\tau$ of a shearon is determined not only by the parameters of the embedded system such as β or $\gamma_{\parallel,\perp}$, but also to a large extent by the external parameters such as the lateral spring constant $\alpha_{\parallel}$. A weaker lateral spring lengthens the lifetime of the shearon, whereas a stiffer spring shortens it. This is consitent with the dependence of the slip relaxation time in stick-slip motion on the spring constant [10].

ACKNOWLEDGMENTS

Financial support from the United States-Israel Binational Science Foundation, the German Israeli Foundation, and DIP and SISITOMAS grants is gratefully acknowledged.

REFERENCES

[1] F.P. Bowden and D. Tabor, *The Friction and Lubrications of Solids* (Claredon Press, Oxford, 1985); B.N.J. Persson, *Sliding Friction, Physical Properties and Applications* (Springer Verlag, Berlin, 1998); S. Granick, Physics Today **52**, 26 (1999).

[2] H. Yoshizawa, P. McGuiggan, and J. Israelachvili, Science **259**, 1305 (1993);S H.-W. Hu, G.A. Carson, and S. Granick, Phys. Rev. Lett. **66**, 2758 (1991); J. Klein and E. Kumacheva, Science **269**, 816 (1995); A.D. Berman, W.A. Drucker, and J. Israelachvili, Langmuir **12**, 4559 (1996).

[3] H. Yoshizawa, P. McGuiggan, and J.N. Israelachvili, Science **259**, 1305 (1993); G. Reiter, L. Demirel, and S. Granick, Science **263**, 1741 (1994); P.A. Thompson, M.O. Robbins, and G.S. Grest, Isr. J. of Chem. **35**, 93 (1995).

[4] G.A. Tomlinson, Phil. Mag. **7**, 905 (1929); P.A. Tompson, M.O. Robbins, and G.S. Grest, Isr. J. Chem. **35**, 93 (1995); J.M. Carlson and A.A. Batista, Phys. Rev. E **53**, 4253 (1996); A.A. Batista and J.M. Carlson, Phys. Rev. E **57**, 4986 (1998); J.P. Gao, W.D. Luedtke, and U. Landman, Phys. Rev. Lett. **79**, 705 (1997); V. Zaloj, M. Urbakh, and J. Klafter, Phys. Rev. Lett. **81**, 1227 (1998); F.-J. Elmer, Phys. Rev. E **57**, R4903 (1998); T. Baumberger and C. Caroli, Eur. Phys. J. B **4**, 13 (1998); J.B. Sokoloff, Phys. Rev. B **51**, 15573 (1995); J.B. Sokoloff and M.S. Tomassone, Phys. Rev. B **57**, 4888 (1998); J. Röder, J.E. Hammerberg, B.L. Holian, and A.R. Bishop, Phys. Rev. B **57**, 2759 (1998); G. He, M.H. Müser, and M.O. Robbins, Science **284**, 1650 (1999).

[5] J.M. Carlson, J.S. Langer, and B.E. Shaw, Rev. Mod. Phys. **66**, 657 (1994).

[6] Y. Braiman, F. Family, and H.G.E. Hentschel, Phys. Rev. E **53**, R3005 (1996) and Phys. Rev. B **55**, 5491 (1997).

[7] B.N.J. Persson, Phys. Rev. B **50**, 4771 (1994); J.M. Carlson and A.A. Batista, Phys. Rev. E **53**, 4153 (1996).

[8] M.G. Rozman, M. Urbakh, and J. Klafter, Phys. Rev. E **54**, 6485 (1996); M.G. Rozman, M. Urbakh, and J. Klafter, Phys. Rev. Lett. **77**, 683 (1996).

[9] A.D. Berman, W.A. Ducker, and J.N. Israelachvili, Langmuir **12**, 4559 (1996).

[10] M.G. Rozman, M. Urbakh, and J. Klafter, Europhys. Lett. **39**, 183 (1997).

[11] M.G. Rozman, M. Urbakh, J. Klafter, and F.-J. Elmer, J. Phys. Chem. **102**, 7924 (1998); V. Zaloj, M. Urbakh, and J. Klafter, Phys. Rev. Lett. **82**, 4823 (1999).

[12] M. Porto, M. Urbakh, and J. Klafter, J. Phys. Chem. B **104**, 3791 (2000); M. Porto, M. Urbakh, and J. Klafter, Europhys. Lett. **50**, 326 (2000). M. Porto, V. Zaloj, M. Urbakh, and J. Klafter, Trib. Lett. (in print).

[13] M. Weiss and F.-J. Elmer, Z. Phys. B **104**, 55 (1997).

[14] E.D. Smith, M.O. Robbins, and M. Cieplak, Phys. Rev. B **54**, 8252 (1996); Y. Braiman, H.G.E. Hentschel, F. Family, C. Mak, and J. Krim, Phys. Rev. E **59**, R4737 (1999); H.G.E. Hentschel, F. Family, and Y. Braiman, Phys. Rev. Lett. **83**, 104 (1999).

[15] H. Yoshizawa and J. Israelachvili, J. Phys. Chem. **97**, 11300 (1993).

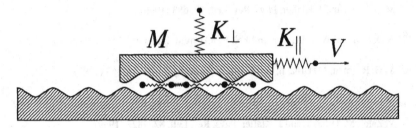

FIG. 1. The top plate of mass M is driven laterally by the spring $K_\parallel$ connected to a stage moving with velocity V, and feels a normal load by the spring $K_\perp$. The particles of mass m (the embedded system) are located between the two plates.

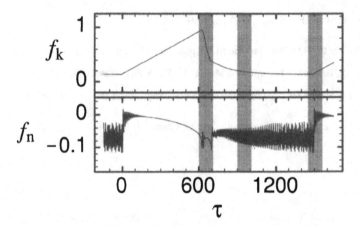

FIG. 2. Plot of the lateral and normal friction forces f_k and f_n vs time τ for parameter values which correspond to stick-slip motion. The model parameters are $\alpha_\parallel = 0.04$, $\alpha_\perp = 0.5$, $\beta = 1$, $\gamma_\parallel = 0.1$, $\gamma_\perp = 0.75$, $\Delta = 0.05$, $\varepsilon = 0.1$, $\lambda = 1$, $\lambda' = 0.05$, $Y_0' = 0$, $N = 15$, and $\nu = 0.13$. The areas shaded in light grey are enlarged in Fig. 3.

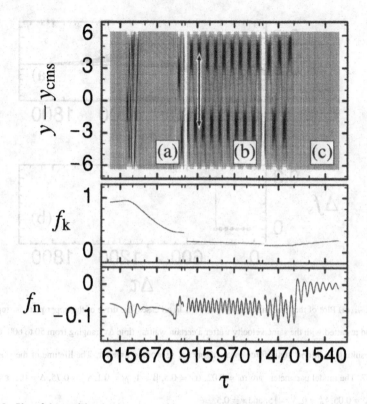

FIG. 3. Plot of the particles density ρ vs position $y - y_{cms}$ and time τ, as well as the corresponding lateral and normal friction forces f_k and f_n vs time τ. In the density plot with σ = 1, the white and black colors indicate low and high density, respectively. The arrow shown in (b) indicate $1/(2q)$. The time shown correspond to the areas shaded in light grey in Fig. 2.

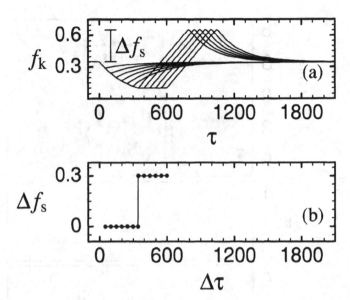

FIG. 4. (a) Plot of the friction force f_k vs time τ. In all cases, the drive of the upper plate is stopped at $\tau = 0$ and restarted with the same velocity v after a certain waiting time $\Delta\tau$, ranging from 50 to 600. (b) Plot of the resulting additional static force (peak hight) Δf_s vs waiting time $\Delta\tau$. The lifetime of the shearon is $\delta\tau \cong 347$. The model parameters are $\alpha_{\parallel} = 0.02$, $\alpha_{\perp} = 0.5$, $\beta = 1$, $\gamma_{\parallel} = 0.1$, $\gamma_{\perp} = 0.75$, $\Delta = 0.1$, $\varepsilon = 0.01$, $\lambda = 1$, $\lambda' = 0.05$, $Y_0' = 0$, $N = 15$, and $v = 0.5$.

Mat. Res. Soc. Symp. Proc. Vol. 651 © 2001 Materials Research Society

Tribological Behavior of Very Thin Confined Films

M. H. Müser
Institut für Physik, WA 331
Johannes Gutenberg-Universität Mainz
55099 Mainz, Germany

ABSTRACT

The tribological properties of two smooth surfaces in the presence of a thin confined film are investigated with a generic model for the interaction between two surfaces and with computer simulations. It is shown that at large normal contact pressures, an ultra thin film automatically leads to static friction between two flat surfaces - even if the surfaces are incommensurate. Commensurability is nevertheless the key quantity to understand the tribological behavior of the contact. Qualitative differences between commensurate and incommensurate contacts remain even in the presence of a thin film. The differences mainly concern the thermal diffusion of the contact and the transition between smooth sliding and stick-slip.

INTRODUCTION

Understanding the dynamics of a system under shear containing a confined fluid is intimately connected with understanding the effects that are invoked through the corrugation of the confining walls. In the case of bare, flat surfaces, not only the degree of corrugation is relevant, but more importantly, the correlation of the corrugation in the upper wall and in the lower wall, i.e., commensurate systems exhibit wearless static friction while incommensurate systems do not [1-3]. This situation is dramatically changed if small amounts of "fluid" are injected into the interface and the fluid atoms do not form covalent bonds with neither surface [4]. Independent of the type of commensurability, static friction can be expected to occur. Here we will discuss why such mobile atoms in the interface lead to friction and address the question whether the differences between commensurate and incommensurate systems are remedied in the presence of a fluid layer.

The studies discussed here address fundamental issues rather than questions of direct, practical use. How does a tiny amount of fluid/contamination between two perfectly flat, crystalline surfaces alter the tribological behavior of the contact and what are the implications of commensurability in the presence of a thin film? Experimentally, it might be impossible to study these effects satisfactorily, because clean surfaces are hard to obtain even in UHV, e.g., the contaminant may reside within the bulk and diffuse to the surfaces only after sliding has been initiated.

THEORY AND COMPUTER SIMULATION MODEL

A simple, idealized model to treat interactions between two flat walls with only atomistic roughness allows a large number of predictions [5]. The main feature of the model is that two surfaces pay a local energy penalty that increases exponentially fast as the distance between the surfaces is decreased or the overlap is increased. The consequences of this model are among others: (i) Commensurate systems show a static friction coefficient μ_s that is independent of the area of contact A_c, (ii) for amorphous crystalline contacts $\mu_s \propto 1/\sqrt{A_c}$ is found, and (iii) for incommensurate contacts $\mu_s = 0$. In this case, sliding is only opposed by a viscous drag force. These predictions are based on the analysis of the Fourier transform of the surface modulation. μ_s can only be independent of A_c if the upper and the lower wall's corrugation is systematically correlated. Atoms that are injected into the interface can easily accommodate the surface modulation of both surfaces simultaneously: The "fluid" atoms sit at positions where the spacing between both walls is maximum. Once the normal pressure becomes large, the atoms are caught in these position and they can only escape via thermal activation, which one can assume to take place on long time scales only. If one tries to initiate relative sliding of the walls, the available free volume of the atoms will decrease. Thus the total energy of the system increases, which results in a force opposing the initiated motion.

However, there is an important difference between commensurate and incommensurate walls if thermal activation and diffusion of the fluid atoms is taken into account. Consider a system that is only subject to thermal noise. In the commensurate case, fluid atoms jump into (more or less) equivalent positions. Therefore, the relative lateral equilibrium position of the top wall does not change even if the fluid atoms undergo diffusion. In other words, there is a well-defined free energy profile of the contact that has the periodicity of the system. Therefore, static friction is an equilibrium phenomenon. In the incommensurate case, the expected situation is strikingly different. The fluid atoms jump into inequivalent positions as they undergo thermally activated diffusion. With each such jump, the lateral, relative equilibrium position of the walls shifts slightly. If we allow the fluid to explore the whole phase space, all relative, lateral positions of the two walls are identical. Therefore the free energy is not a function of the relative, lateral displacement of the two walls unlike the commensurate case. For incommensurate systems, static friction is a nonequilibrium phenomenon. From this discussion, one would expect an exponential slowing down of thermal diffusion with contact area of the whole contact in the commensurate case (at fixed normal pressure), while creep motion would be expected for incommensurate surfaces.

A generic model is emploied in order to analyze the thermal motion of a mechanical contact including a thin confined layer of atoms. It consists of two (111) surfaces whose atoms are harmonically pinned to their ideal lattice sites. Periodic boundary conditions are emploied in the plane of the two walls. The fluid atoms, which are confined between the walls, interact with Lennard Jones potentials among each other and with the walls. For further details of the model, we refer to Refs. [3-5].

RESULTS

We first want to report results for the thermal diffusion of the contact. Commensurate surfaces show the expected exponentially slowing down with increasing area of contact A_c for a given normal pressure $p_\perp$ and temperature T. Of course, a similar slowing down of the diffusion is achieved by either decreasing T at fixed $(A_c, p_\perp)$ or by increasing $p_\perp$ at fixed (A_c, T) [3]. Surprisingly, dramatic size effects are also found for incommensurate surfaces as shown in figure 1.

At small temperatures, the system appears to be pinned elastically during a time window of 10^6 molecular dynamics (MD) time steps. The effective lateral stiffness increases linearly with the area of contact A_c similar to the net friction force in Amontons law, which is proportional to $p_\perp \times A_c$. This generalized Amontons law for elastic pinning has also been observed experimentally [6]. It can be extracted from the small magnitude of the MSD and the decrease of the MSD with the number of atoms N per surface layer (in Lennard Jones units N and A_c are related by a factor

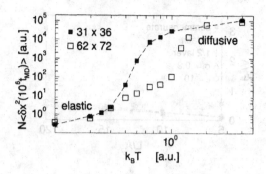

Figure 1. *Mean square displacement of incommensurate top wall times number of atoms N per surface layer after 10^6 MD steps for two system sizes, $N = 31 \times 36$ and $N = 62 \times 72$, as function of temperature.*

close to unity). At large temperatures, the contact can be considered "diffusive". In this regime the same scaling factor, namely N, collapses the data for the two system sizes. However, in between these two regimes, a non-trivial size dependence is observed, where the increase of the system size by a factor of four decreases the MSD of the top wall by more than a factor of hundred. It is important to note that the diffusion of individual fluid atoms is not affected by the size of the upper wall. In all cases, the MSD of individual atoms is larger than N times the MSD of the top wall.

This observation goes beyond the theoretical considerations from the last section. The pinning mechanism seems to be even more effective than anticipated. The large wall still appears to be pinned even when individual fluid atoms have typically diffused over significantly more than hundred lattice constants. We speculate that the indirect long-range interactions between monomers mediated through the top wall's center-of-mass may be the reason for the unexpected slowing down of the equilibration of the top wall. (Remember that all lateral position of an incommensurate top wall are equivalent provided that the fluid is in thermal equilibrium.) Long-range interactions have been shown to frequently slow down equilibration [7].

There is nevertheless a significant difference at what system size commensurate or incommensurate systems appear to be pinned at fixed $(T, p_\perp)$. Other parameters like coverage of the fluid, the strength of the repulsive interactions, etc. play a much less crucial role. In our model, commensurate systems always pin at smaller system sizes as compared to incommensurate systems. This is also reflected in the static friction, which is shown in figure 2. The default system in figure 2 consists of a quarter layer lubricant composed by simple Lennard Jones atoms that interact with $V(r) = 4\epsilon[(\sigma/r)^{12} - (\sigma/r)^6)]$ with the default values for σ and ϵ being unity. Potentials are cut-off such that interactions are purely repulsive. In the absence of free surfaces, the main effect of the long-range adhesive part can be absorbed into the load. We note in passing that the net interaction between two fluid atoms is nevertheless attractive due to their elastic interactions with the confining walls. As long as the two confining walls are incommensurate, the static friction coefficient μ_s, which can be defined as the slope of the curves in figure 2, is relatively independent of the model: The inequality $0.0022 < \mu_s < 0.0035$ holds for the default system, and systems for which one parameter is changed, namely, a half layer lubrication instead of a quarter lubrication, $\epsilon = 10$, and

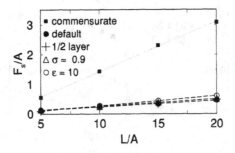

Figure 2. *Shear pressure $p_s = F_s/A$ versus normal pressure $p_\perp = L/A$ for different model systems. The default system consists of two flat incommensurate walls with a quarter layer of fluid confined in between them. The default Lennard Jones interaction parameters are $\sigma = 1$ and $\epsilon = 1$.*

$\sigma = 0.9$. Rotating the walls to make them orientationally perfectly aligned, however, increases μ_s by nearly a factor of five to $\mu_s \approx 0.17$.

While these differences in μ_s are merely quantitative, there is a qualitative difference in the transition from stick-slip to smooth sliding. Here, we calculate the (average) kinetic friction coefficient $\mu_k = F_k/L$ for a system that is pulled with a spring of stiffness k. For small values of k the system shows stick-slip for large values of k smooth sliding at constant velocity is found. The average friction force is shown in figure 3. The commensurate surfaces show similar behavior as dry commensurate surfaces. For a weak spring, the frictional forces are large. In this regime, the top wall shows large scale stick-slip motion. In an intermediate regime, in which atomic scale ratcheting is found, the frictional forces decrease dramatically upon increasing k.

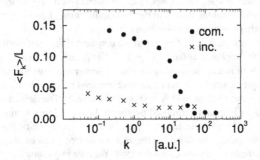

Figure 3. *Average kinetic friction force per load as a function of the stiffness k of the spring that drives the system for commensurate and incommensurate surfaces separated by a quarter layer of fluid atoms.*

At large k, the upper wall goes up and down the free energy surface adiabatically, resulting in small frictional forces. Of course, in the smooth sliding regime, F vanishes all together in the limit $k \rightarrow \infty$ and driving velocity $v \rightarrow 0$. The incommensurate surfaces, however, can not be driven adiabatically resulting in considerably larger friction at large values of k. We want to note in passing that commensurate surfaces show periodic stick-slip events while the incommensurate surfaces with boundary lubrication show much more erratic stick-slip events.

So far, we have only studied the friction between flat surfaces. However, boundary lubrication also effects the friction between curved surfaces. In order to study effects related to the curvature of surfaces, we have studied a Hertzian contact of a curved tip on a flat surface as a function of commensurability and boundary lubrication [8]. Due to the finiteness of the contact, even incommensurate contacts show friction, although it can certainly be considered insignificantly small - as long as the friction remains wear free. The situation changes when a thin layer with mobile atoms is present on at least one of the two surfaces. In the study of such systems, the interaction of the tip with the lower wall and with the mobile atoms is purely repulsive. This choice eliminates adhesive effects such as the so called offset load. All other interactions are chosen to be attractive. Some representative results are shown in figure 4.

Non-adhesive commensurate tips show the expected linear relationship between friction force and load, e.g., $F_s = L^{\alpha}$ with $\alpha = 1$. A dry amorphous tip on a crystalline substrate shows sublinear behavior with $\alpha = 2/3$, which agrees well with friction force microscopy experiments [9]. Dry incommensurate surfaces show the smallest friction forces with an insignificant increase in F_s force with increasing L. As soon as a thin film is present in the interface, the wearless friction increases dramatically. However, unlike for the friction between flat surfaces, sublinear behavior is found. In the present study, the exponent $\alpha = 0.85$ is found, which we do not believe to be universal. Details of the friction-load dependences will depend among other on the wetting properties of the fluid.

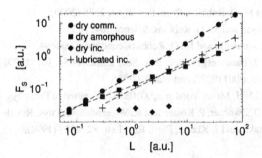

Figure 4. *Static friction force F_s vs load L for different curved tips on a crystalline substrate.*

CONCLUSIONS

We have shown that the commensurability of two walls confining a thin fluid film has systematic implications for the tribological properties despite the presence of the confined film. The differences are both quantitative and qualitative. Static friction is considerably larger between two flat commensurate surfaces than between incommensurate surfaces. In the case of a Hertzian contact, the friction-force load dependence can even be qualitatively different, e.g., incommensurate contacts may have a sublinear dependence on the load. The most striking difference, however, is the transition from stick-slip to smooth sliding. In our studies, commensurate surfaces show a dramatic decrease in kinetic friction from stick-slip to smooth sliding. A rather abrupt decrease in kinetic friction is seen as the slipped distances become small. This abrupt decrease in kinetic friction, which was achieved by increasing the stiffness of the driving device, is absent for incommensurate surfaces. It would be an interesting question, how friction control mechanisms such as proposed in Ref. [10] would be effected by the incommensurability of the surfaces in the presence of a thin confined film.

ACKNOWLEDGMENTS

The author would like to thank K. Binder, M. O. Robbins, and L. Wenning for discussions. Research supported by the Israeli-German D.I.P.-Project No 352-101.

REFERENCES

1. M. Hirano and K. Shinjo, Phys. Rev. B **41**, 11837 (1990); Wear **168**, 121 (1993).

2. M. R. Sørensen, K. W. Jacobsen, and P. Stoltze, Phys Rev. B **53**, 2101 (1996).

3. M. H. Müser and M. O. Robbins, Phys. Rev. B **64**, 2335 (2000).

4. G. He, M. H. Müser, and M. O. Robbins, Science **284**, 1650 (1999).

5. M. H. Müser, L. Wenning, and M. O. Robbins, cond-mat/0004494.

6. P. Berthoud and T. Baumberger, Proc. Roy. Soc. Lond. A **454**, 1615 (1998).

7. C. Tsallis, cond-mat/0011022; cond-mat/9903356.

8. L. Wenning and M. H. Müser, cond-mat/0010396 (submitted to Europhys. Lett.).

9. U.D. Schwarz , O. Zwörner, P. Köster, and R. Wiesendanger, Phys. Rev. B **56**, 6997 (1997).

10. V. Zaloj, M. Urnakh, and J. Klafter, Phys. Rev. Lett. **82**, 4823 (1999).

Mat. Res. Soc. Symp. Proc. Vol. 651 © 2001 Materials Research Society

Mechanical, adhesive and thermodynamic properties of hollow nanoparticles

U. S. Schwarz[1], S. A. Safran and S. Komura[2]
Department of Materials and Interfaces, Weizmann Institute, Rehovot 76100, Israel
[1]Max-Planck-Institute of Colloids and Interfaces, 14424 Potsdam, Germany
[2]Department of Chemistry, Tokyo Metropolitan University, Tokyo 192-0397, Japan

Abstract

When sheets of layered material like C, WS_2 or BN are restricted to finite sizes, they generally form single- and multi-walled hollow nanoparticles in order to avoid dangling bonds. Using continuum approaches to model elastic deformation and van der Waals interactions of spherical nanoparticles, we predict the variation of mechanical stability, adhesive properties and phase behavior with radius R and thickness h. We find that mechanical stability is limited by forces in the nN range and pressures in the GPa range. Adhesion energies scale linearly with R, but depend only weakly on h. Deformation due to van der Waals adhesion occurs for single-walled particles for radii of few nm, but is quickly suppressed for increasing thickness. As R is increased, the gas-liquid coexistence disappears from the phase diagram for particle radii in the range of 1-3 nm (depending on wall thickness) since the interaction range decreases like $1/R$.

Introduction

The prototypical hollow nanoparticle is the buckyball C_{60}, which crystallizes into *fullerite*. Many other morphologies for carbon sheets have been found since the early 1990s, including fullerenes C_n of varying size, multi-walled fullerenes (*carbon onions*), and single- and multi-walled carbon nanotubes. Closed structures of carbon are formed in order to avoid dangling bonds at the edges of finite sized carbon sheets. This mechanism is generic for anisotropic layered material of finite size, and up to now more than 30 other materials have been prepared as hollow nanoparticles (see the reviews [1] and references therein). This includes the metal dichalcogenides MeX_2 (Me = W, Se, X = S, Se), BN, GaAs and CdSe. Usually *inorganic fullerenes* like WS_2 and MoS_2 are multi-walled. In Fig. 1 we show transmission electron micrographs of WS_2-particles which were synthesized by solid-gas reaction. Control of size and shape is rather difficult, as evidenced by the irregularly faceted shapes. However, the methods used for the synthesis of hollow nanoparticles are developing very rapidly, and it is to be expected that control of size, thickness and shape will become much better in the future.

Hollow nanoparticles combine covalent in-plane strength with flexible out-of-plane bending of thin films, which results in high mechanical stability. They interact by van der Waals (vdW) forces, which for example determine the material properties of fullerite [2] and the phase behavior of buckyballs [3]. VdW interactions become even more important for larger fullerenes, carbon onions and inorganic fullerenes; these materials, therefore, feature a large degree of non-specific adhesion. In particular, vdW adhesion can lead to

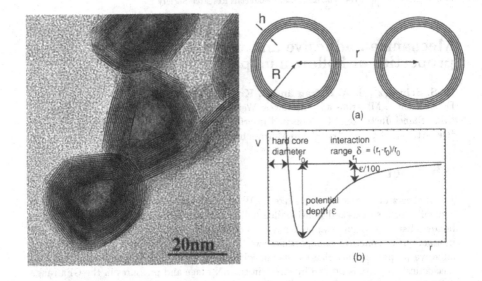

Figure 1: Left: Transmission electron micrographs of hollow WS_2-nanoparticles (courtesy of R. Tenne). Van der Waals interactions leads to strong adhesion. The faceted shape is due to grain boundaries. Right: (a) Model for two spherical nanoparticles of radius R and thickness h each, interacting over a distance r. (b) The effective interaction potential V as a function of distance r is characterized by the hard core diameter $2R$, potential depth $\epsilon \sim R$ and interaction range $\delta \sim 1/R$.

deformation of the hollow nanoparticles. For example, it has been found that carbon nanotubes adhering to each other or to flat surfaces show observable deformations for radii as small as 1 nm [4]. In the presence of local surface features like crystal steps, adhering nanoparticles will have spacially varying electronic properties due to spacially varying deformation.

Hollow nanoparticles may lead to new applications in nanoelectronics and -optics. In contrast to filled nanoparticles, they can also be used for storage and delivery systems. They show superior mechanical properties, like high flexibility, high tensile strength and light weight, which will lead to applications as ultra-strong fibers. Larger fibers can be produced using the cohesiveness provided by vdW interactions. Inorganic fullerenes have also been shown to offer favorable tribological properties [5]. Detailed investigations with the Surface Force Apparatus demonstrated that low friction and wear is caused by material transfer onto the sliding surfaces [6].

For these applications, a good theoretical understanding of the physical properties of hollow nanoparticles is important. Here we discuss structural properties of spherical hollow nanoparticles in a continuum approach that allows us to focus on unusual generic properties which result mainly from geometrical effects. In particular, we address mechanical properties, van der Waals interaction, thermodynamic behavior and deformation by adhesion. In this paper, we focus on the main results of our work; more details can be found elsewhere [7, 8].

Mechanical properties

During recent years, mechanical properties of fullerenes and carbon nanotubes have been extensively studied at the level of molecular calculations, that is by first principles, tight binding and force field techniques. The advantage of molecular models is that they can provide detailed quantitative predictions. For larger systems like multi-walled nanoparticles, their implementation becomes difficult due to computer time requirements. Here we use classical continuum elasticity theory, which is asymptotically correct for large systems and allows us to treat all these systems using the same framework. The large advantage of this approach is that it provides insight into the generic properties of these systems. Molecular calculations for carbon nanotubes have shown that the predictions of continuum elasticity theory persist well into the limit of radii smaller than 1 nm [9]. Recently this observation could be verified also for BN and MoS_2 nanotubes [10].

The layered material discussed here usually has a hexagonal lattice structure, for which elasticity theory [11] predicts that there are only two in-plane elastic constants and the sheet is elastically isotropic. Its deformation energy consists of stretching and bending terms. For single-walled carbon sheets, the in-plane stretching modulus $G \approx 3.6 \times 10^5$ erg/cm^2 and the out-of-plane bending rigidity $\kappa \approx 3 \times 10^{-12}$ erg. For both MoS_2 and WS_2, G is smaller by a factor 4 and κ is larger by a factor 5. For multi-walled nanoparticles, the elastic constants scale with thickness h as $G \sim C_{11}h$ and $\kappa \sim C_{11}h^3$, where C_{11} is the largest in-plane elastic constant of the corresponding layered material [12]. The values for C_{11} are 1060, 238 and 150×10^{10} erg/cm^3 for C, MoS_2 and WS_2, respectively.

Stretching can be avoided completely by bending the sheet into a cylindrical nanotube. However, this is not possible for spherical nanoparticles, which we discuss here. The same holds true for saddle-like structures, and in general for all surfaces with non-zero Gaussian curvature. On a microscopic level, the requirement for non-zero Gaussian curvature also means that a flat sheet cannot be bent into a sphere without introducing defects. Continuum elasticity approaches to fullerenes therefore explicitly treated the 12 pentagonal defects necessary to close a sheet of carbon hexagons [13]. In order to avoid this complication, here we assume that the curvature generating defects are distributed in a homogeneous manner and result in a vanishing internal strain (like for a ping-pong ball). In this case, the hollow nanospheres can be modeled as elastic shells with thickness h and prefered radius R.

Mechanical collapse in tribological applications is likely to be caused by either direct forces or high pressure. In the latter case, an inward buckling instability occurs, since the pressure energy scales with a higher power of indentation H than the restoring elastic energy (H^2 versus $H^{3/2}$) [7]. Mechanical collapse is expected when indentation H becomes of the order of radius R. The critical pressure for collapse can then be estimated to be

$p_c \sim G^{1/4}\kappa^{3/4}/R^{5/2}$. For buckyballs and typical inorganic fullerenes, this yields values of 15 and 2 GPa, respectively. For nanoparticles, the corresponding forces are in the nN-range.

Van der Waals interaction and phase behavior

Since vdW interactions are very difficult to treat on a molecular level, a continuum approach is even more useful in this case. It is well known from colloid science that vdW interactions strongly depend on the geometry of the system under investigation. Geometrical aspects of vdW interactions are well treated in the Hamaker approach, in which an atomic Lennard-Jones interaction is summed in a pairwise manner over the relevant geometry and the molecular details of the interaction are lumped into the effective Lennard-Jones parameters σ and ϵ. For carbon, $\sigma = 3.5\text{Å}$ and $\epsilon = 4.6 \times 10^{-15}$ erg [2]. The corresponding Hamaker constant A is in the order of 10^{-12} erg. For two buckyballs, the Hamaker approach for thin sheets leads to the *Girifalco potential* [2], which is considerably shorter ranged than a Lennard-Jones potential. Since a small range of attraction leads to the disappearance of the gas-liquid coexistence from the phase diagram [14], it has been suggested that buckyballs constitute the first non-colloidal system for which the gas-liquid critical point does not appear in the phase diagram. However, large scale simulations have shown that buckyballs indeed do show a gas-liquid coexistence, which extends roughly from 1900 to 2000°C [3].

For hollow nanoparticles in general, the effect of a finite thickness h becomes important. However, for nanoparticles in a crystal or adhering to a flat substrate or to each other, the distance of closest approach D between interacting particles is on an atomic scale. Under such conditions, the vdW interaction saturates on the atomic scale D with increasing thickness h. For example, the surface energy for the interaction with a halfspace scales as $\epsilon h/D^3$ and ϵ/D^2 for thin films and halfspaces, respectively, with the crossover occuring for $h = D$. Since $h \gtrsim D$ in our case, the surface energy is ϵ/D^2 independently of h and no significant scaling with thickness is expected.

For a more detailed investigation, one has to consider the general case of two nanospheres of finite thickness interacting in a pairwise manner by a Lennard-Jones potential [8]. The interaction energy can be written as $V = V_{R,R} + V_{R-h,R-h} - 2V_{R,R-h}$, where V_{R_1,R_2} is the interaction energy between two filled balls of radii R_1 and R_2, respectively, which can be calculated analytically for a Lennard-Jones potential. In order to predict phase behavior, three quantities characterizing the resulting potential are especially important: the hard core diameter $2R$ sets the density scale, the potential depth sets the temperature scale, and the interaction range determines if a gas-liquid coexistence occurs (compare Fig. 1). Our full analysis has shown that for hollow nanoparticles of radius R and thickness h, the potential depth and interaction range scale linearly and inversely with R. Both quantities depend only weakly on h. In Fig. 2 we show our numerical results for the interaction range as a function of R and h. Recent work for the Double-Yukawa potential suggests that the gas-liquid coexistence disappears for an interaction range smaller than $\delta = 0.4$ [15]. Our analysis predicts that this happens for $R = 12\text{Å}$ and $R = 35\text{Å}$ for single- and multi-walled nanoparticles, respectively. For larger R, the critical point will disappear from the phase diagram.

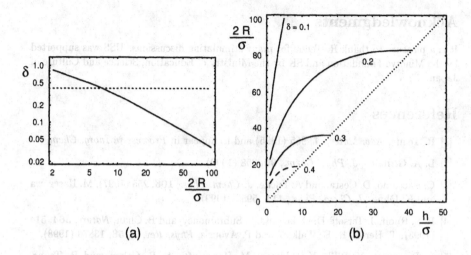

Figure 2: Numerical results for the interaction range δ (a) for single-walled nanoparticles as a function of radius R and (b) for multi-walled nanoparticles as a function of radius R and thickness h. $\sigma = 3.5$ Å is the atomic length of the Lennard-Jones potential. The gas-liquid coexistence is expected to disappear from the phase diagram for $\delta < 0.4$ (dashed lines).

Deformation due to adhesion

We consider the case where a spherical hollow nanoparticle is indented a distance H due to adhesion to a flat substrate. The vdW energy gained on adhesion can be estimated to be $E_{vdW} \sim (A/D^2)DH$ where A is the Hamaker constant and D an atomic cutoff [16]. The energy of deformation can be estimated to be $E_{el} \sim G^{1/2}\kappa^{1/2}H^2/R$. With increasing R, deformation becomes more likely, since both the vdW energy and deformability increase. However, deformability decreases rapidly with increasing thickness h, since $G^{1/2}\kappa^{1/2} \sim h^2$. Solving for the indentation H and using the values given above, it is found that for hollow carbon nanoparticles with few walls, H can be in the nm-range [7]. However, for typical inorganic fullerenes deformations are suppressed by the larger shell thickness h.

Finally we briefly discuss a nanotube adhering to a flat substrate. As explained above, nanotubes can be bent without being stretched. On deformation, one has to pay a bending energy per length of $E_{el}/L \sim \kappa H^2/R^3$. The gain in vdW energy scales as $E_{vdW}/L \sim AH/D^{3/2}R^{1/2}$. It follows that for carbon nanotubes with small thickness h, the indentation H can be in the nm-range for even smaller values of radius R than for hollow carbon nanoparticles.

Acknowledgments

It is a pleasure to thank R. Tenne for many stimulating discussions. USS was supported by the Minerva Foundation and SK by the Ministry of Education, Science and Culture of Japan.

References

[1] R. Tenne, *Adv. Mater.*, **7**, 965 (1995) and to appear in *Progress in Inorg. Chem.*

[2] L. A. Girifalco, *J. Phys. Chem.*, **96**, 858 (1992).

[3] C. Caccamo, D. Costa, and A. Fucile, *J. Chem. Phys.*, **106**, 255 (1997). M. Hasegawa and K. Ohno, *J. Chem. Phys.*, **111**, 5955 (1999).

[4] R. S. Ruoff, J. Tersoff, D. C. Lorents, S. Subramoney, and B. Chan, *Nature*, **364**, 514 (1993). T. Hertel, R. E. Walkup, and P. Avouris, *Phys. Rev. B*, **58**, 13870 (1998).

[5] L. Rapoport, Y. Bilik, Y. Feldman, M. Homyonfer, S. R. Cohen, and R. Tenne, *Nature*, **387**, 791 (1997). M. Chhowalla and G. A. J. Amaratunga, *Nature*, **407**, 164 (2000).

[6] Y. Golan, C. Drummond, M. Homyonfer, Y. Feldman, R. Tenne, and J. Israelachvili, *Adv. Mat.*, **11**, 934 (1999). Y. Golan, C. Drummond, J. Israelachvili, and R. Tenne, *Wear*, **245**, 190 (2000).

[7] U. S. Schwarz, S. Komura, and S. A. Safran, *Europhys. Lett.*, **50**, 762 (2000).

[8] U. S. Schwarz and S. A. Safran, *Phys. Rev. E*, **62**, 6957 (2000).

[9] D. H. Robertson, D. W. Brenner, and J. W. Mintmire, *Phys. Rev. B*, **45**, 12592 (1992). J. P. Lu, *Phys. Rev. Lett.*, **79**, 1297 (1997).

[10] E. Hernandez, C. Goze, P. Bernier, and A. Rubio, *Phys. Rev. Lett.*, **80**, 4502 (1998). G. Seifert, H. Terrones, M. Terrones, G. Jungnickel, and T. Frauenheim, *Phys. Rev. Lett.*, **85**, 146 (2000).

[11] L. D. Landau and E. M. Lifshitz, *Theory of elasticity* (Pergamon Press, 1970).

[12] D. J. Srolovitz, S. A. Safran, and R. Tenne, *Phys. Rev. E*, **49**, 5260 (1994).

[13] J. Tersoff, *Phys. Rev. B*, **46**, 15546 (1992). T. A. Witten and H. Li, *Europhys. Lett.*, **23**, 51 (1993).

[14] A. P. Gast, C. K. Hall, and W. B. Russel, *J. Coll. Inter. Sci.*, **96**, 251 (1983).

[15] C. F. Tejero, A. Daanoun, H. N. W. Lekkerkerker, and M. Baus, *Phys. Rev. Lett.*, **73**, 752 (1994) and *Phys. Rev. E*, **51**, 558 (1995).

[16] In [7], we incorrectly took this energy to be $E_{vdW} \sim (A/D^2)RH$. This overestimates the deformation by adhesion by a factor of R/D, but does not change our general conclusions.

Mat. Res. Soc. Symp. Proc. Vol. 651 © 2001 Materials Research Society

Velocity Distribution of Electrons Escaping from a Potential Well

James P. Lavine
Image Sensor Solutions, Eastman Kodak Company, Rochester, NY 14650-2008, U.S.A.

ABSTRACT

Electron escape over a one-dimensional potential barrier is treated with a Monte Carlo method that incorporates simple models for the electron-phonon interaction. The consequences of these models are considered here through the calculation of the escaping electron velocity distribution and the electron energy distribution before escape. Effective temperatures are derived from both distributions. The numerical results are compared with those from the classical model of thermionic emission.

INTRODUCTION

Particles confined by geometry [1] or by a potential well [2-5] have the opportunity to escape over time. The time scale for escape is important for chemical reactions and semiconductor device operation [5]. A variety of computational methods [1-4] has been used to determine the time scale. In most cases, the methods are based on solving a diffusion equation. The present work considers the velocity distribution of electrons escaping from a confining potential well in one spatial dimension. A Monte Carlo method is used that is based on Newton's equation of motion. Highly evolved three-dimensional Monte Carlo computer programs [6-8] are available for the simulation of semiconductor devices and the programs include sophisticated models of electron-phonon scattering. Electron escape over a potential barrier has been explored with such methods [7,9]. However, these programs are complex, and it is useful to see how models for phonon absorption and emission affect the escape time, the electron velocity distribution, and the electron energy distribution. For example, Ref. [9] has shown that optical phonon scattering introduces spatial oscillations in the energy distribution of the confined electrons.

A series of simple models is developed to learn how electron-phonon scattering influences the details of electron escape from a potential well. Section 2 describes the sequence of models, and section 3 presents the numerical results. The calculated velocity distribution is also compared with that from the classical theory of thermionic emission [10]. The final section contains a concluding discussion.

MODEL PROBLEM

Each electron moves in a constant electric field according to Newton's equation of motion between electron-phonon scattering events. The first model bases the length of the electron's free flight on the τ from the electron's mobility

$$\mu = q\tau/m^* \qquad , \qquad (1)$$

with $m^* = 0.28$ times the free electron mass and q the magnitude of the electron charge. The electric field-dependent mobility model used is that of Scharfetter and Gummel [11] for silicon with an assumed doping density of $3 \times 10^{16}/cm^3$. The free flight lasts Δt

$$\Delta t = -\tau \ln r \qquad \qquad , \qquad (2)$$

where r is a uniformly distributed random number between 0 and 1 that is generated by the computer subroutine ran2 [12]. The electron's velocity is reset before each free flight to the thermal velocity appropriate for 300 K

$$v_0 = (kT/m^*)^{1/2} \qquad \qquad , \qquad (3)$$

with a randomly chosen direction. Here k is Boltzmann's constant and T = 300 K. The phonon energy E_{ph} is assumed to be $kT/2 = 0.013$ eV. This first model corresponds to a literal interpretation of pictures explaining drift velocity such as that in Grove [13].

The second model allows for a variation in the phonon energy from $\beta^2 kT/2$ to $kT/2$. The phonon energy used at a particular time step is selected randomly from between these limits with $\beta^2 = 0.5$ or 0.75.

The third model lets the electron accumulate energy. After each free flight, either a phonon emission or a phonon absorption event is selected at random based on a fixed emission (E) to absorption (A) ratio of E/A. The energy of the phonon involved in absorption is rE_{ph} with r generated by ran2 and $E_{ph} = 0.013$ eV. Emission is treated similarly with rE_{ph} or rE if the electron energy E is less than E_{ph}. The third model uses $\tau = 0.714 \times 10^{-13}$ s, which is comparable to the τ from the mobility for an electric field of 2500 V/cm. The scattering frequency is considered to be independent of the electron's energy. The energy distribution was calculated for selected E/A within a range of E/A with zero electric field and the resultant extracted temperature is plotted in Figure 1. An E/A of about 1.4 corresponds to 300 K. Larger values of E/A are used in the next Section in order to slow the rate of electron escape.

Each electron starts at zero time with v_0 or $0.5v_0$ and moves in a direction chosen at random. In all cases, each electron is followed from its initial position at 0.1 μm until it reaches 1.0 μm, where escape is declared. The time taken and the electron's velocity are then recorded. The electron's energy before each phonon scattering is saved to develop the electron energy distribution. The electron's position at selected times is also recorded. Ten thousand electrons are followed for each case.

RESULTS

The first model yields the dashed curve in Figure 2. The electrons escape over the 0.25 V barrier faster than with the second model, which is represented by the solid curve for $\beta^2 = 0.5$ and the dotted curve for $\beta^2 = 0.75$. It is found that the wider the range of phonon energies, the slower the electrons are to escape. The number of electrons left decays with an exponential in time. The extracted characteristic time τ is 3.3 ns for the first model, and 5.4 and 11.6 ns for β^2 equal to 0.75 and 0.50, respectively. The velocity distribution shifts to lower average velocity as the phonon energy range expands. The single phonon energy model results in 1.21×10^7 cm/s,

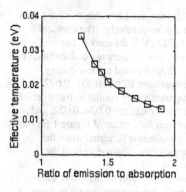

Figure 1. Effective temperature from the energy distribution for electrons moving in zero electric field with E/A varied.

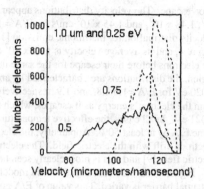

Figure 2. Velocity distributions upon escape. First model = dashed curve, second model = dotted and solid curves.

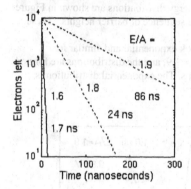

Figure 3. Electrons in the potential well versus time. Solid curve for E/A = 1.6, dashed curve for E/A = 1.8, dotted curve for E/A = 1.9.

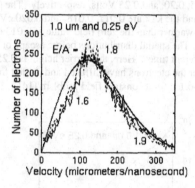

Figure 4. Velocity distributions upon escape for the cases of Fig. 3. The smooth curve is the classical thermionic emission result.

while the second model shows 1.13×10^7 and 1.05×10^7 cm/s for $\beta^2 = 0.75$ and 0.5, respectively. These are all less than the thermal velocity v_o of 1.274×10^7 cm/s.

The next set of results uses the third model for a barrier of 0.25 V. Figure 3 has the decay curves with τ indicated. As the emission to absorption event ratio E/A increases, τ increases. More emission events delay the electron's escape and lower the average velocity of the electrons upon escape. The velocity distributions appear in Figure 4 and the average velocities are 1.56 x 10^7, 1.48 x 10^7, and 1.45 x 10^7 cm/s for E/A = 1.6, 1.8, and 1.9, respectively. The classical velocity distribution for thermionic emission [10] at kT* = 0.022 eV is the smooth curve of Figure 4, and its average velocity is 1.47 x 10^7 cm/s. Figure 5 shows the energy distribution of the electrons before their escape for the same three values of E/A. Beyond the low energy region, the distributions are characterized by an effective temperature kT* of 0.031, 0.022, and 0.020 eV for E/A =1.6,1.8, and 1.9, respectively. These temperatures are similar to those derived from the electron's energy as it escapes, which for the same three cases are 0.024, 0.022, and 0.021 eV. In all cases, the effective temperature exceeds the zero field value of Figure 1 for the same E/A by at least 50%. Apparently, the electron energy distribution is heated when the electrons drift with the electric field. The electrons lose energy as they diffuse against the electric field [9] and this is most clearly seen here for E/A = 1.6.

Figures 6 to 9 are based on the third model with E/A fixed at 1.9 and the height of the potential barrier is varied. This value of E/A gives characteristic decay times similar to those from the Smoluchowski equation for a linear potential [4]. The decay curves of Figure 6 lead to the τ versus barrier height plot of Figure 7, where a roughly exponential increase is seen [2,4]. In fact, the rise in τ is $e^{qV/kT*}$ for barrier height V with kT* = 0.028 eV, which is larger than the effective temperatures extracted above. Figure 8 has the escaping electron velocity distributions and the average velocity is 1.36 x 10^7, 1.37 x 10^7, and 1.45 x 10^7 cm/s, for barrier heights of 0.15, 0.20, and 0.25 Volts, respectively. The pre-escape energy distributions are shown in Figure 9, and the kT* are 0.017, 0.018, and 0.020 eV for the same sequence of barrier heights. These are warmer than the zero field value of 0.013 eV.

The spatial distributions before escape of Figure 10 look exponential and similar for different times. Here, the barrier height is 0.25 Volts, E/A = 1.9, and the distributions are formed after the electrons have diffused and drifted for 10 and 50 ns. The exponential distribution is expected for a constant field problem [14].

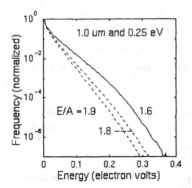

Figure 5. Electron energy distributions before escape for the cases of Fig. 3.

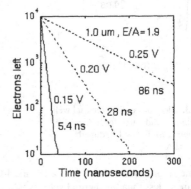

Figure 6. Electrons in the potential well versus time. Potential barrier = 0.15 eV (solid), 0.20 eV (dashed), 0.25 eV (dotted).

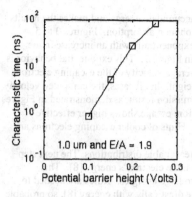

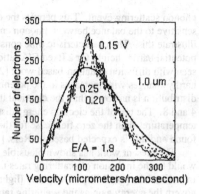

Figure 7. Characteristic time versus potential barrier height for the cases of Fig. 6 with the 0.10 eV case added.

Figure 8. Velocity distributions upon escape for the cases of Fig. 6. The smooth curve is the classical thermionic emission result.

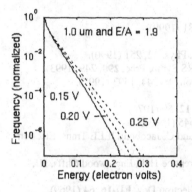

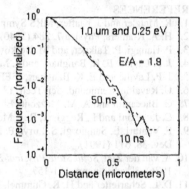

Figure 9. Electron energy distributions before escape for the cases of Fig. 6.

Figure 10. Spatial distributions of the electrons before escape at 10 ns (solid curve) and 50 ns (dashed curve).

DISCUSSION

The previous section contains the numerical results for three simple models of electron-phonon scattering. The first and second models show that the escaping electron's velocity distribution broadens when the range of phonon energies increases. The characteristic time constant for escape also increases as more low-energy phonons are included with the second model. The third model is more realistic in that the electron can retain energy beyond a single phonon scattering event. This permits the electron to accumulate energy and makes the results sensitive to the balance between phonon emission and phonon absorption. Figures 3 to 5 illustrate this. The characteristic time constant grows exponentially with an increase in the potential barrier height for a fixed E/A ratio as shown in Figure 7. This exponential behavior is seen with diffusion equation-based work [2-5]. The average velocity of the escaping electrons hardly changes with increases in the potential barrier height. In all cases, the computed velocity distribution is narrower than the classical thermionic emission result, as demonstrated in Figures 4 and 8. The tail of the electron energy distribution before escape shows higher effective temperatures than the zero field case for the same E/A. Hints of cooler escaping electrons are found and this is expected based on Refs. [9] and [10].

The present work has shown plausible trends and reasonable distributions. The next step would be the inclusion of scattering rates that depend on the electron's energy [8]. This introduces the possibility of long free flights that lead to large energies. Rules are needed to govern the energy gain as the scattering rate can change drastically with energy [8], so probable electron-phonon collisions are ignored. The inclusion of phonon absorption and emission probabilities that depend on temperature is also warranted. Preliminary calculations with such probabilities show extremely rapid electron escape. It appears prudent to invoke the existing more complete Monte Carlo computer programs at this point.

REFERENCES

1. R. Metzler and J. Klafter, MRS Symp. Proc. **543**, 281 (1999).
2. H. A. Kramers, Physica **7**, 284 (1940).
3. P. Hanggi, P. Talkner, and M. Borkovec, Rev. Mod. Phys. **62**, 251 (1990).
4. J. P. Lavine, E. K. Banghart, and J. M. Pimbley, MRS Symp. Proc. **290**, 249 (1993).
5. J. P. Lavine and E. K. Banghart, IEEE Trans. Electron Dev. **44**, 1593 (1997).
6. U. Ravaioli, Semicond. Sci. Tech. **13**, 1 (1998).
7. G. Baccarani and A. M. Mazzone, Electronics Lett. **12**, 59 (1976).
8. C. Jacoboni and L. Reggiani, Rev. Mod. Phys. **55**, 645 (1983).
9. F. Venturi, E. Sangiorgi, S. Luryi, P. Poli, L. Rota, and C. Jacoboni, IEEE Trans. Electron Dev. **38**, 611 (1991).
10. A. van der Ziel, *Solid State Physical Electronics* (Prentice-Hall, Englewood Cliffs, NJ, 1976), 3rd edition, pp. 151-157.
11. D. L. Scharfetter and H. K. Gummel, IEEE Trans. Electron Dev. **ED-16**, 64 (1969).
12. W. H. Press, S. A. Teukolsky, W. T. Vetterling, and B. P. Flannery, *Numerical Recipes in FORTRAN* (Cambridge University Press, New York, 1992), 2nd edition, pp. 271-273.
13. A. S. Grove, *Physics and Technology of Semiconductor Devices* (John Wiley, New York, 1967), p. 107.
14. S. Chandrasekhar, Rev. Mod. Phys. 15, 1 (1943). See especially pages 57-59.

Mat. Res. Soc. Symp. Proc. Vol. 651 © 2001 Materials Research Society

Static Friction between Elastic Solids due to Random Asperities

J. B. Sokoloff
Physics Department
Northeastern University
Boston, MA 02115

ABSTRACT

Several workers have argued that there should be negligible static friction per unit area in the macroscopic solid limit for interfaces between both incommensurate and disordered atomically flat weakly interacting solids. In contrast, the present work argues that in the low load limit, when one considers disorder on the multiasperity scale, the asperities act independently to a good approximation. This can account for the virtual universal occurrence of static friction.

INTRODUCTION

It is well known that in virtually all known circumstances one must apply a minimum force (i.e., static friction) in order to get two solids, which are in contact, to slide relative to each other. It has been argued, however, that there might be no static friction for the interface between two weakly interacting nonmetallic crystalline solids, which are incommensurate with each other. Aubry showed this for the weak potential limit of the one dimensional Frenkel-Kontorova model[1], and recently He, et. al.[2] and Muser and Robbins[3] have shown for clean weakly interacting two dimensional incommensurate interfaces that the force of static friction per unit area falls to zero as $A^{-1/2}$ in the thermodynamic limit, where A is the interface area. Even identical solids are incommensurate if their crystalline axes are rotated with respect to each other. Since this is a potentially technologically important result, it is important to study the circumstances under which it is true. Disorder can pin contacting solids, just as it pins sliding charge density waves[4,5] and vortices in a superconductor[6], provided that the Larkin domains (i.e., domains over which the solids are able to distort to accommodate the disorder at the interface) are small compared to the interface size. Recently, however, it has been shown that Larkin domains for contacting three dimensional elastic solids are enormously large compared to typical solid sizes[7-10], implying that the force of static friction per unit area due to interface disorder should also fall off as $A^{-1/2}$ in the thermodynamic limit. In contrast to Refs. 2 and 3, where it was proposed that the presence of a submonolayer film of mobile molecules at the interface is a requirement for the occurrence of static friction between incommensurate surfaces, it is argued here that disorder that occurs on the multi-micron scale, due to disordered asperities, results in Larkin domains that are much smaller than the interface size, even for clean interfaces, implying that there is static friction for a clean macroscopic surface.

THEORY

Refs. 7-9 seem to imply that weakly interacting disordered surfaces cannot exhibit static friction. We shall see, however, that unlike weak atomic scale defects, for which the elastic interaction between them can dominate over their interaction with the second surface, for contacting asperities that occur when the problem is studied on the multimicron scale, the interaction of two contacting asperities from the two surfaces dominates over the elastic interaction between them. It is suggested here that this could be responsible for the virtual universal occurrence of static friction. This is true because if one attempts to slide two contacting asperities with respect to each other, one must shear the crystalline interface at the region of asperity contact. The resulting shear stress must, therefore, vary considerably when the asperities are sheared over a distance of the order of an atomic spacing. Because of the smallness of an atomic spacing compared to typical inter-asperity distances, almost all of the contacting asperities can sink to their minimum contact potential energy by moving a distance of the order of an atomic spacing, with a negligible cost in elastic energy of the solids. As a consequence, when one attempts to slide the two surfaces with respect to each other, the contributions to the restoring forces from each of the contacting asperities will add together coherently. Thus, when one considers macroscopic surfaces with contacting asperities, one concludes that clean surfaces will always exhibit static friction. The details of this argument are given in Ref. 10.

CONCLUSIONS

In conclusion, when one considers atomically smooth surfaces, arguments based on Larkin domains indicate that the disorder at an interface between two nonmetallic elastic solids in contact will not result in static friction. When one applies such arguments to the distribution of asperities that occur on multimicron length scales, however, one finds that the asperities are virtually always in the "strong pinning regime," in which the the Larkin domains are comparable in size to a single asperity. This accounts for the fact that there is almost always static friction. Muser and Robbins' idea [2,3],however, is not invalidated by this argument. Their result will still apply for a smooth crystalline interface. It will also apply in the present context to the contact region between two asperities, implying that for a clean interface, the shear force between contacting asperities is proportional to the square root of the contact area. The GW model[11] predicts for this case that the average force of friction is proportional to the 0.8 power of the load and is much smaller than it wouldbe for dirty surfaces.

ACKNOWLEDGMENTS

I wish to thank the Department of Energy (Grant DE-FG02-96ER45585).

REFERENCES

1. S. Aubry, in Solitons and Condensed Matter, ed. A. R. Bishop and T. Schneider (Springer, New York, 1978), p. 264.
2. G. He, M. H. Muser and M. O. Robbins, Science 284, 1650 (1999).
3. M. H. Muser and M. O. Robbins, Phys. Rev. B 61, 2335 (2000).
4. H. Fukuyama and P. A. Lee, Phys. Rev. B17, 535 (1977).

5. P. A. Lee and T, M. Rice, Phys. Rev. B19, 3970 (1979).

6. A. I. Larkin and Yu. N. Ovchinikov, J. Low Temp. Phys. 34, 409 (1979).

7. B. N. J. Persson and E. Tosatti in Physics of Sliding Friction, ed. B. N. J. Persson and E. Tosatti (Kluwer Academic Publishers, Boston, 1995), p. 179.

8. V. L. Popov, Phys. Rev. Lett. 83, 1632 (1999).

9. C. Caroli and Ph. Noziere, European Physical Journal B4, 233 (1998).

10. J. B. Sokoloff, Phys. Rev. Lett. 86, 3312 (2001) and J.B. Sokoloff, unpublished.

11. J. A. Greenwood and J. B. P. Williamson, Proc. Roy. Soc. A295, 3000 (1966).

Mat. Res. Soc. Symp. Proc. Vol. 651 © 2001 Materials Research Society

Observation of Laser Speckle Effects in an Elementary Chemical Reaction

Eric Monson and Raoul Kopelman
University of Michigan
Departments of Chemistry and Applied Physics
Ann Arbor, MI 48109-1055 USA

ABSTRACT

An experimental demonstration is shown for non-classical reaction kinetics in a homogeneous system with an elementary reaction, $A+B \rightarrow C$. Sensitivity to the initial distribution of reactants is observed, along with a new reaction-kinetics regime which is a direct consequence of speckles in the laser beam. The long-time regime gives the first experimental demonstration of the asymptotic self-segregation ("Zeldovich") effect, in spite of the non-random, speckled initial distribution of reactant B. Monte-Carlo simulation results are consistent with the experiments, and spatial analysis of these results correlates the excess of long-wavelength components in the initial reactant distribution with an anomalous slowing of the reaction progress.

INTRODUCTION

The field of "non-classical" reaction kinetics has been established in the past 20 years [1, 2], and is distinguished from "classical" kinetics in that the latter, which is taught in many chemistry and physics textbooks, is based on a mean-field result. This type of solution to the kinetics problem includes the built-in assumption of a constantly randomized system, so that the reactants always have optimal access to each other. Out of this treatment comes relationships which do not depend on the dimensionality of the system or the initial reactant distribution. In "non-classical" kinetics, on the other hand, diffusion is the only transport mechanism for the reactants, and because the characteristics of diffusive motion are both dimension-dependent and inefficient in exploring space (when compared to convection or other types of mixing), many very interesting and important results are revealed which deviate dramatically from the "textbook" case.

The well known classical result for an $A+B \rightarrow C$ reaction (where the product C is inert and irreversible) gives

$$\rho_A = \rho_B \sim t^{-1},$$

whereas the non-classical result for reactants which are initially randomly distributed has been shown with Monte-Carlo simulation and analytical results to be

$$\rho_A = \rho_B \sim t^{-d/4}$$

(where d is the dimensionality of the system; $d \leq 4$) [3, 4]. This can equivalently be expressed, partly for historical convenience, in terms of the "reaction progress" as

$$(1/\rho - 1/\rho_0) = t^{d/4}.$$

When viewed on a log-log plot, this asymptotic result gives a straight line with a slope of $d/4$, referred to as the Ovchinnikov-Zeldovich rate. This result gives a dramatically slower reaction progress than the classical case, due entirely to the inefficiency of the diffusive process when compared to convection or other types of stirring. In the diffusion-limited case, any fluctuations in the initial reactant distribution persist and lead to segregated regions of reactants, forcing the reaction to proceed only at the interfaces between these regions. In contrast, the classical case assumes that any fluctuations in the distributions, whether present initially or formed by the reaction process itself, are instantly smoothed out, giving the reactants optimal access to each other throughout the process. Even in the three-dimensional, non-classical case, the reaction is slowed due to fluctuations in the reactant distributions which can not be eliminated by the diffusive motion.

Recently, theoretical work has also been done on more initially correlated systems. Lindenberg and coworkers [5, 6] showed a strong slowing of the reaction progress in a system where the reactants are initially distributed in a fractal pattern on a standard lattice. This distribution is characterized by an extreme excess of long wavelength components in the initial distribution when compared to the random case, which leads to even more dramatic slowing of the reaction progress.

Experiments have been done in the $A+B \rightarrow C$ case on initially separated reactants, but work has been absent for both the initially random case and for more correlated initial conditions (besides stirred cases verifying the classical result, see e.g. [7]). We present here the first results showing the extreme sensitivity of these simple reactions to the initial reactant distribution and the dramatic slowing of the reaction progress in the absence of convection and stirring [8].

EXPERIMENTAL METHODS

One of the reasons for the lack of experimental work on these initially random systems is the challenge of finding and suitable chemical system. An observable reaction is required. For a diffusion limited, initially random system, the reactants must be quickly mixed to a known distribution, then the mixing and convection must be stopped and a well defined "time zero" established before the study can even begin. But, the "geminate" case must also be avoided (where a product is dissociated into its reactant pair by some sudden perturbation such as light or heat), since this produces small initial reactant separations and strong pair correlations resulting in a reaction which is much more controlled by the A and B interactions rather than diffusive motion.

In our work, a system was chosen, as shown schematically in Figure 1, where the "A" particles (a calcium sensitive fluorophore – Calcium Green-1 from Molecular Probes) are distributed randomly, and the "B" particles (calcium ions) are initially bound to a "cage" molecule (DM-Nitrophen from CalBiotech). This allows solution preparation without reaction, but when the sample is hit with a pulse of near-UV light, some of the cages are photolyzed, releasing calcium ions which are now free to diffuse with the dye and react, forming a highly fluorescent product which can be monitored experimentally with high signal to noise ratio.

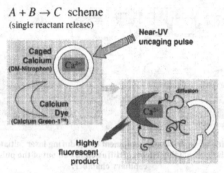

$A + B \rightarrow C$ scheme
(single reactant release)

Figure 1: Experimental chemical scheme for A+B→C reaction kinetics.

Convection was avoided in our system by confining the samples to small-diameter, fused-silica capillaries (50-75 µm). The polyimide coating was removed with fuming sulfuric acid and samples were injected into the capillaries after preparation and immediately before experiments were performed.

Two different light sources were used for the experiments. A xenon flash lamp (Hamamatsu) created an incoherent source for uncaging after the addition of an efficient condenser and optical filtering system. As a more powerful, coherent source, a wavelength tripled pulse from a 10 Hz, 10 ns Nd:YAG laser was guided to the sample chamber with a 500 µm fused-silica optical fiber (Polymicro Technologies).

The chemical system was optimized for tighter binding and against some experimental artifacts. BAPTA, which is the parent compound of the dye, has binding strength which increases with lowered ionic strength[9]. EGTA, related to the cage, on the other hand, has increased affinity for calcium ions with increasing pH (ibid.). So, for the laser initiated experiments, a Tris buffer with pH 8.5 and 10 mM ionic strength (including 2 mM potassium chloride) was used, and for the flash lamp experiments a carbonate buffer with pH 10 and 10 mM ionic strength (2 mM KCl) was used. P-phenylene diamine, a strong anti-fade agent, was included at a level of 1 mg/ml in regular buffer, and 2 mg/ml when used in 50% wt:wt glycerol, to eliminate some reversible photobleaching observed in the dye after each laser pulse.

A schematic of one experimental equipment setup is shown in figure 2A. During each experiment, a single flash of near-UV light is used to uncage some of the calcium ions, and the change in sample fluorescence (due to product formation) is followed by placing the samples on an inverted optical microscope in a standard epi-illumination fluorescence configuration. The region monitored for kinetics was always kept at about half the size of the flashed region for uncaging (as shown in figure 2B) to guard against effects of reactant/product diffusion in and out of the monitored space. A Hamamatsu PMT detects the light signals, and then, after current to voltage conversion, the signal is digitized by a Tektronix oscilloscope. The data is then transferred to a computer, where a logarithmic smoothing is performed in the same way as in ref [10], resulting in evenly spaced data on a logarithmic plot in time.

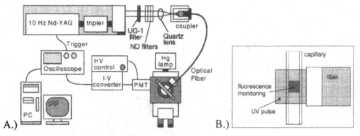

Figure 2: (A) Schematic of experimental equipment setup during laser initiated kinetics measurements, along with (B) the method of avoiding effects of diffusion in and out of the pulsed region of the sample capillary chamber.

To compare with past theoretical and simulation work, it is preferable to conduct the experiments with equal initial A and B global densities. But, it is difficult to determine the correct amount of UV energy to use for each pulse in the experiment, so an appropriate range of energies was determined empirically by hitting the sample with multiple pulses and observing the number of flashes necessary to saturate the calcium sensitive dye. The correct number of pulses can be closely determined by knowing the amounts of cage and dye in the solution. Nonetheless, during each experiment, a range of pulse energies was used in order to include the equal initial A and B population point.

Simulations were conducted using the standard methods [11]. In brief, each particle was landed on a lattice using no excluded volume conditions, and every particle of each type was moved randomly once during each time step before checking for reaction. Speckled initial conditions were obtained by landing all A particles randomly, and all B particles according to a speckled probability map (e.g. X position vs. Y position vs. probability in two dimensional simulations) generated with Matlab software using a method similar to Fugii [12].

Wavelet-based spatial analysis was conducted on some of the simulation results. Algorithms were modified from Numerical Recipes in C [13]. The simple Haar wavelet was used as the basis function for the transform, but routines were modified to perform a complete convolution with each wavelet (rather than using the efficient "pyramidal" algorithm supplied in the text, which is too sensitive to spatial shifts in the data). A wavelet transform results in a data set of amplitude vs. spatial scale vs. spatial position along the data set. The average square amplitude of the transforms were taken over the spatial position dimension (motivated by Scargle [14]), resulting in the wavelet equivalent of a Fourier power spectrum.

RESULTS

Figure 3 shows the results of two sets of experiments conducted on the same basic chemical systems. The chemical reaction data set which rises smoothly and faster in time (circular symbols) has been initiated with the xenon flash lamp, and the second set (square symbols) is a laser initiated reaction. The flash lamp data looks quite as would be expected with random initial distributions of both A and B, the product increasing smoothly over time and then flattening off as the system moves toward equilibrium. (Note that the downturn in the curves at very late times is due to the unavoidable presence of unphotolyzed, initially unbound cage molecules taking up some of the calcium ions at equilibrium.) The laser initiated reaction, on the other hand, exhibits features not expected for a random initial distribution A+B→C reaction, whether classical or non-classical.

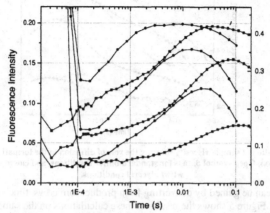

Figure 3: Experimental data showing fluorescence (reaction product) increase in time. Circular symbols show flash-lamp initiated reactions. Square symbols represent laser initiated reactions. The reactions were conducted in 50% wt:wt glycerol solutions as an experimental convenience to slow the reactions. (There are three curves in each data set because reaction initiating pulses are varied in energy to include the equal A and B initial global density point.)

The laser initiated product increase with time begins with a fast portion (before about 200 µs), but then slows down (between ~200 – 2000 µs) before increasing in rate once more (~3 ms – 30 ms). The mid-time slow region of these curves is completely unexpected if the reactants are initially randomly distributed. To test whether the features of these curves are really related to the diffusion and reaction behavior, the experiments were performed in two different viscosity solutions. One was a standard buffer solution, and the other a 50% wt:wt buffer in glycerol solution. Figure 4 shows that in these experiments, all of the main product increase with time features are reproduced in both cases, just shifted in time by the viscosity difference of the preparations.

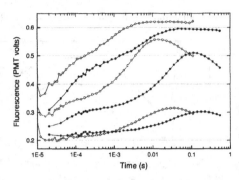

Figure 4: Variation of sample viscosity showing the role of diffusion in the formation of kinetic features in product formation. Open symbol data is for standard buffer conditions, and closed symbols are for 50% wt:wt glycerol conditions.

More insight can be gained by converting from product formed vs. time to a plot of reaction progress vs. time. Figure 5 shows the results of these calculations on the same data shown in figure 3. (Larger errors are introduced here because the conversion from product density to reactant density involves subtractions of the product fluorescence intensity from the maximal and minimal possible values for each experimental run. It is errors in the estimation of these F_{max} and F_{min} numbers which accounts for most of the error associated with each point.) Examination of the resultant curves reveals behavior in the flash-lamp initiated reactions consistent with the non-classical theoretical results. Dotted lines are included on this log-log plot as a guide to the eye, indicating the 3/4 power law slope predicted (for this three dimensional situation) by the Ovchinnikov-Zeldovich analytic result. In the laser initiated experiments, there is also a short portion of the curve, during the later-time region, where the data parallels this Zeldovich 3/4 power law slope. Another interesting point to note is that the slow, middle-time portion of the curve has a slope around 1/8, a much slower reaction progress than would be expected even in the 1-dimensional, initially random non-classical case.

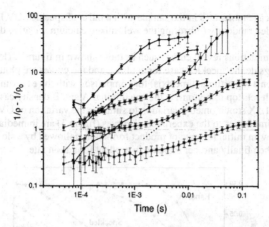

Time (s)

Figure 5: Experimental reaction progress. Circular symbols show flash-lamp initiated reactions, and square symbols show laser-initiated reactions. Dotted lines are included as a guide to the eye showing the 3/4 power law slope predicted by the "Zeldovich" asymptotic result (3-dimensional, initially random).

If the reactants are both starting at about the same global density, then some sort of inhomogeneity in the initial reactant distribution is the only likely explanation for the difference between the flash lamp and laser initiated reactions. When a CCD camera was set up to image the end of the laser delivery fiber, the pattern shown in figure 6(A) was observed. This laser speckle pattern has the characteristic, triangular Fourier power spectrum [15, 16]. Simulations were necessary, though, to test whether this pattern in the light from the UV laser pulse (which governs where the B particles in the A+B→C reaction are created) is sufficient to reproduce the types of features seen in the experimental curves.

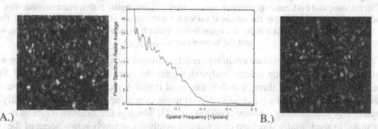

A.)   Spatial Frequency [1/pixels]   B.)

Figure 6: (A) Experimental image of laser speckle at the end of the UV pulse delivery fiber. The spatial intensity pattern has the characteristic triangular Fourier power spectrum. (B) Matlab generated speckles for use as a probability map for reactant landing during A+B→0 reaction kinetics Monte-Carlo simulations.

Figure 6(B) shows an artificial speckle pattern which was calculated similarly to the method of Fugii [12] using the software package Matlab. Briefly, the calculation simulates a plane wave scattered off a slightly smoothed gaussian random surface, and then propagates this light through space to a region far away where this image is formed. Patterns such as this were used as probability maps for the generation of B particles on 1D or 2D lattices (simulating the release of calcium ions from cage molecules in the sample with a pulse of speckled UV laser light), with the A particles landed randomly (mirroring the well-mixed calcium sensitive dye in the experiments).

Monte-Carlo simulations reveal curves such as those shown in figure 7. Here the random case is shown alongside the speckled case for contrast, and the curves are plotted as product increase vs. time (on a logarithmic time axis) for comparison with the experimental results. The center curve of each set (top to bottom) is the equal initial A and B density case, and the other curves represent a 1.5X stoichiometry variation to simulate the variation of UV pulse energies in the experiments. Similarities to the experimental curves (figure 3) are immediately obvious. The speckled cases show an initial fast rate of product formation, followed by a slowdown during the mid-time regime, then finally another increase in product formation rate.

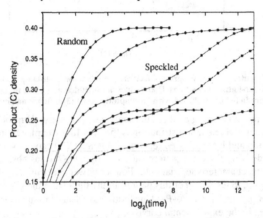

Figure 7: Simulated product formation vs. log time (base 2), showing the results of random initial A distributions, and both random (circles) and speckled (square) initial B distributions. Stoichiometry variations are included to mirror the effects of varied UV pulse energies in the experiments, with the center curve (top to bottom) in each case being the equal initial A and B density case. In the top curve $\rho_{B0} = 1.5*\rho_{A0}$, and the bottom curve $\rho_{B0} = \rho_{A0}/1.5$.

Converting these values once again into reaction progress vs. time, and plotting on a log-log scale (base 2 for convenience in later analysis), we see the curves presented in figure 8. Focusing on the center of each set of three, which is the equal initial A and B global density case, we see the random initial conditions (circular symbols) producing a curve which rises smoothly, asymptotically approaching a power law slope of 1/2, which is the Zeldovich rate in two-dimensions. In the initially speckled case (square symbols), strongly reminiscent of the experimental laser initiated results, the reaction progress starts out with a steep slope, then moves through a time period of very shallow slope, followed by a region of steeper slope. This late-time regime is seen to exhibit a power law slope close to the Zeldovich asymptotic prediction for the initially random case. To show that this is not just a coincidence for this two-dimensional simulation, figure 9 shows initially speckled and random simulations on both one and two-dimensional lattices, along with some variation in speckle size.

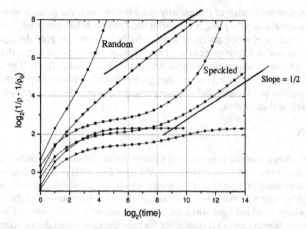

Figure 8: Simulated reaction progress showing both initially random (circles) and speckled (square) reactant distributions. Stoichiometry variations are also included to mirror the effects of varied UV pulse energies in the experiments.

It is clear that the speckle size controls the time at which the reaction progress begins to leave the mid-time, slow regime and climb into the Zeldovich-like behavior. There is a factor of two size difference between the two speckle "sizes" used for the simulations, so the factor of four time shift between the size 64 and 128 curves suggests (with this limited data set) that this behavior goes with the typical Einstein diffusion relation, where $\langle r^2 \rangle \sim t$.

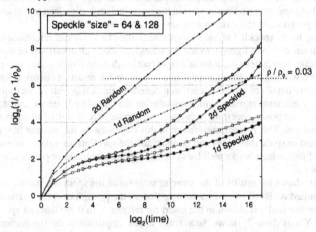

Figure 9: Simulation results showing the effects of dimensional and speckle size variations. Both 1D and 2D speckled results are shown, along with initially random results to serve as a comparison in the asymptotic slopes of the reaction progress.

Another interesting feature of the initially speckled curves is the point at which (especially in the two-dimensional case) they begin to exhibit the Zeldovich-like behavior. It has been shown in simulations that the reaction progress in the initially random case goes through a crossover to the Zeldovich regime at a consistent ratio of the reactant density to the initial density (ρ_A/ρ_{A0}) [17]. In two dimensions, this value is about 0.03, and is indicated in figure 9 by a horizontal dotted line. The initially speckled curves, though, have made the transition to this behavior regime much earlier, presumably due to some pre-segregation of the reactants which is built-in to the speckled initial distribution.

DISCUSSION

It has been shown that the experimental kinetics for this elementary, diffusion-limited chemical reaction have a very sensitive dependence on the initial reactant distribution which has been predicted by analytic results and computer simulations. Consistent results were also obtained when comparing simulations with speckled initial distributions of the B reactant with experiments where speckled light was used to release one of the reactants and initiate the reaction. The reaction progress vs. time plots for this speckled initial distribution show some interesting features which can be understood in more detail by looking at a spatial analysis of the reactants over the time of the simulations.

Lindenberg studied a couple types of correlated initial conditions, including simulations of a fractal initial distribution on a one-dimensional lattice [5, 6]. A random distribution, by definition, has fluctuations in the local reactant densities whose representation is equal on all spatial scales (i.e. a flat distribution of fluctuations vs. spatial scale). The fractal distribution that Lindenberg studied has a power law increase in fluctuation representation with increasing spatial scale which goes on forever in principle (because a fractal is self-similar on all scales). This increase in long-wavelength fluctuations in the reactant distribution lead to a dramatic slowing of the reaction progress, in this case to a power-law slope around 1/10.

Returning to the speckled initial distribution studied here, we can refer back to the Fourier power spectrum shown in figure 6 to see the triangular shape characteristic of laser speckle. Unlike a fractal, this pattern has a finite extent (notice that the plot in figure 6 is on a linear scale), and we will see that this has a definite impact on the resultant chemical kinetics.

To see more details of how the reaction proceeds in time, it is instructive to look at the distribution of reactants at many time points during the (simulated) reaction. Instead of the more traditional Fourier power spectrum technique, we choose here to utilize an equivalent form of a wavelet transform. This is more convenient in this case because the wavelet basis functions are more localized in space than sines and cosines, and the wavelets are related to each other in size by powers of two, which works well for systems like these which have power law behaviors both in space and time.

Figure 10 shows the results of the wavelet analysis on the speckled initial condition, diffusion-limited A+B→0 reaction at many points in the process over time. The reaction progress (log-log scale) is shown in the insert for reference. In these wavelet spectra, the logarithmic Y axis (base 2) shows the average square amplitude of the fluctuations present at each spatial scale, and the logarithmic X axis (base 2) shows spatial scale, which is similar to a wavelength. The top curve is the initial distribution of the A and B reactants on the one-dimensional lattice, and each curve below is generated at a time step during the course of the reaction which is a power of four (i.e. simulation time step = 0, 1, 4, 16...).

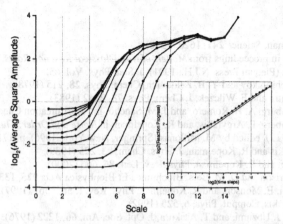

Figure 10: Wavelet spectra of the speckled reaction simulations. The initial distribution is shown at the top, revealing the excess of long wavelength fluctuations present in the distribution (when compared to a flat, random distribution). As the reaction progresses, a loss of information can be seen at small spatial scales, followed by a slowing of the reaction while it proceeds through the spatial scale of the speckles (as seen in the reaction progress plot inset – straight line with a power law slope of 1/4), followed, finally, by an approach to the Zeldovich random result as the distribution flattens out again.

The top curve, then, is the equivalent of the triangular power spectrum, but now plotted on a log-log graph, emphasizing the limited extent of the distribution (in contrast to the infinite, idealized fractal case mentioned earlier). At the smallest spatial scales (scales 1 – 4), we have a flat distribution, representing the randomness of the reactants themselves on scales smaller than the speckles. Then, the curve rises (scales 4 – 7), showing the increase in long wavelength fluctuations present due to the speckle pattern. Finally, at even larger spatial extents (scales 8 – 14), the speckles have a flat distribution again because of the finite extent of the pattern. As the reaction begins, we see an immediate drop-off in information at the smallest spatial scales as reactants annihilate each other, while fluctuations on larger scales persist – again, because diffusion is not an efficient enough transport mechanism (in these low dimensions) to redistribute, or randomize, the reactants. At later times, the progress is seen to slow while the reaction proceeds through the spatial scales of the speckles. In the later stages, the reactant groupings with fluctuation scales in the large, flat region of the initial curve (similar to a random distribution) begin to be affected by the reaction-diffusion process and the reaction progress is seen to exhibit the Zeldovich-like behavior (slope of 1/4 shown in the inset plot).

So, even in this case of a non-ideal, real-life reaction system, it is clear that the lack of stirring (by convection or other means) creates situation where the reaction system is very sensitive to the initial distribution of reactants because of the non-classical behavior of a diffusion limited reaction. It has also been shown that the details of the reaction progress with time, in this case of a chemical reaction initiated with a speckled laser beam, can be correlated with the spatial fluctuations present in the initial distribution of reactants.

We gratefully acknowledge funding from NSF Grant No. DMR-9900434.

REFERENCES

1. R. Kopelman, Science **241**, 1620 (1988).
2. G. Weiss, in proceedings from *Models of Non-Classical Reaction Rates*, edited by J. L. Lebowitz (Plenum Press, N.I.H., 1991), J. Stat. Phys. Vol. 65.
3. A. A. Ovchinnikov and Y. B. Zeldovich, Chem. Phys. **28**, 215 (1978).
4. D. Toussaint and F. Wilczek, J. Chem. Phys. **78**, 2642 (1983).
5. K. Lindenberg, A. H. Romero, and J. M. Sancho, Int. J. Bifurcation Chaos **8**, 853 (1998).
6. K. Lindenberg, P. Argyrakis, and R. Kopelman, in *Noise and Order: The New Synthesis (Chapter 12)*, edited by M. Millonas (Springer, New York, 1996), p. 171.
7. P. Argyrakis and R. Kopelman, J. Phys. Chem. **93**, 225 (1989).
8. E. Monson and R. Kopelman, Phys. Rev. Lett. **85**, 666 (2000).
9. S. M. Harrison and D. M. Bers, Biochimica Et Biophysica Acta **925**, 133 (1987).
10. A. L. Lin, E. Monson, and R. Kopelman, Phys. Rev. E **56**, 1561 (1997).
11. P. Argyrakis, Comput. Phys. **6**, 525 (1992).
12. H. Fujii, J. Uozumi, and T. Asakura, J. Opt. Soc. Am. **66**, 1222 (1976).
13. W. H. Press, S. A. Teukolsky, W. T. Vetterling, *et al.*, in *Numerical Recipes in C: The art of scientific computing* (Cambridge University Press, 1992), p. 591.
14. J. D. Scargle, T. Steimancameron, K. Young, *et al.*, Astrophys. J. **411**, L91 (1993).
15. T. S. McKechnie, Optik **39**, 258 (1974).
16. J. C. Dainty, Optica Acta **17**, 761 (1970).
17. P. Argyrakis, R. Kopelman, and K. Lindenberg, Chem. Phys. **177**, 693 (1993).

Mat. Res. Soc. Symp. Proc. Vol. 651 © 2001 Materials Research Society

Mechanical measurements at a submicrometric scale : viscoelastic materials

Charlotte Basire and Christian Fretigny
ESPCI/LPQ, CNRS ESA 7069
Systèmes Interfaciaux à l'Echelle Nanométrique
10 rue Vauquelin
F-75231 Paris cedex 05, France

ABSTRACT

Adhesive and tribological properties of the tip of an AFM on viscoelastic samples are studied.
The kinetics of the indentation process is shown to be governed by *bulk* rather than by contact
edge dissipations. It is shown that the transition from static to dynamic friction regimes takes
place at a critical strain. During the static friction regime, the contact size remains nearly
constant. Owing to this property, viscoelastic moduli is measured in this regime. Obtained
results, characteristic of a micrometric domain are in a very good agreement with the same
properties measured on centimetric samples using dynamical mechanical analysis. Finally, a
stick-slip friction regime is observed in a range of velocities.

INTRODUCTION

Despite their importance in many application fields, adhesive and tribological properties
of elastomers are not fully understood: Their description is complicated by viscoelastic effects.
Indeed, though equilibrium adhesive static contact between elastic materials is now well
established [1,2], it has early been realized that the kinetics of the adhesion process can be very
slow on polymers, even on those that possess a low glass transition temperature [3-6].
Mathematical description of the kinetics of adhesion is rather complicated and no complete
theory is presently available [7]. Similarly, break of the adhesion through shear solicitation,
which corresponds to the sliding transition, has been theoretically investigated on elastic
materials [8,9] but viscoelastic effects have not yet been taken into account. In both cases, the
physical phenomena can be related to the problem of fracture propagation in viscoelastic media,
which has not been completely solved.

In contact mechanics experiments, the domain of the samples which determines the
mechanical response has a characteristic size comparable to the contact size itself. Small size
contact properties reflect thus the mechanical properties of the sample near its surface. It also
allows the analysis of deposited thin films without perturbations from the substrate provided that
their thickness is larger than several times the contact size. On an other hand, when the size of
the contacts become less than a characteristic size of the roughness, experimental studies lend
themselves to simpler comparison to existing theories of contact mechanics or friction: The
averages necessary to interpret the macroscopic experiments are then avoided.

The measurements of the interaction forces between the surface of a sample and a small
tip are at the very principle of atomic force microscopy (AFM). Mechanical interactions are used
to form images related to the local topography, friction forces or contact stiffness. For more
quantitative studies, however, some difficulties inherent to the AFM technique have to be
overcome. Firstly, the shape of the tip end is usually poorly known: Rather sophisticated
characterization methods are needed in order to describe it (typical radius of curvature are in the

range of 10 to 30 nm). Another difficulty originates in the impossibility to directly measure the contact radius while the experiments are performed. Finally, cantilever stiffness (usually in the range of .01 to 10 N/m) is frequently comparable to the contact stiffness. Then, contact mechanics experiments using AFM cannot be made under fixed grip nor fixed load conditions. This complicates their interpretation, specially when viscoelastic properties are involved.

Using AFM, local mechanical information is usually obtained through the measurement of the stiffness of the contact under imposed displacement (position modulation). Several studies have also been performed by direct application of an external force to the tip. A discussion of these techniques can be found in ref. 10. Normal and lateral stiffness of a contact linearly scale with the contact radius and with the Young modulus of the sample. For both excitation directions, the amplitudes should remain sufficiently small in order that the contact size can be considered constant. In the following, a study of adhesion and friction of the tip on soft viscoelastic samples is presented. Lateral modulation position of the sample is used in order to monitor the adhesion kinetics and the transition from static to sliding friction. From their analysis viscoelastic moduli are deduced. Finally, the sliding friction regime is analyzed.

Samples are styrene-butadiene or styrene-isoprene copolymers which have been described elsewhere [11,12]. Their glass transition (at 1 Hz) is below the room temperature (resp. –2 and 9°C) and they are partially cross-linked. Experiments are performed under ambient conditions, the samples are then in a true viscoelastic state. We use a commercial microscope (Multimode Nanoscope IIIa, Digital Instruments, Santa Barbara). Cantilevers (NanoProbe, Digital Instruments) have nominal stiffness of 0.12 and 0.6 N/m and tips are silicon nitride square based pyramids. Experiments presented here involve large indentation depth, making them few sensitive to the geometry of the tip end. Moreover, in these soft materials, contact stiffness is small as compared to the cantilever lateral stiffness allowing thus an interpretation of the experiments as a fixed grip one.

ADHESION KINETICS

A large scale image is first realized in contact mode in order to choose the point where adhesion kinetics is tested. Scan size is suddenly reduced, centered on this point : a small amplitude scan (+/- 1.25 nm at 20.3 Hz) is kept perpendicularly to the cantilever axis [13]. This experiment is performed under zero applied load : the servo-loop maintains the cantilever in a non-normally-deflected position. We have verified that the presence of a small lateral modulation of the position of the sample does not modify the observed height kinetics. Lateral force is recorded during the experiment ; it is pseudo-periodic : the shape of the cyclic response is constant while its amplitude increases (figure 1). To analyze these data, a reference cycle is computed from the average of the ten last recorded cycles. The amplitude factor is then calculated by least square fitting of the successive periods of the signal (an offset value being allowed to account for a possible slow drift of the lateral force). This amplitude is plotted in figure 2 together with the recorded height of the sample. Assuming that the contact remains in a static friction regime (see below), the amplitude of this lateral force is a measurement of the contact stiffness and scales thus with the contact size.

Due to interfacial forces, the tip penetrates into the material. The difference between the instantaneous height value and its initial value is the indentation depth under zero applied load. Adhesion on viscoelastic materials is a rather complicated problem. For the longer times, however, the experimental data can be simply interpreted by considering the sample modulus as nearly relaxed. The material behaves then as an elastic sample with a compliance D_∞.

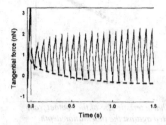

Figure 1. *Lateral force recorded at the very beginning of the indentation when a +/- 1.25 nm modulation is applied to the sample position at 20.3 Hz.*

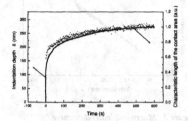

Figure 2. *Indentation depth and characteristic size of the contact during the zero load indentation experiment.*

Adhesive equilibrium state can thus be described using the Johnson, Kendall and Roberts theory (JKR) [14]. Due to geometrical similarities, one can consider that the pyramidal shape of the tip can be well approximated with a conical punch [15-17]. Under this assumption, generalized JKR theory for axisymmetric punches [18-21] is used. Assuming that the equivalent cone has a half angle $\beta = 45°$, the thermodynamic work of adhesion, w, can be deduced from the long times indentation depth, δ_∞ : $\delta_\infty = 6wD_\infty/\tan\beta$. Experimental data ($D_\infty \approx 5$Mpa^{-1}, $\delta_\infty \approx 300$ nm) yield $w \approx$ 10 mJ/m^2. Though this value is small by a factor of about 3, it is in a correct order of magnitude agreement with the expected value.

To analyze the kinetics part of the data, a comparison is presented in figure 3 of the macroscopically determined compliance function of the material, $D(t)$, to the indentation depth, $\delta(t)$, taken at the same times. Taking into account the long term equilibrium relation given above, the observed proportionality shows that a similar equilibrium relationship holds during the indentation process: $\delta(t) = 6wD(t)/\tan\beta$. Another insight on the adhesion kinetics can be obtained from figure 4 where the contact size deduced from figure 2 is plotted against the indentation depth. If a small offset of the contact size at the initial time is neglected, a proportionality is observed between these parameters. Taking into account the long times expected JKR equilibrium relation $\delta_\infty = \pi/4 \, a_\infty \, \tan\beta$, one can deduce that a similar instantaneous relation exists between the instantaneous values of the indentation depth and of the contact radius $a(t)$: $\delta(t) = \pi/4 \, a(t) \tan\beta$. ($a_\infty$ is the long times contact radius). From both experimentally obtained relations, the adhesion kinetics can then be described as the equilibrium adhesive contact of the tip on an elastic material which compliance function is the instantaneous value of the compliance function.

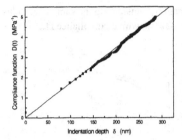

Figure 3. *Macroscopically determined compliance function plotted against the indentation depth. The straight line is a linear fit to the data through the origin.*

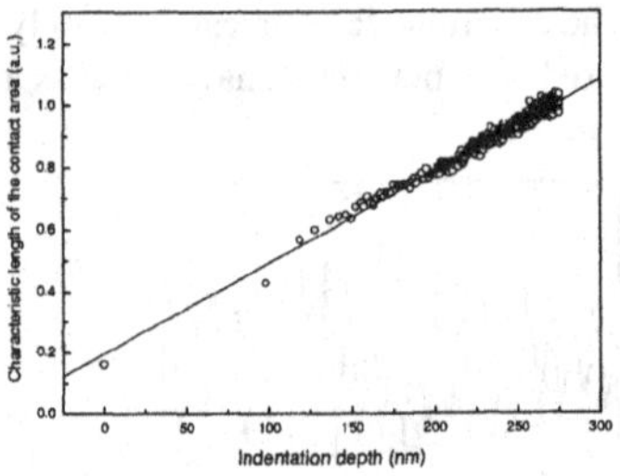

Figure 4. *Characteristic size of the contact from figure 2 plotted against the indentation depth. The straight line is a linear fit to the data.*

Several descriptions of adhesion on viscoelastic materials recently appeared in the literature [7,22-24]. In a fracture mechanics frame, it has been proposed that the dependence of the stress intensity factor on the crack velocity could be a characteristic of the material [25]. Johnson [7] extends the Maugis-Dugdale theory of adhesion in remarking that the time scales for fracture and bulk dissipation are different. In this model, however, an hertzian bulk stress field applies, the compliance being the relaxed compliance of the material. This is clearly not realized in our experience where the characteristic times of the material are comparable to the experiment duration. Obtained pseudo-equilibrium relations suggest that kinetics is dominated by bulk dissipation rather than by fracture propagation effects. In contrast, this model has been successfully used to extract the work of adhesion from indentation experiments at intermediate loading rates [26]. Recent theoretical investigations of the adhesion between viscoelastic material [23,24] may help to analyze the presented results.

RECOVERY OF THE IMPRINT

After an indentation experiment, the tip is extracted from the sample by large scan imaging [27]. In figure 5, the imprint left by the tip in the sample is seen to disappear slowly. Sections of the images through the imprint are presented in figure 6-a). Once normalized, all these profiles are similar (see figure 6-b): recovery is affine. This is indeed what is expected from a simple analysis of the recovery of a deformation at the surface of a viscoelastic sample. Depth of the imprint follows the variations of a corresponding homogeneous creep-recovery experiment : both kinetics are compared in figure 7. By imaging the imprint, this simple experiment allows, to obtain the variations of the polymer compliance function.

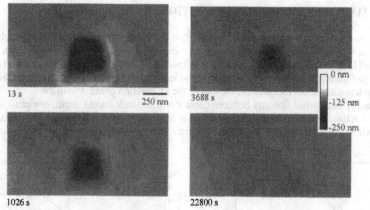

13 s   250 nm   3688 s

1026 s   22800 s

0 nm
-125 nm
-250 nm

Figure 5. *Recovery of the imprint after an indentation experiment. Indicated times correspond to the times elapsed after the extraction of the tip from the sample*

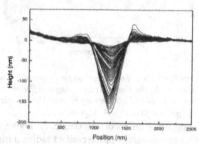

Figure 6-a). *Section of the imprints for times between 13 and 5000 s.*

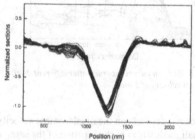

Figure 6-b). *Section of figure 6-a), normalized to a given depth. 36 sections are superimposed corresponding to an affinity factor of about 13.*

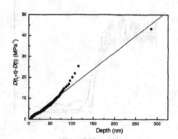

Figure 7. *Comparison of the time dependence of the imprint depth with recovery which could be measured in a creep-recovery experiment. The straight line is a linear fit of the data through the origin*

STATIC TO DYNAMIC FRICTION TRANSITION

Static to dynamic friction transition is studied by recording the *stiction curve*, which corresponds to the variations of the tangential force during a constant velocity displacement of the sample [12]. The experiment begins after a given dwell time (figure 8), where adhesion develops. In figure 9, it is shown that, for a given velocity, the position of the transition scales with the indentation depth measured at the end of the indentation phase. Using the experimentally established linearity between contact size and indentation depth, one can conclude that the static friction regime terminates when the displacement is about 2.7 times the contact radius. For a given dwell time, it is clear from figure 10 that the transition thresholds

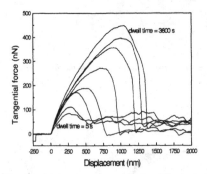

Figure 8. *Stiction curves recorded after different dwell times, at a velocity of 5 nm/s.*

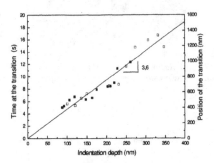

Figure 9. *Position of the transition for different indentation depths. Data corresponding to different cantilever (0.12 and 0.6 N/m) are plotted. The straight line is a linear fit to data through the origin.*

does not depends on the displacement velocity in a range of 4 orders of magnitude. As above, a small modulation of the position of the sample gives a measurement of the contact radius during the experiment. Figure 11 shows that the contact size remains nearly constant during the static phase and suddenly drops at the transition position. The way the contact shrinks appears to be almost identical in the range of tested velocities.

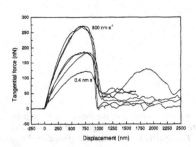

Figure 10. *Stiction curves recorded for different displacement velocities (0.4 nm/s < v < 800 nm/s) after a dwell time of 30 minutes.*

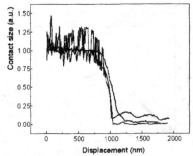

Figure 11. *Characteristic size of the contact during stiction experiments at .01, 0.1, 1 and 10 μm/s.*

From the above results, it can be conclude that static contact is rather abruptly broken at a given characteristic strain. Indeed, on this viscoelastic sample, stress effects should be highly velocity dependant. An estimation of this strain is obtained from the ratio of the displacement threshold to the initial contact size. Transition occurs at a deformation of about 270 % whatever the velocity is. Such a high strain value suggests that it may be related to the onset of non-linear elasticity. It can be noticed that, on similar partially cross-linked materials, tack experiments indicate also that final normal break of adhesion takes place at a given strain, independent on the strain rate [28].

VISCOELASTIC MODULI MEASUREMENTS

As we have verified that the contact radius remains constant during most of the static friction phase and knowing the threshold position it is possible to measure the viscoelastic properties of the sample through a constant velocity or a periodic displacement.

The displacement y of a disk of the surface of an elastic material requires a tangential force

$$T = \frac{8a}{3(2-\upsilon)} Ey$$, E is the Young modulus, υ the Poisson ratio and a the contact radius. Since the

radius of the contact is constant, this relation can be generalized to viscoelastic materials using the correspondence principle [29].

In a constant velocity displacement of the sample from a static contact position, the tangential force $T(t)$ is related to the relaxation modulus $E(t)$ by the expression [12,30] :

$$T(t) = \frac{8av}{3(2-\upsilon)} \int_0^t E(t')dt'$$

In figure 12-a) stiction curves normalized to the velocity and to the initial contact radius deduced from the analysis of the adhesion kinetics (see above) are plotted. The static parts of the data lie on a single curve as expected from the above expression. A differentiation of this curve gives the experimentally determined relaxation function which is very similar to the same function macroscopically determined (figure 12-b).

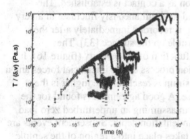

Figure 12-a). Stiction curves normalized to the contact radius and to the displacement velocity (0.4 nm/s < v < 4μm/s).

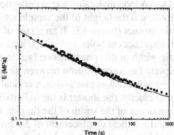

Figure 12-b). Relaxation function obtained by differentiation of the master curve of figure 9-a) (dots) compared to the macroscopically determined relaxation function (line).

On the same way, a periodic modulation of the position allows for the determination of the complex modulus [11,12,30]. Linearity of the response to a saw-tooth position modulation is confirmed in figure 13-a) where a Lissajous representation of the tangential force response normalized to the excitation amplitude is seen to be invariant. Fourier analysis of the harmonics of the response for different excitation frequencies gives the complex modulus [30]. Figure 13-b) compares the determined values to the same quantities macroscopically measured. Here again, a good agreement is found.

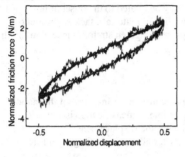 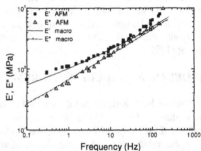

Figure 13-a). *Lissajous representations of the static friction response normalized to the saw-tooth excitation amplitude in the range 0.5 to 200 nm.*

Figure 13-b). *Complex modulus deduced from Fourier analysis of the periodic static friction response compared to the same property macroscopically determined.*

STICK-SLIP REGIME

An analysis of the friction loops [31] of the tip on the polymer reveals a linear dependence of an *averaged value* of the tangential force with the normal load (figure 14-a). The friction coefficient is a decreasing function of the velocity (figure 14-b). This behavior is characteristic of an unstable sliding friction regime [32]. In the following, a characterization of the underlying stick-slip regime is presented. Experiments are performed under constant velocity displacement, zero normal load and zero dwell time [12]: the tip is approached of the moving surface and servo-loop maintains a zero deflection as soon as a contact is established. The tangential force and the height of the sample are observed to simultaneously fluctuate in a very characteristic manner (figure 15). If an image of the zone is recorded immediately after the experiment, sequences of holes and bumps are clearly visible on the surface [33]. The characteristic width of this phenomenon is a decreasing function of the velocity (figure 16). It can be interpreted as a competition between an indentation process due to interfacial forces and a sliding process which takes place when a critical lateral strain is reached. The higher is the displacement velocity, the shorter is the indentation phase. A good agreement is found for the velocity dependence of the width of the first stick phase in assuming an unperturbed zero load indentation and a critical displacement of 3.6 times the indentation depth. Subsequent events are shorter, probably because the second indentation phase takes place into a region of the sample which have already undergone normal stress. Periodic plastic deformations of polymers under an AFM tip have been observed by several groups [34-37]. Presented experiments may help understanding these effects.

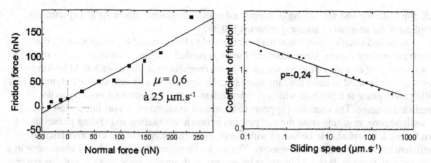

Figure 14. a) *Averaged friction law deduced from friction loops.* b) *Velocity dependence of the apparent friction coefficient.*

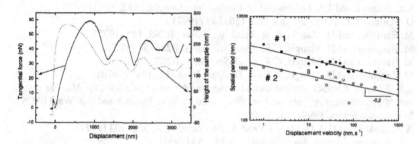

Figure 15. *Height of the sample and tangential force during a constant velocity experiment. Stick slip behavior is observed on both of this curves.*

Figure 16. *Width of the first two stick-slip events for different displacement velocities.*

CONCLUDING REMARKS

Adhesive and tribological properties of viscoelastic polymer surfaces have been investigated using an AFM. An important issue concerns the size of the volume of the material participating to the response. Due to the regularity of the equations governing contact mechanics, one expect that the participating domain has a characteristic size which represents several times the contact size, where boundary conditions apply. The spatial resolution of the presented determination of the modulus, for example, is about one micrometer. This size could be reduced by lateral displacement modulation after shorter dwell times. In this case, however, precision of the determination should be lower, due to shorter acquisition times. On stiffer materials, resolution is expected to be higher but moduli determination is complicated by the reduced static friction domain and by the low stiffness of the AFM cantilevers which prevents the description of the experiments as a fixed grip one. When sliding, the contact radius on soft materials is very much reduced. One can expect that mechanical testing of the sliding contact could reveal the mechanical properties of the material with a good resolution. However, the presented complex stick-slip behavior and the mixing of lateral and normal displacements in AFM experiments complicates the analysis of such experiments [10,38].

Adhesion kinetics under zero load is observed to be governed by dissipation on the volume covered by the tip. More theoretical work is needed in order to understand this result. As a consequence of this slow process, we have shown elsewhere that the so-called AFM-force-curves cannot be used to determine the mechanical properties of soft viscoelastic polymers [13]. Sliding takes place at a position which is independent on the displacement velocity in the experimental range. The transition appears to be related to the onset of non-linear elasticity. A stick-slip regime, resulting from the competition between indentation and sliding phases has been found. It is probably to be linked with the decreasing behavior of the effective friction coefficient with the displacement velocity. The stick-slip regime has been clearly observable in a given velocity domain. It should however be present, with smaller amplitudes out of this range. Correlation analysis of the sliding friction force may reveal it.

REFERENCES

1. K.L. Johnson and J.A. Greenwood, *J. Colloid Interface Sci.*, **192**, 326 (1997).
2. D. Maugis, *J. Colloid Interface Sci.*, **150**, 243 (1992).
3. M. Barquins and D. Maugis, *C. R. Acad. Sci. Paris* , **B285**, 125 (1977).
4. M. Barquins and D. Maugis, *C.R. Acad. Sci. Paris*, **B286**, 57 (1978).
5. M. Barquins and D. Maugis, *C.R. Acad. Sci. Paris*, **B287**, 49 (1978).
6. M. Barquins and D. Maugis, *J. Phys. D: Appl. Phys.*, **11**, 1989 (1978).
7. K.L. Johnson, Contact Mechanics and Adhesion of Viscoelastic Spheres, *Microstructure and Microtribology of Polymer Surfaces*, edited by V.V. Tsukruk and K.J. Wahl (ACS Symposium Series, 1999), pp. 24-41.
8. A.R. Savkor and G.A.D. Briggs, *Proc. R. Soc. London, A*, **356**, 103 (1977).
9. K.L. Johnson, *Proc. R. Soc. London, A*, **453**, 163 (1997).
10. S.A. Syed Asif, R.J. Colton, and K.J. Wahl, Nanoscale surface mechanical property measurements: force modulation techniques applied to nanoindentation, *Interfacial Properties on the Submicron Scale*, edited by J. Frommer and R. Overney (ACS, *in press*).
11. C. Fretigny, C. Basire, and V. Granier, *J. Appl. Phys.*, **82**, 43 (1997).
12. C. Basire and C. Fretigny, Experimental study of the friction regimes on viscoelastic materials, *Microstructure and Tribology of Polymer Surfaces*, edited by V. Tsukruk and K. Wahl (ACS Symposium Series, 1999), pp. 239-257.
13. C. Basire and C. Fretigny, *Tribology Lett.*, *in press*.
14. K.L. Johnson, K. Kendall, and A.D. Roberts, *Proc. R. Soc. London, B*, **A324**, 301 (1971).
15. K.L. Johnson, *J. Mech. Phys. Solids*, **18**, 115 (1970).
16. J.L. Loubet, J.M. Georges, O. Marchesini, and G. Milelle, *J. Tribol.*, **106**, 43 (1984).
17. R.B. King, *Int. J. Solids Structures*, **23**, 1657 (1987).
18. D. Maugis and M. Barquins, *J. Phys. Lett.*, **42**, L95 (1981).
19. M. Barquins and D. Maugis, *J. Mec. Theor. Appl.*, **1**, 331 (1982).
20. D. Maugis and M. Barquins, *J. Phys. D: Appl. Phys.*, **16**, 1843 (1983).
21. C. Basire and C. Fretigny, *C. R. Academie. Sci., Ser. II* , **B326**, 273 (1998).
22. A. Falsafi, P. Deprez, F.S. Bates, and M. Tirrel, *J. Rheol.*, **41**, 1349 (1997).
23. Y.Y. Lin, C.-Y. Hui, and J.M. Baney, *J. Phys. D: Appl. Phys.*, **32**, 2250 (1999).

24. Y.Y. Lin, C.-Y. Hui, and A. Jagota, (*to be published*).
25. C.-Y. Hui, J.M. Baney, and E.J. Kramer, *Langmuir*, **14**, 6570 (1998).
26. M. Giri, D. B. Bousfield, and W.N. Unertl, (*to be published*).
27. C. Basire and C. Fretigny, *C. R. Academie. Sci. Paris* , **B325**, 211 (1997).
28. J. Hooker and C. Creton, (*to be published*).
29. G.A.C. Graham, *Q. Appl. Math.*, **26**, 167 (1968).
30. C. Basire and C. Fretigny, *Eur. Phys. J. AP* , **6**, 323 (1999).
31. C.M. Mate, G.M. McClelland, R. Erlandsson, and S. Chiang, *Phys. Rev. Lett.*, **59** (17), 1942 (1987).
32. E. Rabinowicz, Friction Fluctuations, *Fundamentals of Friction : Macroscopic and Microscopic Processes*, edited by I.L. Singer and H.M. Pollock (NATO ASI Series 220, 1992), pp. 25-34.
33. O.M. Leung and M.C. Goh, *Science*, **255**, 64 (1992).
34. G.F. Meyers, B.M. DeKoven, and J.T. Seitz, *Langmuir*, **8**, 2330 (1992).
35. X. Jin and W.N. Unertl, *Appl. Phys. Lett.*, **61**, 657 (1992).
36. Z. Elkaakour, J.P. Aime, T. Bouhacina, C. Odin, and T. Masuda, *Phys. Rev. Lett.*, **73**, 3231 (1994).
37. C. Fretigny and D. Michel, (*to be published*).

Mat. Res. Soc. Symp. Proc. Vol. 651 © 2001 Materials Research Society

Adsorption in ordered porous silicon : a reconsideration of the origin of the hysteresis phenomenon in the light of new experimental observations

B. Coasne, A. Grosman, N. Dupont-Pavlovsky*, C. Ortega and M. Simon
Groupe de Physique des Solides, UMR 7588, Universités Paris 7&6,
2 Place Jussieu 75251 Paris Cedex 05 France
* Laboratoire de Chimie du Solide Minéral, UMR 7555 CNRS,
BP 239, 54520 Vandoeuvre les Nancy Cedex France

ABSTRACT

Porous silicon formed in p+ single crystal silicon is an interesting material for the study of the behavior of a fluid confined in mesoporous media, because it is a simple system of non interconnected straight pores, all perpendicular to the substrate and it can be well characterized by different methods. Moreover, the pores may be closed at one end when the porous layer is supported by the substrate or opened at both ends when the layer is removed from the substrate. Surprising results have been obtained. A hysteresis loop of type H2 (IUPAC classification), which is generally obtained for highly interconnected porous material, is also observed. Furthermore, a hysteresis of same type is also observed in pores closed at one end which is in contradiction with the Cohan model. The steep desorption process does not account for the large pore size distribution extracted from transmission electronic microscopy study. It is believed that, during the desorption process, the presence of an adsorbed layer on the external surface of the pores and/or the coupling between the pores via the silicon walls must be taken into account.

INTRODUCTION

The behavior of a fluid confined in mesoporous media which exhibits pore size in the range 2-50 nm, still aroused a great interest since it concerns materials which find more and more applications in industrial processes. From a fundamental point of view, it concerns the understanding of fluid properties in small confined geometries. These physics are generally studied by carrying out adsorption isotherms. A typical adsorption isotherm performed on a mesoporous solid exhibits two important features : (i) at a pressure lower than the bulk saturating vapor pressure P_0, a sharp increase of the adsorbed amount is observed which is assigned to the capillary condensation of the fluid in the pores and (ii) when the pressure is decreased from P_0, the evaporation process fails to retrace the condensation, so that a wide hysteresis loop appears revealing the irreversibility of the phenomenon. The origin of the hysteresis has been the subject of many discussions from which two ideas have emerged.

(1) The irreversibility is an intrinsic property of a fluid confined in a single pore.
In an earlier work, Cohan [1], on the basis of a Foster suggestion, describes different scenarii for the filling and the emptying of a cylindrical pore opened at both ends. While the filling consists in overlaying the surface of the pore by the dense phase, during the emptying, a hemispherical meniscus appears, provided a perfect wetting. The condensation and the evaporation occur at different pressures which are related to the corresponding curvature of the meniscus (cylindrical during the adsorption and hemispherical during the desorption) via the macroscopic Kelvin equation. On the contrary, in the case of a cylindrical pore closed at one end, this model predicts that the phase transition should be reversible since both the

condensation and evaporation proceed through the same hemispherical meniscus. More recently, microscopic calculations based on the Density Functional Theory (DFT) [2] as well as Monte Carlo simulations in the Grand Canonical ensemble (GCMC) [3] have also predicted a hysteresis loop in adsorption isotherm for a single pore opened at both ends, but related to the existence of gas-like and/or liquid-like metastable states. Calculations in the case of a pore closed at one end are lacking.

(2) The irreversibility is ascribed to the so-called pore network effect.
By a generalization of the earlier work by Everett and Barker on the pore blocking effect, Mason [4] has shown that, in a network of pores interconnected to the others by constrictions, the hysteresis loop could be due to a percolation transition : while the pressure at which the condensation occurs is characteristic of both the pore and constriction sizes, the emptying can only occur when one of the constrictions, which isolate the pore from the gas, empties, so that evaporation pressure is only characteristic of the constriction sizes.

Most of the mesoporous solids used in adsorption studies exhibit a highly interconnected and complex morphology which is in addition poorly characterized. The hysteresis loop observed for such materials could thus be the sum of single pore effects and pore network effects, so that it is quite difficult to check the validity of the above models.

We present here adsorption measurements performed on highly boron doped porous silicon. This porous solid appears as an interesting medium for adsorption studies, because it is a simple system with non interconnected straight pores, perpendicular to the silicon substrate. It can be well characterized by methods such as Transmission Electron Microscopy (TEM) and nuclear microanalysis [5]. Moreover, we can obtain pores opened at one or at both ends, whether the porous layer is supported or come off the substrate. In this paper, we check, for the first time, the validity of the Cohan model by comparing adsorption on porous layers with pores opened at one or at both ends. We next discuss the position on the pressure axis and the shape of the hysteresis in the light of the existing single pore models.

EXPERIMENTAL

Porous silicon layers are obtained by an electrochemical etching of a highly boron doped single crystal silicon in a HF solution. We can vary the porosity, which depends on the current density, from ~ 30% up to ~ 90% and the thickness of the layer from ~ 1 up to a few hundreds micrometers, according to the duration of the anodic dissolution. Porosity, thickness and pore volume are determined with a high precision by weighing the sample. We have studied the morphology of the porous layers by TEM (figure 1). It consists in a honeycomb-like structure of straight parallel pores of polygonal shapes which are separated by silicon walls of a constant thickness about 5 nm. From nuclear techniques, we have shown that the deviation between the pore direction and the [100] axis of the silicon substrate is less than 0.1º [6]. From a numerical treatment of the plane view which consists in adjusting the threshold to obtain a binary contrast in order to reproduce the porosity and the constancy of the wall thickness, we extract information such as the perimeter and the surface of each pore. The Pore Size Distribution (PSD), shown in figure 2, corresponds to cylindrical pores having the same surface area as the polygonal pores. Pores have a mean dimension about 13 nm and the pore size dispersion is large (± 6 nm). A transmission electronic diffraction study reveals that the porous material conserves the single crystal character of the silicon substrate. We have experimentally proved [5] the total absence of lateral interconnection between the pores. The experiment consists in depositing aluminum cap on a part of the porous layer, followed by a thermal oxidation of the sample in oxygen highly enriched in ^{18}O. From the nuclear reaction $^{18}O(p, \alpha)^{15}N$, we measure the ^{18}O content under and beside the cap. While we obviously found ^{18}O atoms in the porous part beside the cap, no ^{18}O atom has been found under the cap which shows that this porous medium is composed of non-interconnected pores, in contact with the gas reservoir during the adsorption experiment only by their ends.

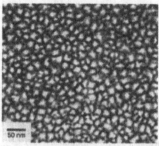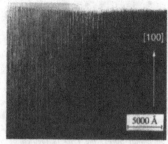

Figure 1. *Bright field TEM images in plane (left) and cross section (right) views of a porous silicon layer with ~ 50% porosity prepared by an electrochemical etching of a p^+ type (~ 3 10^{-3} Ω.cm) [100] Si substrate under 50 mA/cm². These views show a honeycomb-like structure with pores (white) separated by silicon walls (black). The pore density is about ~ $10^{11}/cm^2$.*

At the end of the electrochemical etching, we obtain a porous silicon layer supported by the silicon substrate, so that pores are opened only at one end. It is possible by a sharp increase of the current density to go to an electropolishing regime during which the anodic dissolution occurs only at the bottom of the pores. By the application of such a process during few seconds, the porous layer comes off the substrate and we obtain a porous silicon membrane with pores opened at both ends.

Sorption measurements were made by a volumetric method, using a commercial apparatus (ASAP 2010 Micromeritics) under a secondary vacuum about 10^{-6} Torr. Nitrogen adsorption isotherms were performed at pressures up to the bulk saturating vapor pressure P_0 and at 77 K, using a dewar in which the level of liquid nitrogen is kept constant.

RESULTS AND DISCUSSION

Figure 3 shows a N_2 adsorption isotherm carried out on a porous silicon membrane. This curve exhibits a wide and asymmetrical hysteresis loop which corresponds to type H2 with reference to the IUPAC classification (1985). The adsorbed amount of nitrogen sharply increases during the capillary condensation which occurs from ~ 0.55 P_0 up to ~ 0.80 P_0 and then, reaches a plateau region where all the pores are filled by the dense phase. The steep desorption process occurs on a pressure range about three times narrower than that of the adsorption. In figure 4, we compare N_2 adsorption isotherms corresponding to a porous layer supported by the substrate i.e. with pores closed at one end, and to a porous membrane with pores opened at both ends. The two hysteresis loops have similar shape.

The irreversibility of the adsorption-desorption cycle observed in the case of pores closed at one end, is not predicted by the Cohan model. Such a result implies that at least one of the two processes (condensation or evaporation) cannot be described in the framework of this model whether pores are opened at one or at both ends.

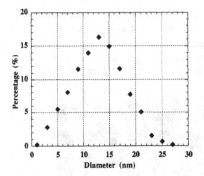

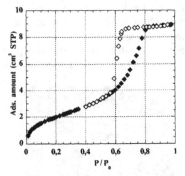

Figure 2. The pore size distribution of a porous silicon layer, of 50% porosity, estimated from a numerical treatment of the TEM plane view shown in figure 1. It corresponds to cylindrical pores having the same surface area as the polygonal pores; the PSD is located at ~ 13 nm ± 6 nm.

Figure 3. Nitrogen adsorption isotherm measured at 77 K in a porous silicon membrane with pores opened at both ends, 20 μm long, the morphology of which is shown in figure 1. The hysteresis loop is of type H2 with reference to the IUPAC classification (1985).

To go further in the analysis of this result, it is then necessary to discuss in more details, 1) the position on the pressure axis and 2) the shape of the hysteresis loop, with regard to the existing models. These two points will be discussed in the case of pores opened at both ends since no microscopic calculations have been performed in a cylindrical pore closed at one end.

1) We have analyzed the hysteresis loop by the Barrett, Joyner and Halenda method (BJH) [7], based on the modified Kelvin equation which predicts that, for a pore of radius r, the condensation pressure is related to r-t where t is the thickness of the adsorbed film prior to the condensation. By using the classical t-plot proposed either by Halsey [8] or by Harkins and Jura, such analyses of both the adsorption and the desorption curves yield a PSD centered on about 5 nm which is a mean value about 2.5 times less than that we extracted from TEM (fig. 2). Density functional calculations performed by Ball and Evans [2] in cylindrical pores having a gaussian diameter distribution centered on 7.2 nm yield a hysteresis loop (fig. 5) for which the position on the pressure axis (between 0.5 P_0 and 0.9 P_0), similar to that observed in figure 4, is close to that predicted by the modified Kelvin equation for such a mean value.

Hence, it seems that neither the macroscopic Cohan model nor the microscopic calculations can account for the position on the pressure axis of the hysteresis loop we observe. However, such a large discrepancy may be due to the use of a cylindrical geometry in the calculations which do not account for the numerous acute angles of the polygonal pores and their surface irregularities [9]. GCMC simulations and calculations based on the DFT are in progress.

2) The other important result is that the asymmetrical shape (type H2) of the hysteresis loops shown in figure 4 is not explained by the single pore models. Indeed, for a given size distribution of cylindrical pores, the macroscopic Cohan model predicts that the adsorption branch should be steeper than the desorption branch exactly in contradiction with what we experimentally observe. Moreover, as shown in figure 5, the hysteresis shape predicted by the DFT calculations is more symmetric with similar slopes for the two branches.

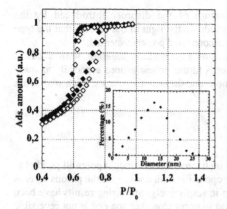

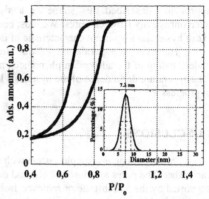

Figure 4. Nitrogen adsorption isotherms at 77 K in a porous Si membrane with pores opened at both ends (opened diamonds) and in a porous Si layer with pores closed at one end (filled diamonds). The two layers have a porosity of 50%.

Figure 5. Xenon adsorption isotherm at 174 K in Vycor glass obtained from density functional calculations for an assembly of independent pores opened at both ends. The authors [2] used a gaussian distribution of pore diameters centered on about 7.2 ± 1.5 nm.

At this point, we would like to focus on the steepness of the desorption process which is surprising according to the large width of the PSD we extracted from TEM. The analysis of the desorption branch using BJH method [7] yields a PSD with a width about 0.5 nm which is very far from that (± 6 nm) of the actual PSD extracted by TEM (fig. 2). Such a result rules out that the evaporation process occurs at coexistence of the low and high dense phases. It must be noted that the incompatibility of the evaporation process with the Cohan description supports the first conclusion we made from the observation of the presence of a hysteresis loop for pores closed at one end. Microscopic calculations [2] can no more explain the steepness of this process. Indeed, while the PSD we extracted from TEM study (fig. 2) is 5 times larger than that used by Ball and Evans (7.2 ± 1.5 nm), the desorption process we observe (fig. 4) is even steeper than the calculated one (fig. 5).

From these analyses, we must conclude that no single pore model, neither the macroscopic Cohan model nor the microscopic calculations describe the desorption process we observe.

As there is no lateral interconnection in the porous Si layers, we must believe that this process should result from another coupling between the pores. In such an idea, we would like to stress the fact that all the models never consider the external surface of the pores. We believe that the film adsorbed on this surface, at high pressure, should connect the pores and hence plays a key role during their emptying. Another and/or complementary explanation should be that a coupling of the pores could be supplied by pore-pore correlation effects via the silicon walls. Indeed, inter-pore adsorbate interactions, through adsorption strains, might exist for materials with closely spaced pores such as the porous silicon which exhibits few nanometers thick partitions between the pores. The emptying of one pore would lead to different strains on both sides of the Si walls which separate this pore to the neighboring filled pores.

These assumptions may explain the fact that surprisingly, although porous silicon exhibits non-interconnected pores, we observe the same hysteresis shape (type H2) as those systematically observed for highly interconnected mesoporous solids such as controlled pore glass or Vycor. From a theoretical point of view, both macroscopic [4] and microscopic [2] approaches described this shape as a signature of the disorder introduced by the interconnectivity between pores. We must conclude, in the light of our results, that the type H2 of hysteresis loop is not characteristic of interconnected porous materials.

However, the validation of our previous assumptions together with a better understanding of the adsorption phenomena needs further investigations especially to study the temperature dependence of the hysteresis loops and the scanning curves. Such studies are in progress.

CONCLUSIONS

Porous silicon is a simple system with non interconnected straight pores all parallel to each other. The pores are closed at one end or opened at both ends when the porous layer is supported by the Si substrate or removed from it, respectively. Surprising results have been obtained in this porous material. The adsorption in pores closed at one end is not reversible on the contrary to the Cohan predictions. The position ont the pressure axis of the hysteresis loops is too low according to the actual pore size distribution extracted from TEM study and the single pore models. Whereas the pores are not interconnected, the hysteresis loops observed whatever pores are opened at one or at both ends are asymmetrical (type H2). This shape is found and predicted only in porous solids highly interconnected. We think that the desorption process depends, in addition, on physical parameters which have never been taken into account in the existing models : the presence of an adsorbed layer on the external surface of the pores and/or the coupling between pores via the silicon walls.

ACKNOWLEDGMENTS

The authors would like to thank Pr. Xavier Duval for enthusiast and fruitful discussions held in Nancy, France. We thanks Pr. R Evans and Dr. P. C. Ball for authorizing us to use data published in [2].

REFERENCES

[1] L. H. Cohan, *J. Am. Chem. Soc.*, **60,** 433 (1938).
[2] P. C. Ball and R. Evans, *Langmuir*, **5**, 714 (1989).
[3] A. Papadopoulou, F. Van Swol and U. Marini Bettolo Marconi, *J. Chem. Phys.*, **97**, 6942 (1992).
[4] G. Mason, *J. Colloid Interface Sci.*, **88**, 36 (1982).
[5] B. Coasne, A. Grosman, N. Dupont-Pavlovsky, C. Ortega and M. Simon, To be published.
[6] G. Amsel, E. d Artemare, G. Battistig, V. Morazzani and C. Ortega, *Nucl. Instrum. Meth. in Physics Research*, **B122**, 99 (1997).
[7] E. P. Barrett, L. G. Joyner and P. H. Halenda, *J. Am. Chem. Soc.*,**73**, 373 (1951).
[8] G. Halsey, *J. Chem. Phys.*, **16** (10), 931 (1948).
[9] B. Coasne, A. Grosman, N. Dupont-Pavlovsky, C. Ortega and M. Simon, submitted in *Physical Chemistry and Chemical Physics*.

Mat. Res. Soc. Symp. Proc. Vol. 651 © 2001 Materials Research Society

Electrokinetic phenomena in montmorillonite

V. Marry, J.-F. Dufrêche, O. Bernard, P.Turq

ANDRA and Laboratoire Liquides Ioniques et Interfaces Chargées, case courrier 51, Université P. et M. Curie, 4 place Jussieu, F-75252 Paris Cedex 05, France.

Abstract

Clays present remarkable electrokinetic features since they exist from very dilute colloidal state to nanoporous materials, according to the ration water/clay. The case of low volume fraction V_{water}/V_{tot} which corresponds to compact systems is examined. The ionic distributions have been evaluated from Poisson-Boltzmann like models and compared to discrete solvent simulations. Electroosmosis and conductance have been calculated, in the framework of the Mean Spherical Approximation (MSA) introduced in the Fuoss-Onsager transport theory.

1 Introduction

Clays are formed by large platelets composed of stacks of elementary sheets. In the case of montmorillonite, a sheet is made up of a layer of octahedral oxides (Al^{3+}, Mg^{2+}, etc...) between two layers of tetrahedral oxides (Si^{4+}, Al^{3+}, etc...). As some of the cations are substituted by other cations of lower valency, clay sheets are negatively charged. Counterions which are localized between sheets are partly responsible for the swelling behavior of montmorillonite in the presence of water. In this article, the motion study will be reduced to the transport between two parallel sheets of clay.

Monte-Carlo and molecular dynamics simulations have already been undertaken to describe the equilibrium and dynamic properties of this system: several kinds of counterions as Li^+ (Boek [1], Chang [2]), Na^+ (Boek [1], Delville [3], Karaborni [4], Siquiera [5], Skipper [6]), K^+ (Boek [1], Chang [7], Delville [8]) and Cs^+ (Shroll [9], Smith [10]) were studied. These simulations use microscopic descriptions of the system: water and clay sheets are considered to be discrete sets of atoms. In this article, we offer to develop a quite simple mesoscopic model to describe ions' motions in confined water. Clay sheets are considered as uniformly charged plans and water as a continuum. Such a model, based on the Poisson-Boltzmann equation, was already used to calculate equilibrium quantities as osmotic pressures (Delville [11]). Ionic distributions between two clay sheets were compared with those obtained by continuous solvent Monte-Carlo simulations (Delville [12]).Donnan effects deduced from Poisson-Boltzmann for didodecyldimethylammonium bromide, which is a similar swollen lamellar medium, were in good agreement with experiments of atomic adsorption (Dubois [13]). It was shown that the Poisson-Boltzmann equation was satisfactory for medium and large interlayer spacings.

Here, ionic distributions obtained from the Poisson-Boltzmann equation are compared to those obtained by discrete solvent Monte-Carlo simulations. Then they are used to determine transport properties for several salt concentrations in the interlayer spacing (electroosmosis and conductivities). A similar theory was previously used for Nafion membranes and succeeded in reproducing experimental conductivity data (Lehmani [14]).

2 Equilibrium properties: microscopic and mesoscopic approaches

2.1 Ionic distributions

In the adopted model, clay sheets are considered as parallel infinite plans. They are uniformly charged and the surface charge density is $\sigma = 0.0161/2e$ Å^{-2}. The water inside the interlayer spacing is a continuum. Its viscosity and dielectric constant are bulk water's ones: at 298K, $\eta = 0.8910^{-3}$ SI and $\epsilon_r = 78.3$ SI. Counterions are Na$^+$ and the added salt NaCl. The electrolyte is simply treated by taking an average radius.

If $c_i(r)$ is the concentration of the ion i at a distance r from a central particle the Poisson-Boltzmann treatment is obtained from the Boltzmann statistics:

$$c_i(r) = M_i \exp(-\frac{V_i(r)}{k_B T}) \tag{1}$$

where $V_i(r)$ can identify with the electrostatic energy $V_i(r) = e_i\psi(r)$. e_i is the charge of i and $\psi(r)$ the electrostatic potential. In the case of two horizontal charged plans separated by a distance L, ψ and c_i only depend on z. Then M_i can be obtained by integrating $c_i(z)$. By replacing in the Poisson equation $\Delta\psi = -\sum_i e_i c_i(r)/\epsilon_0\epsilon_r$, and setting $\phi = e\psi/k_B T$, we get:

$$\Delta\phi = -4\pi L_B \sum_i z_i M_i e^{-z_i\phi} \tag{2}$$

where $L_B = e^2/4\pi\epsilon_0\epsilon_r k_B T$ is the Bjerrum length. The symmetry of the system implies $d\phi/dz(z=0) = 0$ as one of the boundary conditions. The counterionic concentration if there is no added salt is given by:

$$c(z) = \frac{1}{2\pi Z^2 L_B} \frac{\alpha^2}{\cos^2(\alpha z)} \tag{3}$$

with $\alpha\tan(\alpha L/2) = 2\pi Z L_B \sigma/e$. In the presence of added salt, $\psi(z)$ and $c_i(z)$ are numerically determined.

The ionic distributions of this system calculated by the Poisson-Boltzmann equation have already been validated by Monte-Carlo simulations in which water is considered as a continuum (Delville [12]). At low distances L (< 10 Å), discrete solvent simulations showed that distributions presented oscillations, indeed even peaks, that are characteristic of the discrete nature of ions and water molecules. As we assumed that this feature would die down as L increases, we decided to compare Poisson-Boltzmann distributions with those obtained with a discrete solvent Monte-Carlo program for larger values of the interlayer spacing (> 20 Å).

The structure of the montmorillonite used in the microscopic simulation was taken in literature from X-ray diffraction studies (Maegdefrau [15], Brindley [16]). Its formula is: $Na_{0,75}[Si_{7,75}Al_{0,25}]$ $(Al_{3,5}Mg_{0,5})O_{20}(OH)_4$ The simulation box is formed by two half-sheets containing eight clay units. The atom charges are Skipper's ones (Skipper [17]). Van der Waals interactions are reduced to Lennard-Jones potentials, whose parameters are given by Smith (Smith [10]). The solvent is SPC/E water. Monte-Carlo equilibration have been performed in the (N, P, T) ensemble. with 300 water molecules. The average height of the box evaluated after equilibration was 35.5 Å. The counterionic distributions and their integrals obtained by the Poisson-Boltzmann equation and discrete solvent simulation when no salt is added are given on Fig. 1 and 2. We notice the presence of oscillations on molecular dynamics curves due to the discrete description of water, as explained before. It just shows that the mesoscopic model will be more acceptable for rather large interlayer spacings ($L > 20$ Å) even if in this article transport properties calculations have been made until $L = 10$ Å.

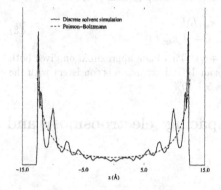

Figure 1: Counterionic distributions by discrete solvent simulation and Poisson-Boltzmann continuous solvent calculation. The distributions are symmetrical between 0 and 12.3 Å.

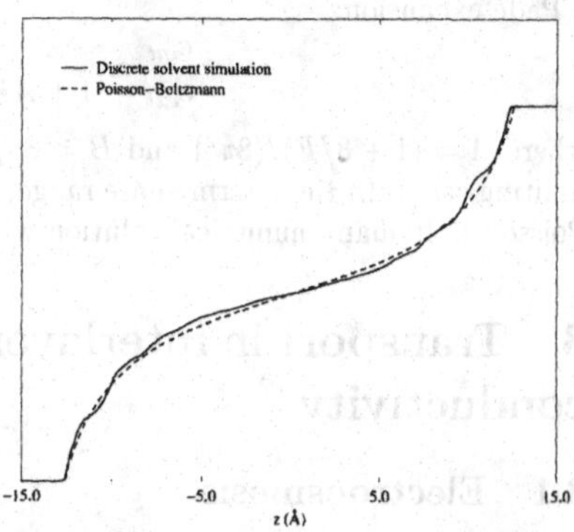

Figure 2: Integrated values of the counterionic distributions. The distributions are symmetrical between 0 and 12.3 Å. This corresponds to the half size of the box minus the thickness of the clay sheet (6.45 Å /2) and the radius of Na$^+$ (2.1 Å).

2.2 Donnan effect

When a salt, NaCl for example, is added in the water in contact with the clay, some salt penetrates between the sheets. It is well-known that $c_{int} \neq c_{ext}$ where c_{int} and c_{ext} are respectively the concentrations of salt inside and outside the pore: this is the Donnan effect. Between clay sheets, the average concentration of the cation (which is the counterion too) c_1^0 and the average concentration of the anion c_2^0 are given by:

$$c_1^0 = \frac{2\sigma}{LZe} + c_{int} \qquad (4)$$

$$c_2^0 = c_{int} \qquad (5)$$

The relation between c_{int} and c_{ext} is found by equalizing the chemical potentials of the salt inside and outside the pore: $\mu_{ext}^s = \mu_{int}^s$.

In the case of low salinities or interlayer spacings ($\kappa_{ext}L \ll 1$), ϕ is not disrupted by the presence of salt. For a highly charged clay: $F = \pi L Z L_B \sigma / e \geq 1$, a development of the analytical solution of the Poisson-Boltzmann equation without salt leads to (Dubois [13])

$$\frac{c_{int}}{c_{ext}} = \frac{\kappa_{ext}^2 L^2}{8\pi^2} \left(1 + \frac{3}{F}\right) \qquad (6)$$

In the case of high salinities or interlayer spacings ($\kappa_{ext}L \gg 1$), the effects of both charged plans can be decoupled, so that a superposition approximation can be used. It gives

$$\frac{c_{int}}{c_{ext}} = 1 - \frac{\gamma}{1 + \gamma} \frac{8}{\kappa_{ext}L} \tag{7}$$

with $\gamma = \sqrt{1 + 1/ZE} - 1/ZE$ and $E = 2\pi L_B \sigma/\kappa_{ext}e$.

In the whole concentration range, the Donnan ratio c_{int}/c_{ext} can be approximated by a Padé expansion:

$$\frac{c_{int}}{c_{ext}} \approx \frac{A(\kappa_{ext}L)^2}{1 + AB\kappa_{ext}L + A(\kappa_{ext}L)^2} \tag{8}$$

where $A = (1 + 3/F)/(8\pi^2)$ and $B = 8\gamma/(1 + \gamma)$. This Padé approximation gives both limiting cases. In the intermediate range, this analytical expansion is consistent with the Poisson-Boltzmann numerical solution with a $5 - 10\%$ accuracy.

3 Transport in interlayer spacing: electroosmosis and conductivity

3.1 Electroosmosis

We took an adaptation of the classical Smoluchowski method for electroosmosis to porous systems.

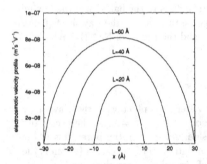

Figure 3: Electroosmotic velocity profiles for different interlayer spacings without added salt.

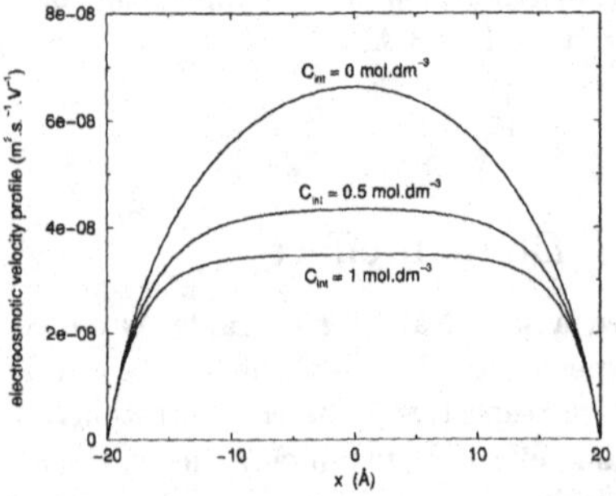

Figure 4: Electroosmotic velocity profiles for different added salt concentrations. The interlayer spacing is 40 Å. The velocity pattern tends rapidly to a plateau value.

The electroosmotic velocity (or the solvent velocity) $\mathbf{v}_s$ verifies the Stokes equation:

$$\eta \Delta \mathbf{v}_s + \mathbf{F}_v - \nabla p = 0 \tag{9}$$

with $\nabla . \mathbf{v}_s = 0$ if the solvent is incompressible. We obtain after integration

$$\mathbf{v}_s(z) = \frac{e}{4\pi \eta L_B} [\phi(z) - \phi(L/2)] E \mathbf{e}_y \tag{10}$$

In the case of no added salt, we get:

$$\mathbf{v}_s(z) = \frac{e}{2\pi\eta Z L_B} \ln \frac{\cos \alpha z}{\cos \alpha L/2} E\mathbf{e}_y \qquad (11)$$

The electroosmotic mobility profiles without added salt, for several interlayer spacings, are represented on Fig. 3. In the case of an added salt, electroosmotic velocities are calculated numerically. The variation of electroosmotic velocity profile is represented for an interlayer distance of 40 Å, for various concentrations (Fig. 4).

3.2 Conductivity

The conductivity has several contributions whose order of magnitude varies with increasing concentration: Without added salt the conductivity is only due to the intrinsic contribution of counterions and to the electroosmotic flow.

$$\sigma = \sigma_{int} + \sigma_{e.o.} \qquad (12)$$

The ionic atmosphere around cations is not distorted by the opposite moving of possible anions. In our model, the intrinsic contribution of counterions does not involve any relaxation nor electrophoretic effect, and follows strictly the Nernst-Einstein relation. This intrinsic conductivity of counterions will is our reference value, by respect to which any other effect will be evaluated. In the case of no added salt, the electroosmotic conductivity can be directly evaluated by averaging the electroosmotic velocity over the interlayer spacing:

$$\sigma_{e.o.} = \frac{e^2\alpha^2}{4\pi^2 Z^2 L_B^2 \eta} \left[\frac{\tan(\alpha L/2)}{\alpha L/2} - 1 \right] \qquad (13)$$

In the case of added salt a new contribution appears, which has to be added to the intrinsic and electroosmotic ones: the salt conductivity. We have then:

$$\sigma = \sigma_{int} + \sigma_{salt} + \sigma_{e.o.} \qquad (14)$$

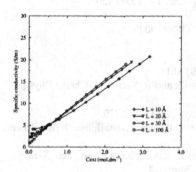

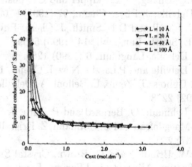

Figure 5: Total specific conductivity (taking the intrinsic conductivity as zero reference value) as a function of the external concentration.

Figure 6: Total equivalent conductivity (taken from the total specific conductivity without the intrinsic contribution) versus external concentration for different interlayer spacings.

The electroosmotic specific conductivity is higher for thiner interlayer spacing and decreases slightly with increasing external salt concentration. The MSA treatment of the specific conductivity of the salt gives σ_{salt} by using the Fuoss-Onsager equations for the non equilibrium pair distribution functions. We used an analytical theory that gives extended laws for the variations with concentration of transport coefficients of strong and associated electrolytes (Bernard [18-20].) The total specific conductivity (taking as zero value the intrinsic contribution to the counterion conductivity) can then be represented as a function of the external concentration. The results are presented in Fig. 5.

Those results can be also expressed in terms of equivalent conductivity

$$\Lambda = \frac{\sigma}{Z c_{ext}} \tag{15}$$

In any case (large or small interlayer spacing), the total equivalent conductance represented on Fig. 6 exhibits a rapid decrease with increasing external concentration. It should be noticed that Onsager's limiting laws for equivalent conductance are not valid with the external concentration, on the grounds of the Donnan effect. For exemple :

$$\Lambda_{salt} = \frac{\sigma_{salt}}{Z c_{ext}} = A c_{ext} - B c_{ext}^2 \tag{16}$$

The total equivalent conductivity varies in c_{ext} at low salinities.

REFERENCES

1. E.S. Boek, P.V. Coveney and N.T. Skipper, J. Am. Chem. Soc., 117 (1995) 12608.
2. F-R.C. Chang, N.T. Skipper and G. Sposito, Langmuir, 13 (1997) 2074.
3. A. Delville, Langmuir, 8 (1992) 1796.
4. S. Karaborni, B. Smit, W. Heidug, J. Urai and E. van Oort, Science, 271 (1996) 1102.
5. A.V.C de Siquiera, N.T. Skipper, P.V. Coveney and E.S. Boek, Molecular Physics, 92 (1997) 1.
6. N.T. Skipper, G. Sposito and F-R.C. Chang, Clays Clay Miner., 43 (1995) 294.
7. F-R.C. Chang, N.T. Skipper and G. Sposito, Langmuir, 14 (1998) 1201.
8. A. Delville, J. Phys. Chem., 97 (1993) 9703.
9. R.M. Shroll and D.E. Smith, J. Chem. Phys., 111 (1999) 9025.
10. D.E. Smith, Langmuir, 14 (1998) 5959.
11. A. Delville, Langmuir, 6 (1990) 1289.
12. A. Delville and P.Laszlo, New J. Chem., 13 (1989) 481.
13. M. Dubois, T. Zemb, L. Belloni, A. Delville, P. Levitz and R. Setton, J. Chem. Phys., 96 (1992) 2278.
14. A. Lehmani, O. Bernard and P. Turq, J. Stat. Phys., 89 (1997) 379.
15. E. Maegdefrau and U. Hofmann, Z. Kristallogr. Kristallgeom. Kristallphys. Kristallchem, 98 (1937) 299.
16. G.W. Brindley and G. Brown, Crystal Structures of Clay Minerals and their X-ray Identification, Mineralogical Society, London, 1980, Chapter 3.
17. N.T. Skipper, F-R.C. Chang and G. Sposito, Clays and Clay Miner., 43 (1995) 285.
18. O. Bernard, W. Kunz, P.Turq and L. Blum, J. Phys. Chem. 95 (1991) 9508
19. P. Turq, L. Blum, O. Bernard and W. Kunz, J. Phys. Chem. 99 (1995) 822
20. A. Chhih, P. Turq, O. Bernard, J.M.G. Barthel and L. Blum, Ber. Bunsenges Phys. Chem. 12 (1994) 1516

Mat. Res. Soc. Symp. Proc. Vol. 651 © 2001 Materials Research Society

Spectroscopic studies of liquid crystals confined in sol-gel matrices

Carlos Fehr, Philippe Dieudonne, Christophe Goze Bac, Philippe Gaveau, Jean-Louis Sauvajol, Eric Anglaret
Groupe de dynamique des Phases Condensées, UMR CNRS 5581, Université Montpellier II, Montpellier, France

ABSTRACT

Thermotropic liquid crystals (5CB and 8CB) were confined in silica porous matrices (xerogels and xero-aerogels) with different pore sizes. The structure and dynamics of confined liquid crystals were studied by Raman spectroscopy and Nuclear Magnetic Resonance. In Raman, the frequency of the CN stretching peak is a good probe of the smectic_A-crystal phase transition. The CN peak is observed to split upon quenching. This suggests a coexistence of crystalline and supercooled liquid phases for confined LC which was not observed in the bulk. In NMR, strong differences in both chemical shifts and linewidths are observed in confinement with respect to the bulk. We present a model which analyses the changes in the spectra in terms of changes of the order parameter and molecular dynamics for the confined LC.

INTRODUCTION

The effect of confinement on the properties of liquid crystals has been a matter of intense research [1]. More generally, liquid crystals are good candidates to probe the effect of confinement on the properties of fluids because of their rich and original phase diagrams. Phase transitions and structural properties of confined LC were so far essentially studied by calorimetry and diffraction experiments [1]. Only a few studies were carried out using spectroscopic techniques [2],[3],[4] although spectroscopy may provide useful and complementary informations on the structure and dynamics of confined fluids.

Here, we report on Raman scattering and Nuclear Magnetic Resonance (NMR) studies of LC confined in model porous matrices. Monolithic porous silica glass with tailored pore size were prepared using an original process based on a 2-steps drying procedure of silica gels. We describe the materials in the first part of the paper. In a second part, we present a Raman study of the low-temperature phases of confined LC. We compare the results obtained by slow cooling down and quenching from the smectic A phase. The temperature-dependence of the CN stretching mode is used as a probe to study the liquid crystal-solid phase transition and to discuss the nature of the solid phases. In a third part, we discuss the changes in the NMR spectra (chemical shifts and linewidths) of confined LC with respect to the bulk. We show that the analysis of the spectra allows us to estimate the order parameter of the LC phases as well as the timescale of the molecular motions.

MATERIALS

Silica gels were prepared by hydrolysis and polycondensation of tetramethoxysilane (TMOS) in ethanol in basic conditions (0.016M NH_4OH). The volume fractions of alcohol, TMOS and H_2O in the solution are 0.4, 0.4 and 0.2, respectively. Gels were then submitted to a two-steps drying procedure.

In a first step, gels were partially dried by classical solvent evaporation in an oven (2h at 90 °C). Because of capillary stress and weak mechanical properties, the gel undergoes a volumic shrinkage which affects essentially the largest pores and results in a narrowing of the pore size distribution. In the second drying step, supercritical drying allows to obtain a porous dried material without further modifications of the porous gel texture. This two-steps procedure leads to a monolithic rod of so-called xero-aerogels [5]. These samples are characterized by rather narrow distributions of pore diameters. For a given gel, the mean-pore diameter of the xero-aerogel is fixed by the duration of the first step i.e. by the volumic shrinkage at the end of the first drying step [5]. The dried matrices were then characterized by N_2 adsorption-desorption. Pore volume and pore size distribution were evaluated from desorption curves. The results are summarized in table I.

Sample	Mean pore diameter (nm)	Porous volume (cm^3/g)	Volumic fraction of pores (%)
XA22	22.0	2.86	86
XA19	18.6	1.29	80
XA10	10.0	1.18	71
XA8	7.6	0.52	51

Table I. Compared textural properties of the samples

For the Raman and NMR experiments, the samples were cut in thin slices of about 1 mm. Gas desorption was achieved by heating the sample at 180 °C during 18 hours in a vacuum chamber. The matrices were then filled with LC in the isotropic phase (at 60 °C) under helium flow to avoid formation of bubbles inside the pores. In the bulk, the sequence of phase transitions is *crystal* (**K**)-21.5 °C- *smectic_A* (**S**)-33.7 °C- *nematic* (**N**)-40.5 °C- *isotropic* (**I**) for 8CB and **K**-24 °C- **N**-35.3 °C- **I** for 5CB.

RAMAN SPECTROSCOPY

A few Raman spectroscopy studies of confined LC were reported so far. Shao *et al* used Raman scattering to probe the nematic-crystal phase transition [2] and found that the transition temperature is proportional to the inverse of the mean-pore diameter. A detailed study of the pore-diameter dependence of the isotropic-nematic and smectic_A-crystal transitions of confined 8CB will be reported elsewhere [6]. Here, we use Raman spectroscopy in the frequency range of the CN stretching mode to study the effect of quenching on the low-temperature solid phases of confined 8CB.

The Raman experiments were carried out with a 64000 Jobin-Yvon using the 514.5 nm line of an Ar ion laser. Measurements were achieved over the temperature range 100K-330K in the following thermal cycle : i) quenching from room temperature (smectic A phase) down to 100K, ii) slow heating up to 330K, iii) slow cooling down to 100K. Selected typical spectra for bulk and confined 8CB are presented in figure 1 for various temperatures during the heating cycle (ii).

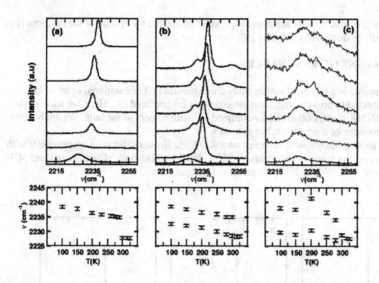

Figure 1. Raman spectra in the range of the CN stretching mode (top) and peak maxima (bottom) for bulk 8CB (a), confined 8CB in xero-aerogels with a mean-pore diameter of 19 nm (b) and 8 nm (c). Temperatures from top to bottom : 100K, 150K, 200K, 250K, 300K, 320K.

A single peak is measured all along the thermal cycle for bulk 8CB. This is also the case for confined LC during the slow cooling cycle (not shown). By contrast, a splitting of the peak is observed for quenched samples. In the bottom part of figure 1, we plot the temperature-dependence of the frequency of the maximum (maxima) of the CN peak(s) upon the heating cycle (ii). Each peak was fit to a Voigt lineshape in order to achieve a good overall fitting, as detailed in reference [6]. For bulk LC, the brutal shift observed around 300 K is the signature of the crystal-smectic A transition.

For LC confined in pores of 18 nm, one recovers the signature of this crystalline phase in the low-temperature spectra. In addition, a second intense peak is measured. Its temperature dependence extends at low temperatures that measured in the liquid phase. Consequently, it is tempting to assign the splitting of the CN peak upon quenching to a coexistence of two solid phases in the confined samples : the crystalline bulk phase and a supercooled liquid phase. Note that in specific conditions of quenching (from the isotropic phase), such a coexistence was already suggested for bulk 8CB [7]. We will discuss in detail differences and agreements between our results and those of reference [7] elsewhere. Finally, for LC confined in pores of 8 nm, a comparable splitting is observed, but the intensity of the low-frequency peak is found to increase to the expense of the high-frequency one. In our interpretation, this corresponds to an increase of the fraction of supercooled liquid in the confined LC. The study of the relation between the volumic fraction of each low-temperature phase and the texture of the matrices is in progress and will be discussed elsewhere [6].

NUCLEAR MAGNETIC RESONANCE

The NMR technique is a powerful tool to study the dynamics and orientational order of LC [3],[4]. It is expected to provide important informations for confined LC. Here, we address the changes in the NMR spectra observed for confined LC with respect to the bulk. We analyse the data in the framework of a simple exchange model.

The ^{13}C NMR measurements were achieved on a 400 MHz Bruker solid-state spectrometer with ^{1}H decoupling over the range of temperatures 250K-340K for both bulk 5CB and confined 5CB. Spectra measured at 320K and 298K are presented in figure 2.

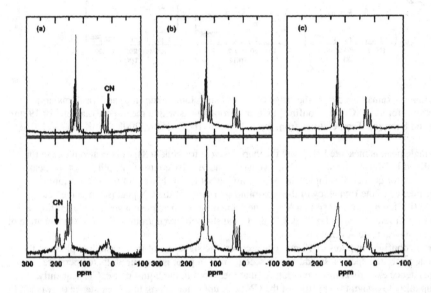

Figure 2. NMR spectra of 5CB bulk (a), confined in a xero-aerogel with a mean-pore diameter of 20 nm (b) and confined in a xerogel with a mean-pore diameter of 10 nm (c), at 320K (top) and 298K (bottom).

For bulk 5CB in the isotropic phase (figure 3a, top), two bunches of lines are observed in the ranges 10-40 ppm and 110-150 ppm which correspond to carbons of the aliphatic tail and aromatic cycles, respectively [3],[4]. The spectrum of bulk 5CB in the nematic phase (figure 3a, bottom) is featured by an upshift and broadening of the two bunches, indicating orientational order and slowing down of the dynamics. The most spectacular change is the large upshift from about 20 to about 200 ppm of the chemical shift of the carbon involved in the CN group (arrows in figure 3a).

At 320K, the spectra of confined LC are very close to those of the bulk. The main difference is a weak broadening of the lines due to a matrix-induced magnetic field inhomogeneity. By contrast, at 298K, drastic changes in the lineshape of the spectra of confined LC are observed with respect to the bulk. This is first due to the appearance of an orientation of the LC in the cavities, which is the signature for confined LC of the isotropic-nematic transition in the bulk. Note that in confinement, the small size of the LC domains in the porous cavities makes the magnetic field insufficient to align the nematic domains by contrast with the bulk where one gets well-oriented monodomains [4]. Therefore, the LC oriented domains are misoriented one from each others so that the resulting spectrum is a powder-like average. In addition, diffusion of the molecules from one domain to another is slowed down at low temperatures which also modifies the NMR lineshapes. In order to study the order parameter and the molecule dynamics, we developed an exchange model for the analysis of the lineshapes [6].

In this model, two parameters are considered : the order parameter S and a correlation time t corresponding to the characteristic timescale for the molecules to diffuse from one cavity to another.

Calculated spectra for various order parameters and correlation times are compared to the experimental data in figure 3.

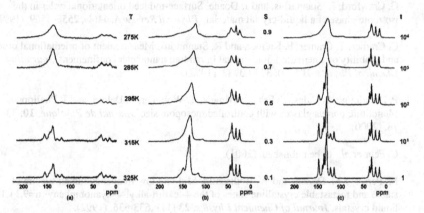

Figure 3. Temperature-dependence of the NMR spectra of 5CB confined in a xerogel (a), as compared to calculations (see text) for fixed correlation time ($t=10^4$) and various order parameters (b) and for fixed order parameter (S=0.4) and various correlation times (c).

The general profile of the calculated spectra is in good agreement with the data. Furthermore, the broadening of the lines is well recovered for an increasing order parameter and/or an increasing correlation time. The quantitative analysis of the data and the correlation with Raman data are in progress and will be presented elsewhere [6].

CONCLUSIONS

Spectroscopic techniques were used to probe the structure and dynamics of confined LC. The smectic_A-crystal phase transition is featured by a brutal shift of the frequency of the CN stretching mode in the Raman spectra. A splitting of the CN peak was observed for samples quenched to low temperatures and assigned to a coexistence of a crystalline and a supercooled liquid phase in the confined LC. The profile of the NMR spectra around the isotropic-nematic phase transition are very different for confined materials with respect to the bulk. For decreasing temperatures, broadening of the lines indicates an increase of the order parameter and/or a slow down of the molecular dynamics. A simple model was developed which allows to get a good qualitative description of the experimental spectra. Quantitative analysis and correlations between Raman and NMR results are now in progress.

REFERENCES

1. G. Crawford and S. Zumer, editors. *Liquid Crystals in complex Geometries*. (Taylor & Francis, 1998).

2. Y. Shao and T. Zerda. Phase transitions of liquid crystals PAA in confined geometries. *Journal of Physical Chemistry B*, **102**, 3387-3394, (1998).

3. G. Crawford, R. Stannarius, and J. Doane. Surface-induced orientational order in the isotropic phase of a liquid-crystal material. *Physical Review A*, **44**(4), 2558-2569, (1991).

4. C. Cramer, T. Cramer, F. Kremer, and R. Stannarius. Measurement of orientational order and mobility of a nematic liquid crystal in random nanometer confinement. *Journal of Chemical Physics*, **106** (9), 3730-3783, (1997).

5. P. Dieudonne, S. Calas, C. Fehr, J. Primera, T. Woignier, P. Delord, and J. Phalippou. Monolithic porous glasses with controlled mesopore size. *Journal de Physique*, **10**, 73-78, (2000).

6. C. Fehr *et al.* to be published. (2001).

5. M. Perrot, J-M. DeZen, and W. Rothschild. Mid-and low-frequency raman spectra of stable and metastable crystalline states of the 4-textitn-alkyl-4'-cyanobiphenyl(n=9,11,12) liquid crystals. *Journal of Chemical Physics*, **23** (11) ,633-636, (1992).

Mat. Res. Soc. Symp. Proc. Vol. 651 © 2001 Materials Research Society

SURFACE AND VOLUME DIFFUSION OF WATER AND OIL IN POROUS MEDIA BY FIELD CYCLING NUCLEAR RELAXATION AND PGSE NMR

S. Godefroy[1,2], J.-P. Korb[1], D. Petit[1], M. Fleury[2] and R. G. Bryant[3]

[1]Laboratoire de Physique de la Matière Condensée, UMR 7643 CNRS, École Polytechnique, 91128 Palaiseau, France
[2]Institut Français du Pétrole, 92852 Rueil-Malmaison, France
[3]Department of Chemistry, University of Virginia, Charlottesville, Virginia 22901

ABSTRACT

The microdynamics of water and oil in macroporous media with SiO_2 or $CaCO_3$ surfaces has been probed at various temperatures by magnetic field-cycling measurements of the spin-lattice relaxation rates. These measurements and an original theory of surface diffusion allowed us to obtain surface dynamical parameters, such as a coefficient of surface affinity of the liquid molecules and the surface diffusion coefficient. The water surface diffusion coefficients are compared to the volume self-diffusion coefficients of water in pores, measured by PGSE method, the latter values being more than an order of magnitude higher than the surface ones. Complementary information on the nature of the solid-liquid interface was given by NMR chemical shift experiments at high magnetic field.

INTRODUCTION

The interest of Nuclear Magnetic Resonance for oil recovery comes mainly from the use of the technique for *in-situ* well logging applications and laboratory characterization of oil-bearing rocks [1]. The technique relies on the measurement of proton nuclear relaxation at low magnetic fields, and gives various petrophysical properties, such as porosity, saturation, permeability, pore size distribution and wettability. NMR proton relaxation times of fluids in pores are enhanced by dipolar interaction with the paramagnetic impurities at the pore surface [2]. However, a better interpretation of the data requires the understanding of the molecular surface dynamics. A question then arises: how is it possible to obtain structural and dynamical information on liquids at the pore surface by nuclear spin-relaxation methods? Probing directly the molecular dynamics at the pore surface by standard NMR relaxation methods is difficult, in particular because the fast exchange of the low fraction of molecules in the surface layer with the bulk averages the spectral properties. Measurements using standard techniques have been previously reported [3,4]. For example, surface dynamics were studied by pulsed-field gradient spin-echo (PGSE) technique on partially saturated samples [3]. Recently, non-standard nuclear magnetic relaxation dispersion (NMRD) experiments were proposed to point out the microdynamics of liquids at the surface of microporous media [5,6].

Our aim here is to understand the surface nuclear relaxation processes, and the dependence of the nuclear spin relaxation on the nature of the solid-liquid interface. We report NMR experiments performed on water and oil saturating homogeneous macroporous media representative of some oil-bearing rock surface properties, with SiO_2 or $CaCO_3$ surfaces. The longitudinal relaxation rates were measured as a function of the magnetic field strength and the temperature, using NMRD technique. The data provide evidence for a two-dimensional surface diffusion of the adsorbed proton species in close proximity to paramagnetic impurities in the surface. We deduce fundamental dynamical surface parameters, such as correlation times of surface dynamics τ_m, residence times τ_s, effective diffusion coefficients D_{eff}, and a coefficient of affinity τ_s/τ_m of the liquid for the pore surface. PGSE-NMR experiments were performed on the water-saturated samples to measure the restricted self-diffusion in the pore volume D_r, which was found to be about an order of magnitude larger than D_{eff}. Complementary information on the nature of the proton species at the surface is given by NMR chemical shifts at high magnetic field.

SYSTEMS AND EXPERIMENTS

We used two types of macroporous samples, which are model systems of the two major groups of oil-bearing rocks (sandstone and carbonate) but which had narrow pore size distributions to allow a simple and non-ambiguous analysis of relaxation times. However, the results presented here do also apply on more complex natural porous media with large pore size distribution. The samples were saturated with water or oil (dodecane) using vacuum line techniques. One type of sample simulated a porous system by packing size-calibrated grains. Nonporous silicon carbide, SiC, grains were packed using different grain diameters varying in the range 8-150 µm. This procedure provided a series of porous media with a porosity of about 45%, and with pore sizes approximately the same as the grain diameters. X-ray photoelectron spectroscopy (XPS) [7] have shown that SiO_2 covers 25% of the grain surface. The observed quadratic relation between the permeability and the pore diameter [8] showed the reproducibility of the structure of the different porous media. The narrow pore size distribution is evidenced by the observed narrow transverse relaxation times (T_2) distribution of the water saturated sample measured at 2.2 MHz [8]. Due to the grain preparation, the surfaces contain paramagnetic impurities that originate either from the grinding process or derived from the silicon carbide synthesis. We removed the grinding impurities by applying a continuous flux of hydrochloric acid into the porous media for a few weeks. Electron Spin resonance (ESR) experiments performed on the dry grains after cleaning showed that ferric ions (Fe^{3+}) remained in the grains. The second type of sample is a limestone rock. The narrow pore size distribution of the rock is evidenced by the narrow T_2 distribution of the water saturated sample measured at 2.2 MHz [8]. ESR measurements performed on limestone rock have shown the presence of Mn^{2+} paramagnetic impurities.

To investigate the surface dynamical properties, we performed Nuclear Magnetic Relaxation Dispersion (NMRD) experiments on the 8 µm grain pack and limestone rock, saturated either with water or oil. We measured the variation of the proton spin-lattice relaxation rates ($1/T_1$), as a function of the magnetic field (corresponding to Larmor frequencies in the range 0.01-25 MHz), from 15 to 45°C, using a field cycling instrument of the Redfield design [9]. The NMRD curves of water or oil in grain packing are reported in Fig. 1. Relaxation rates are much higher for water than for oil due to the difference of affinity of these liquids for the pore surface. Water dispersion curves vary logarithmically with the Larmor frequency, and terminate by a plateau below a frequency cutoff $\omega_{lco} \approx 0.1 \times 2\pi \ 10^6$ rad/s. Oil NMRD curves vary logarithmically over the whole frequency range studied and the slope is much smaller than that observed for water. The dependence of the water spin-lattice relaxation rates on temperature is opposite to that observed for the oil. For water at low frequency, $1/T_1$ increases with temperature but the curves converge at high frequency. For oil, the $1/T_1$ decrease weakly with increasing temperature. The NMRD curves of water or oil saturating the limestone cores are presented in Fig. 2. Again, the proton spin-lattice relaxation rates are much higher for water than for the oil and the water relaxation rates vary linearly with the logarithm of the Larmor frequency and display a low field plateau below $\omega_{lco} \approx 0.03 \times 2\pi \ 10^6$ rad/s. For the limestone samples, the spin-lattice relaxation rates of the oil show a plateau at higher field strengths than the water corresponding to a Larmor frequency of $\omega_{lco} \approx 1 \times 2\pi \ 10^6$ rad/s. The spin-lattice relaxation rates for both the water and oil protons increase with decreasing temperature.

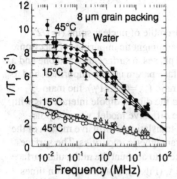

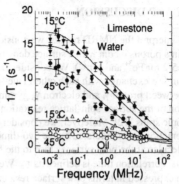

Figure 1. NMRD data of water and oil in 8 μm grain packing. The continuous lines represent the best fit obtained with Eq. (1)

Figure 2. NMRD data of water and oil in limestone rock. The continuous lines represent the best fit obtained with Eq. (1)

Analysis of the NMRD curves provides a characterization of the liquid microdynamics at the pore surface. We now want to probe the diffusive motion of the liquid molecules inside the pores. Pulsed-Gradient Spin-Echo (PGSE) experiments were performed on water saturating the 8 μm grain packing and limestone rock using the echo stimulated Stejskal and Tanner sequence [10] to determine the water self-diffusion coefficient D within the pores. The experiment applied on bulk water leads to $D \approx 2.2 \ 10^{-5} \ cm^2/s$. Fig. 3 shows the variation of the diffusion coefficient of water inside the grain packing as a function of the diffusion time between the gradient pulses of the sequence for gradient strengths of 5 G/cm. One observes a variation of D from $1.6 \ 10^{-5}$ to $1.1 \ 10^{-5} \ cm^2/s$. At short diffusion times, one obtains information on the surface to volume ratio of the porous media [11], which is beyond the scope of this paper. At long diffusion times, D reaches the asymptotic value of the restricted diffusion D_r due to the structural characteristics of the porous media. We use the value of D_r to compare surface and volume dynamics. Fig. 4 displays the diffusion coefficients of water inside the limestone rock as a function of the diffusion time for gradient strengths of 8 G/cm. D varies from $1.2 \ 10^{-5} \ cm^2/s$ to the restricted diffusion $D_r \approx 0.6 \ 10^{-5} \ cm^2/s$.

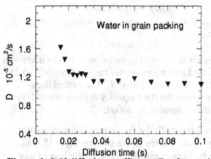

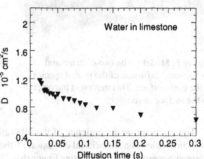

Figure 3. Self-diffusion coefficient, D, of water in packed grains obtained using the PGSE method with magnetic field gradients of 5 G/cm, at 2.2 MHz and 25 °C, for diffusion times varying from 0.015 to 0.1 s (▼). D reaches an asymptotic value of $D_r=1.1 \ 10^{-5} \ cm^2/s$.

Figure 4. Self-diffusion coefficient D of water in limestone rock obtained using the PGSE method with magnetic field gradients of 8 G/cm, at 2.2 MHz and 25 °C, for diffusion times varying from 0.015 to 0.5 s (▼). D reaches an asymptotic value of $D_r=0.6 \ 10^{-5} \ cm^2/s$.

THEORY AND DISCUSSION

To interpret the NMRD results, we consider an ensemble of nuclear spins (I=1/2), diffusing in a porous media with a surface density σ_S of paramagnetic impurities of electronic spin S (S=5/2 for Fe^{3+} and Mn^{2+}). We consider two liquid phases, a surface-affected one and a bulk one, in fast exchange with each other. Due to the surface paramagnetic impurities, and to the ratio between proton γ_I and electron γ_S gyromagnetic ratio ($\gamma_S = 658.21\gamma_I$), the main contribution of the surface spin-lattice relaxation is due to the dipole-dipole interactions with the electronic spins [12]. Moreover, we consider a surface diffusive model, where dipole-dipole interactions are modulated by the two-dimensional translational diffusive motion of the liquid molecules at the pore surface close to the fixed paramagnetic impurities. The surface translational correlation time is defined as τ_m. We also take into account the molecular surface desorption expressed in terms of a surface residence time τ_s (Fig. 5). Both correlation times follow exponential activated laws. The longitudinal relaxation rates of liquids in pores is then given by the expression [5,6]:

$$\frac{1}{T_1(\omega_I)} = \frac{1}{T_{1B}} + \frac{N_S}{N}\frac{\pi}{20}\frac{\sigma_S}{\delta^4}(\gamma_I\gamma_S\hbar)^2 S(S+1)\tau_m \ln\left(\frac{1+\omega_I^2\tau_m^2}{(\tau_m/\tau_s)^2 + \omega_I^2\tau_m^2}\right) \quad (1)$$

Here, Ns/N is the probability that a proton is at the pore surface, which can be estimated as the ratio between the surface volume to total volume of liquid. δ is the distance of minimal approach between I and S spins. In Fig. 6 are displayed the theoretical frequency dependences of Eq. (1) for frequencies ranging from 0.01 to 25 MHz, and different temperatures.

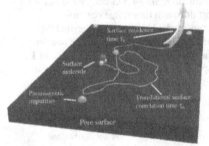

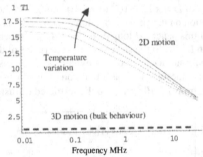

Figure 5. Model of the two-dimensional translational diffusion of the proton species at the pore surface. This motion is interrupted by the surface desorption.

Figure 6. Theoretical variation of $1/T_1$ as a function of the Larmor frequencies ranging from 0.01 to 25 MHz, and from 15 to 45°C. The dashed lines represents the flat frequency dependence of $1/T_1$ for a 3 dimensional bulk motion.

The logarithmic dependence of the relaxation rates with the frequency is the signature of a two-dimensional translational diffusion of the proton species in the surface layer [5,6,8]. The plateau observed for frequencies lower than the frequency cut-off is due to the loss of any dipolar correlation between I and S after the surface desorption. A direct analysis with Eq. (1) of the experimental dispersion curves yields to the evaluation of τ_m and τ_s.

In Fig.1, the best fits of the dispersion curves measured in packed grains with Eq. (1) are shown as a continuous line. The observed logarithmic variation of $1/T_1$ with the frequency evidences a two-dimensional surface diffusion. For water, the correlation time τ_s is directly estimated from the value of the cut-off frequency ω_{lco}, leading to $\tau_s = 1/\omega_{lco} \approx 1.6$ µs, at 25°C (Fig. 1). τ_m is determined from the slope of the logarithmic variation and the value of the plateau, $1/T_{1plateau}$, according to an expression deduced from Eq. (1) :

$$\frac{1}{T_1} = \frac{1}{2T_{1\,plateau}} \ln\left(1 + \frac{1}{(\omega_1 \tau_m)^2}\right) / \ln\left[1/(\omega_{lco}\tau_m)\right]. \qquad (2)$$

This yields $\tau_m \approx 0.6$ ns, at 25°C. Due to the very long surface residence time τ_s compared to τ_m, one can estimate the effective surface diffusion coefficient $D_{eff} \approx \varepsilon^2/(4\tau_m) \approx 6\ 10^{-7}$ cm²/s, where ε is of the order of the molecular size. This surface diffusion coefficient is more than one order of magnitude smaller than the restricted diffusion coefficient D_r measured by PGSE-NMR ($D_r/D_{eff} \approx 22$), showing that the diffusion has a value hindered at the surface. We define a coefficient of surface affinity, τ_s/τ_m, as the number of surface motion events before desorption. For water on the silica surface one finds $\tau_s/\tau_m \approx 2700$. Both ratios ($D_r/D_{eff}$ and τ_s/τ_m) show that water has a high affinity for the surface, which is responsible for the enhancement of the relaxation of water in the pores. All these surface parameters are summarized in Table 1. The anomalous temperature dependence observed for water (Fig. 1) may be explained by the surface diffusion of the protons, limited by the exchange with the local bulk. The NMRD profile of dodecane is typical of pure two-dimensional diffusive process [5,6].

In Fig. 2, the best fits obtained with Eq. (1) for the limestone rocks are presented as continuous lines. From the water dispersion curves, we directly access to the value of the residence time $\tau_s = 1/\omega_{lco} \approx 5.3$ µs, at 25°C. From Eq. (2), we deduce the surface correlation time $\tau_m \approx 1.3$ns, at 25°C, and the effective surface diffusion coefficient $D_{eff} = 2.6\ 10^{-7}$ cm²/s. This coefficient is much lower than the restricted diffusion measured by PGSE-NMR, leading to $D_r/D_{eff} \approx 23$. The coefficient of surface affinity of water on CaCO$_3$ surface is $\tau_s/\tau_m \approx 4100$. The surface parameters are summarized in Table 1. As for silica surface, the values of the ratios D_r/D_{eff} and τ_s/τ_m show that water has a high affinity for the surface. Considering oil dispersion the cut-off frequency occurs at much higher value $\omega_{lco} \approx 1 \times 2\pi\ 10^6$ rad/s than in the water case. The low affinity coefficient $\tau_s/\tau_m \approx 20$ obtained from Eq.2 is consistent with the lower affinity of dodecane molecules for the CaCO$_3$ surface, which is expected.

Table 1. Surface parameters of water obtained from data fitting

	τ_m	τ_s	τ_s/τ_m	D_{eff}	D_r/D_{eff}
Water on SiO$_2$	0.6 ns	1.6 µs	2500	$5\ 10^{-7}$ cm²/s	22
Water on CaCO$_3$	1.3 ns	5.3 µs	4100	$2.6\ 10^{-7}$ cm²/s	23

To investigate the origin of the affinity differences observed for water and oil on silica or calcium carbonate surfaces, we performed inversion-recovery and chemical shift experiments, at high magnetic field (360 MHz). Fig 7 shows the evolution of the NMR spectra during an inversion-recovery sequence performed on packed grains fully saturated with water, for recovery times varying from 0.01 to 2 s. Owing to the very low fraction of surface molecules and the efficiency of the fast exchange with the bulk water, we find a single water population (Lorentzian-like spectra), with a single $T_1 = 0.56$ s. The similar experiment performed on dodecane saturated packed grains yields the same behaviour, but with a longer

$T_1=1.2$ s. In order to observe the effects of the solid-liquid interactions at the pore surface, we performed inversion-recovery experiment on a limestone rock partially saturated with water (layer of water at the surface). Fig. 8 shows the NMR spectra measured at the longer recovery time. We observe two narrow Lorentzian peaks corresponding to water and hydrated species at proximity of the surface with corresponding T_1 of 50 and 780 ms, respectively. This result is consistent with the hindered diffusion observed by NMRD at the pore surface. We also observe a large peak with a Voigt profile and an overall linewidth about 20kHz. It corresponds to liquid proton adsorbed at the surface. On the other hand, for oil, one observes a single Lorentzian lineshape with a single $T_1 \approx 1$ s. This is consistent with the low affinity of dodecane for the pore surface

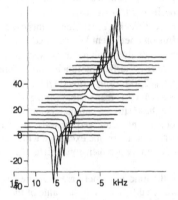

Figure 7. Inversion-recovery experiments of water in grain packing, at 360 MHz and 25 °C. One find a single relaxation time of 0.56 s.

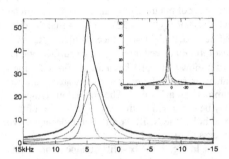

Figure 8. NMR spectra of limestone rock, partially saturated with water, at 360 MHz and 25°C. The two narrow Lorentzian peaks correspond to water and hydrated species at proximity of the surface. The large peak is a Voigt profile of overall linewidth 20KHz.

CONCLUSION

On the basis of experimental and theoretical nuclear relaxation studies at various conditions of magnetic field and temperature, we directly access to surface microdynamics of liquids in macroporous media. Field cycling experiments were performed on water or oil saturated porous media of SiO_2 and $CaCO_3$ surface. Surface correlation times and diffusion coefficients are thus estimated, as well as a coefficient of affinity of liquids for the pore surface. We compare the water surface diffusion and volume diffusion coefficients measured by PGSE-NMR technique, and find that the latter are more than an order of magnitude larger than the former. High-resolution NMR spectra of liquids in the porous media compliment the relaxation studies and demonstrate that the liquid samples more than one environment.

REFERENCES
1. R. L. Kleinberg, Experimental methods in the Physical Sciences: Methods of the Physics of Porous Media, P.-Zen Wong, ed. (1999).
2. J. Korringa, D.O. Seevers and H.C. Torrey, Phys. Rev. **127**, 1143 (1962).
3. F. D'Orazio, S. Bhattacharja, W. P. Halperin, K. Eguchi, T. Mizusaki, Phys. Rev. **42**, 9810 (1990).
4. J.-P. Korb, Shu Xu, and J. Jonas, J. Chem. Phys. **98**, 2411 (1993).
5. J.-P. Korb, M. Whaley-Hodges, and R. G. Bryant, Phys. Rev. E, **56**, 1934 (1997).
6. J.-P. Korb, M. Whaley-Hodges, Th. Gobron, and R. G. Bryant, Phys. Rev. E, **60**, 3097 (1999).

7. V. Médout-Marère, A. El Ghzaoui, C. Charnay, J. M. Douillard, G. Chauveteau, and S. Partyka, J. Colloid Interface Sc. **223**, 205 (2000).
8. S. Godefroy, J.-P. Korb, M. Fleury and R. G. Bryant, submitted to Phys. Rev. E.
9. A. G. Redfield, W. Fite, and H. E. Bleich, Rev. Sci. Instrum. **39**, 710 (1968).
10. J. E. Tanner, E. O. Stejskal, J. Chem. Phys, **49**, 1768 (1968).
11. P. P. Mitra, P. N. Sen, L. M. Schwartz, and P. Le Doussal, Phys. Rev. Let., **65**, 3555 (1992).
12. A. Abragam 1961, *The Principles of Nuclear Magnetism* (Clarendon, Oxford, 1961).

Mat. Res. Soc. Symp. Proc. Vol. 651 © 2001 Materials Research Society

CAYLEY TREE RANDOM WALK DYNAMICS

Dimitrios Katsoulis and Panos Argyrakis
Department of Physics
University of Thessaloniki
54006 Thessaloniki Greece

Alexander Pimenov and Alexei Vitukhnovsky
Lebedev Physical Institute
Russian Academy of Sciences
53 Leninsky prospect
117924 Moscow Russia

Abstract

We investigate diffusion on newly synthesized molecules with dendrimer structures. We model these structures with geometrical Cayley trees. We focus on diffusion properties, such as the excursion distance, the mean square displacement of the diffusing particles, and the area probed, as given by the walk parameter S_N, the number of the distinct sites visited, on different coordination number, z, and different generation order g of a dendrimer structure. We simulate the trapping kinetics curves for randomly distributed traps on these structures, and compare the finite and the infinite system cases, and also with the cases of regular dimensionality lattices. For small dendrimer structures, S_N approaches the overall number of the dendrimer nodes, while for large trees it grows linearly with time. The average displacement R also grows linearly with time. We find that the random walk on Cayley trees, due to the nature ot these structures, is indeed a type of a "biased" walk. Finally we find that the finite-size effects are particularly important in these structures.

Introduction

Diffusion properties in dendrimer structures recently attracted considerable interest, because it has been hypothesized that one might be able to control the energy transport mechanism by varying the generation order, g, of the structure , and the coordination number, z. The chemical synthesis of various such species in the last decade, albeit the difficulties encountered of extremely low yields, opened up this direction, and thus it is important to elucidate the mechanism of particle diffusion, before direct comparison with experimental data on the dynamics of such systems can be made [1-4]. In all considerations it is expected that these properties will strongly depend on the particular geometry of the prepared molecules. It is highly possible that structures that have minute only differences that they might exhibit completely different behavior in their transport properties. For example, the "compact" and "extended" series in repeated phenylacetylene units exhibit enhanced rates, the first one towards the core, but the second towards the opposite direction, the molecular periphery, even though both systems are composed of identical single phenylacetylene chains, but with slightly different molecular layouts [5].

By the nature of the Cayley tree structure, the transport properties are highly asymmetric. Transfer towards the periphery has a higher overall probability, when compared to transfer towards the core of the structure, for any number of emanating branches. In a corresponding random walk model, which is frequently used for calculations, this type of diffusion is analogous to "biased" or directional diffusion. In a biased walk, say in one dimension, the probability for moving to the right is not the same as the probability to move to the left, as is the case for a symmetric walk, where the two probabilities are 0.5 each. When this asymmetry occurs, then it is expected that this will lead to a ballistic-type motion, with the average maximum excursion being proportional to N rather than $N^{1/2}$ (N is the number of steps, or time). On the other hand this natural bias factor can be taken out if we assign a ½ probability for the motion towards the core, and ½ for motion towards the periphery collectively for all bonds in this direction. We thus have two models, one with natural bias, and one with no bias in the motion at all.

In the present paper we deal in detail with the notion of the effective bias present in the dendrimer structures by investigating the average displacement of the walk, <R>. Since motion is ballistic it is not necessary to look at <R^2> (which would be proportional to N^2), but it is suffices to investigate <R>, which is expected to be proportional to N. Additionally, for such a calculation, one does need to keep track on information about the sites (nodes), (e.g. whether a site has been visited or not), making it possible to avoid the finite-size effect. We have thus devised a new algorithm that keeps track of information only for the initial site, the final site, and the number of steps that it has taken to move from one to another. Then the algorithm calculates the distance between these two points to derive the R vs. time relation. This relation is expected to be linear, rather than having a $N^{1/2}$ dependence, as a normal random walk would have. With this algorithm we can use very large size systems, practically infinite.

Additional properties of interest include the area sampled by the random walk, as given by S_N, which is the number of sites visited as a function of time, and which is also expected to behave ballistically, due to the inherent "bias" in the particle motion. This means that $S_N \sim N$, similar to the behavior of 3-dimensional lattices, and this is indeed the case [6]. This has been also shown analytically for the asymptotic limit (in size and time), and it is has the same dependence as R. Analytically [7], we have that:

$$S_N = \frac{z-2}{z-1}N \qquad\qquad N \rightarrow \infty$$

Another peculiarity in the dendrimer structures is that finite size effects are very important and play a dominant role in dendrimer systems, much more important than customary lattice systems. The reason for this is that even for very large structures, the largest possible in use today in computer calculations, any given point in the structure is quite close to the boundaries. For example, a Cayley tree of g=22, and z=3, has about 10 million sites, but no site is further away than 11 units away from the boundaries. Thus, if it is necessary to use computational models where the entire structure must be defined and kept in memory, then it is practically impossible to avoid finite size effects. This, in turn, makes it impossible to compare with existing asymptotic trends, which imply infinite systems. On the other hand, this very point of finite size, might be the key to observed unusual experimental behavior, of real molecular systems and, therefore, it is worth investigating in detail. We should keep in mind that typical dendrimer species utilized today in experiments have a g value of g=6 or 7.

Finally, we examine in detail the effect of trapping on dendrimer structures, and attempt to compare it with trapping on regular lattices, a process that is still poorly understood. We do this by calculating the survival probability, Ψ_N, as a function of time, on dendrimers of different sizes, with and without the bias factor. In order to simulate the trapping process we designated a small part (say, 10^{-3}) of the nodes as traps. When a particle reaches a trap it is irreversibly trapped, and the time to trapping was recorded. The survival probability is deduced from the population of the remainder (untrapped) particles. It is also expected here that the boundary conditions will be important in the decay profile of the survival probability. The trapping intensity curves for the infinite Cayley tree should be monoexponential curves. Here we expect a very complex decay profile, which, to a first approximation could be treated as two-exponential decay: the first exponential corresponding to the fast decay with the rate equal to the infinite case, while the second exponent would be a decay characterized by a decay time several orders higher.

The model of calculations

Computer simulations of random walks on dednrimers are performed in a similar fashion as in discrete lattice systems. We already discussed above the difficulties of finite sizes. Our calculations are thus separated in two models. First the model where no structure is kept in memory, for the R calculation, and then the case where the structure is kept in memory for the S_N and the trapping calculations.

The first algorithm for calculating the <R> distance is the following: We assume that the starting point is some random point in the structure, where a particle is placed. Before choosing the random site we first choose its g. Thus we randomly choose a g value, and assume that the starting point will be a site with this chosen g. The value of g of this point is known and recorded. Next we build a 1-dimensional array with g number of elements, which take random values in the interval 1 to k=z-1. This array gives the exact position of the starting point in the structure, and it is kept intact for the entire duration of the run. Then the random walk is started, and the particle moves to one of the nearest neighbor sites only, which all have the same probability. A second 1-dimensional array is now constructed, which at this point is identical to the first one. When a step is made at random, during the random walk, the second array changes by adding or subtracting one row, according to whether g increases by 1 or decreases by 1, respectively. No other information is necessary, just these two arrays, the first one at time t=0, and the second one, which dynamically changes during the walk. At the end of the process, we compare the two arrays. We will notice that the top parts in both arrays are identical, while their bottom parts are different. In each array of the two that we have, we subtract the common top part, and finally R will be the sum of the remaining two bottom parts. For example, if the 2 arrays have sizes of 15, and 100, respectively, and the top 10 rows are identical in both arrays, then R=(15-10)+(100-10)=95.

The advantage of this method is that it is not necessary to find and monitor the value of the index of the current site, which is time consuming and limited by the size of arrays, but we need to monitor just the value of g, and two 1-dimensional arrays.

Finally, the second model for the S_N and the trapping calculations is constructed in the usual way, and the calculations are performed using routine procedures [8].

Results

In Figure 1 we show the behavior of <R> as a function of time for several different z values. Here R is considered to be the excursion distance from the point of origin, and it is measured in terms of structure bonds traversed during the random walk. This distance R is the shortest distance between any two points on the Cayley tree, and as such no bond is traversed more than once during the random walk. We see, as expected, that <R> grows linearly with time, for any z. This is the opposite of what we have in lattice systems, where <R>=0, say for a 1-dimensional lattice.

One would then expect for the mean square displacement that $<R^2>=N^2$, as the motion is ballistic rather than stochastic, as it is on normal lattices. The same assertion is expected to hold for the area sampled by diffusing particles, as it is depicted by S_N, the number of distinct sites visited during the walk. For lattices, the well known results are that for 1-dimensional systems $S_N \propto N^{1/2}$, for 2-dimensional $S_N \propto N/\log N$, and for 3-dimensional $S_N \propto N$. One here would expect for a dendrimer structures that the 3-dimensional behavior would hold, i.e. S_N would be directly proportional to time.

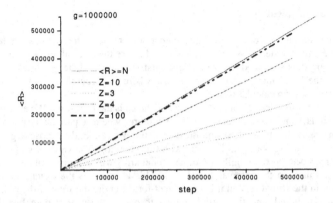

Figure 1: The average distance <R> as a function of time, for random walks on dendrimer structures, for several different z values. The top straight line is <R>=x

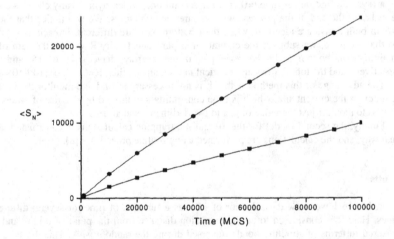

Figure 2: S_N as a function of N for a random walk on a dendrimer with g=20 and z=3. The upper curve is for normal random walk, i.e. all branches have equal probabilities. The lower curve is for the model of no bias, i.e. equal probability to go up or down. In both cases the curves correspond to the model of reflective boundary conditions at the core, but cyclic boundary condition at the periphery. The points correspond to the model of the cyclic boundary conditions at the core and at the periphery.

The coordination number z affects the behavior only at the constant of proportionality, but this holds for any z value. Thus, for z=3, we have that S_N=N/2. The computer simulations at this point suffer from the finite size effect discussed above, so that with a direct algorithm it is rather impossible to reach large values of g and z. Our values for S_N are in the range of 0.1 – 0.2. Figure 2 shows some results for S_N. Since the stochastic walk on a dendrimer is effectively a biased walk, it is of interest to consider a model on which this bias is negated. This can simply be accomplished by assigning a ½ probability for the particle to go up towards the core along the single bond, and also ½ together to the (z-1) bonds to go down towards the periphery. If z=3, then the latter probability would be ¼ along each bond. As we see in Figure 2 the "bias" model shows a higher S_N behavior rather than the unbiased case. This is true for every biased walk on any lattice. The effect of different boundary conditions at the core is not important at all. If the R property were investigated with the "unbiased walk" model, then the result would be trivially <R>=0, because of the equal probabilities to go up and down. In Figure 3 we compare the S_N on the dendrimer with that in the classical 3 dimensionalities. We see that the dendrimer case is closer to the 2-dimensional lattice, rather than the 1-dimensional one. But, of course, this is due to finite size effects, while the asymptotic one should resemble the results in the 3-dimensional lattices.

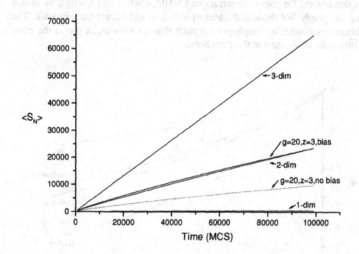

Figure 3: S_N as a function of time N. The random walks are on a dendrimer with g=20 and z=3, using two models: One with normal random walk (effectively biased), and one with a motion of no bias (equal probability to go up or down). Boundary conditions are such that we have reflection at the core, but cyclic motion at the periphery. For comparison purposes, we also show the S_N behavior for 1-dim, 2-dim and 3-dim lattices.

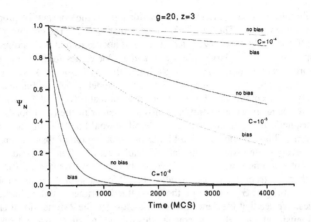

Figure 4: Survival Probability Ψ_N, as a function of time, for random walk on a dendrimer, for trap concentrations $C=10^{-2}$, $C=10^{-3}$ and $C=10^{-4}$, as shown on the graph. We show both cases of unbiased and biased random walk. The boundary conditions employed are such that we have reflection at the core, but cyclic conditions at the periphery.

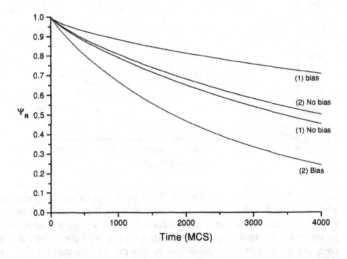

Figure 5: Survival probability for different sets of boundary conditions, as marked: (1) Reflection at the core, so that when the particle reaches the periphery it jumps to the neighbor site (left or right). (2) Reflection at the core, so that when the particle reaches the periphery, it goes up to the core.

Finally, a related property in the same class of random walk properties is that of trapping. The problem of trapping has been known for a long time [9], but a general finite-time solution is still lacking, with the exception of 1-dimensional lattices [9]. An asymptotic solution due to Donsker and Varadhan [9] is valid for any dimensionality d, but this solution is valid for time t→ ∞, so that nobody has ever numerically observed the Donsker-Varadhan behavior. In Figures 4 and 5 we show the survival probability as a function of time for different trap concentrations, and different sets of boundary conditions, both for the case of a biased walk and an unbiased walk. We observe that typically particles survive longer in the case of no bias, under cyclic boundary conditions. However, when at the periphery there is interactions between sites that are adjacent, but belong to different branches, then this trend is reversed. These interactions are included by allowing branch-to-branch hopping, and thus we conclude that these interactions are very important in Cayley tree dynamics.

Conclusions

Summarizing, we have performed computer simulations on dendrimer structures, modeled by geometrical Cayley trees, and investigated in detail well known diffusion properties, especially the ones that have been heavily studied in the past on discete lattices. We point out the character of the "bias" of the random walk towards the periphery of the system, something that is shown in the results of the walk excursion, R, and the number of sited visited, S_N. Due to finite size effects we can not recover the exact analytical solution of Hughes-Sahimi. However, our solution has the same form as the analytical solution, i.e. it is linear, with the difference being in the prefactor.

Acknowledgements: This work was supported by NATO SfP grant 971940

References

(1) D. A. Tomalia, A. M. Naylor, W. A. Goddard, Angew. Chem., Int. Ed. Engl., 29,138(1990).

(2) R. Kopelman, M. R. Shortreed, Z. Y. Shi, W. Tan, Z. Wu, J. S. Moore, A. Bar-Haim, and J. Klafter, Phys. Rev. Lett., 78,1239(1997).

(3) M. R. Shortreed, S. F. Swallen, Z. Y. Shi, W. Tan, Z. Xu, C.Devadoss, J. S. Moore, R. Kopelman, J. Phys. Chem., 101,6318(1997).

(4) A. Bar-Haim, J. Klafter, and R. Kopelman, J. Am. Chem. Soc., 119,6197(1997).

(5) S. F. Swallen, R. Kopelman, J. S. Moore, and C.Devadoss, J. Molec. Struct., 485-486,585(1999).

(6) P. Argyrakis and R. Kopleman, Chem. Phys., 261,391(2000).

(7) B. D. Hughes and M. Sahimi, J. Stat. Phys., 29,781(1982).

(8) P. Argyrakis, Comp. Phys., 6,525(1992).

(9) G. H. Weiss, Aspects and Applications of the Random Walk, North-Holland, Amsterdam (1994).

Mat. Res. Soc. Symp. Proc. Vol. 651 © 2001 Materials Research Society

Thermodynamics and kinetics of shear induced melting of a thin lubrication film trapped between solids

V.L. Popov[1] and B.N.J. Persson
IFF, FZ Jülich, D-52425 Jülich, Germany
[1] IFF, FZ Jülich, D-52425 Jülich, Germany
and
University of Paderborn, Dept. of Theoretical Physics
33098 Paderborn, Germany

ABSTRACT

Shear melting of a thin layer trapped between two crystalline solids is considered in the frame of the Landau's theory of phase transitions. Kinetics of melting and solidifying by static and periodic loading is analyzed.

INTRODUCTION

If two solid bodies are brought together, the thickness of the lubrication layer will decrease monotonically until it reaches a thickness of about 10 molecular layers. The remaining layer often shows elastic properties and remains trapped between the surfaces, provided the applied pressure is not too high [1]. This layer has a tendency to form a layered structure in the direction perpendicular to the confining surfaces and a long range crystalline order in the layer plane (experimental investigations of these aspects can be found in [2,3], and molecular dynamics simulations in [1,4]).

Computer simulations of sliding of solids separated by a thin lubrication layer show that the sliding often is accompanied by shear melting of the intermediate layer [1,4]. The "melting" can occur both in the whole volume of the lubricant and in single layers [4]. The melting can also be achieved through a sufficient increase of temperature [1].

In the present paper we propose a theoretical description of the shear melting transition of thin boundary lubrication layers in the frame of the Landau's theory of phase transformations [5]. We show that the experimentally observed kinetics of shear melting can be understood if we assume that the melting occurs (by increasing the elastic deformation) as a phase transformation of the second order. The possibility of the melting transition being a second order transition does not contradict to the Landau's conclusion that melting is always a first order transition [6], because the latter refers only to unbounded media, and assumes a transition from a state with a discrete symmetry group, into an isotropic state, invariant with respect to rotations. In the present case the invariance with respect to rotations is broken – even in the liquid state – due to the presence of the confining surfaces and to the presence of a chosen direction in the plane of the layer (direction of deformation). Below we suppose that the shear melting occurs as a second order transition.

TEMPERATURE AND SHEAR INDUCED PHASE TRANSFORMATIONS

The starting point for the Landau's thermodynamic theory of phase transitions is the choice of an order parameter φ which characterizes the qualitative change in the state of a medium at the phase transition point. In the case of melting, the qualitative change can be characterized by the shear modulus μ, which is nonzero in the solid phase and zero in the liquid phase.

In the Landau's phenomenological theory of phase transitions the free energy of a medium is expanded as a series in powers of the order parameter. This assumes that the order parameter can take any value near zero. Since the shear modulus is always nonnegative, it cannot be taken as the order parameter itself but may be represented as the square of this parameter:

$$\mu = \varphi^2. \tag{1}$$

At temperatures near the phase transition point, the expansion of the free energy density in series of the order parameter has (in the case of a second-order transition) the form:

$$f_{therm} = \alpha \cdot (T - T_c)\varphi^2 + \frac{b}{2}\varphi^4, \tag{2}$$

where a and b are functions of the temperature and pressure.

The function (2) describes the system in the minimum free-energy state, i.e., in the elastically undeformed state. In the crystalline phase, as a result of the non-vanishing transverse rigidity of the system, the solid film may be transferred to a deformed metastable state. In order to describe this state, we need to add the free energy of elastic deformation $\mu\varepsilon_{el}^2 / 2$ to the free energy (2). Thus, the total free energy density can (with account of (1)) be written as

$$f = \alpha(T - T_c)\varphi^2 + \frac{b}{2}\varphi^4 + \frac{1}{2}\varphi^2\varepsilon_{el}^2, \tag{3}$$

where ε_{el} is elastic deformation.

Elastic stresses in the layer are defined as

$$\sigma_{el} = \frac{\partial f}{\partial \varepsilon_{el}} = \varphi^2 \varepsilon_{el} = \mu\varepsilon_{el}. \tag{4}$$

For $T < T_c$ and $\varepsilon_{el} = 0$ the coefficient of the second power of the order parameter is negative which implies that the substance is in the solid state with shear modulus

$$\mu = \varphi^2 = \frac{\alpha(T_c - T)}{b} \neq 0. \tag{5}$$

As the elastic distortions increase, the modulus of the coefficient in the second-order terms becomes smaller. Thus, the order parameter and the shear modulus decrease:

$$\mu = \varphi^2 = \frac{\alpha(T_c - T) - \varepsilon_{el}^2 / 2}{b}. \tag{6}$$

For the specific strain

$$\varepsilon_{el,c} = \sqrt{2\alpha(T_c - T)} \tag{7}$$

the transverse rigidity of the medium becomes zero.

Elastic stress as function of elastic deformation has the form

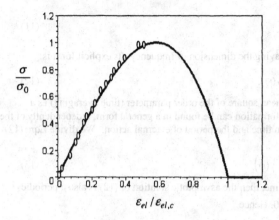

Figure 1. *Dependence of the shear stress on the elastic deformation in a layer undergoing the melting transition.. The theoretical curve was plotted accordingly to Eqn. (8). The experimental data have been taken from the paper [Reiter G., Demirel A.L., Peanasky J. et al. In: Physics of Sliding Friction. Eds. B.N.J. Persson and E. Tosatti, p.119-138].*

$$\sigma_{el}(\varepsilon_{el}) = \mu\varepsilon_{el} = \begin{cases} \dfrac{\alpha(T_c - T) - \varepsilon_{el}^2/2}{b} \cdot \varepsilon_{el}, & \varepsilon_{el} < \varepsilon_{el,c} \\ 0, & \varepsilon_{el} > \varepsilon_{el,c} \end{cases} \tag{8}$$

For $\varepsilon_{el} = \varepsilon_{el}^*$, with

$$\varepsilon_{el}^{*2} = \frac{2}{3}\alpha(T_c - T) = \frac{1}{3}\varepsilon_{el,c}^2 \tag{9}$$

it reaches the maximum value

$$\sigma_0 = \frac{1}{b}\left(\frac{2}{3}\alpha(T_c - T)\right)^{3/2}. \tag{10}$$

When the stress reaches the critical value (10), the layer loses its shear stability. The theoretical dependence (8) is in Fig.1 compared with experimental data obtained in [2].

KINETICS OF THE ORDER PARAMETER

If the temperature or deformation of the layer are changed abruptly, some time is necessary until the order parameter will reach the new equilibrium value. The kinetics of the order parameter can be determined as follows. The derivative $-\partial f/\partial \varphi$ defines the generalized thermodynamic force, tending to move the order parameter to its equilibrium value. Near the phase transition, this derivative is small and we can write down a linear (in thermodynamic force) kinetic equation

$$\dot{\varphi} = -\gamma \frac{\partial f}{\partial \varphi}, \tag{11}$$

where γ is a kinetic coefficient having the dimension of frequency. Its explicit form is

$$\dot{\varphi} = -\gamma \left(2\alpha(T - T_c)\varphi + 2b\varphi^3 + \varphi \varepsilon_{el}^2(t) \right). \tag{12}$$

Let us show that the equilibrium mean square of the order parameter (time averaging) as a function of mean square elastic deformation can be found in a general form, independently of the interrelation between the relaxation time and the period of external action. We divide Eqn. (12) by φ and rewrite it in the form

$$\frac{\partial \ln \varphi}{\partial t} = -\gamma \left(2\alpha(T - T_c) + 2b\varphi^2 + \varepsilon_{el}^2(t) \right) \tag{13}$$

If $\varepsilon_{el}(t)$ is a periodic function of time, then the asymptotic solution of (12) is also a periodic function of time and $\langle \partial \ln \varphi / \partial t \rangle = 0$. Hence

$$\langle \varphi^2 \rangle = \frac{\alpha(T_c - T) - \langle \varepsilon_{el}^2 \rangle / 2}{b}. \tag{14}$$

For $\langle \varepsilon_{el}^2 \rangle$ larger than the critical value $2\alpha|T_c - T|$ the mean value $\langle \varphi^2 \rangle \equiv 0$. The Eqn. (14) is

analogous to Eqn. (6), but relates the mean values of $\langle \varphi^2 \rangle$ and $\langle \varepsilon_{el}^2 \rangle$. We see that, under

periodic action, the layer undergoes a phase transition similar to the case of stationary action. Note that the amplitude value of ε_{el} at the transition point exceeds the critical value $\varepsilon_{el,c}$ for static transition. There were just these non-stationary phase transformations which have been observed in experimental work [2].

RELAXATION OF THE ORDER PARAMETER AT A GIVEN STRESS

Often it is not the deflection amplitude but the stress, which is kept constant in experiment. From the condition that the stress remains constant

$$\sigma = \varphi^2 \varepsilon_{el} = const \tag{15}$$

we can express the elastic deformation as a function of order parameter and substitute it into the kinetic equation (12). In this way we come to the kinetic equation

$$\dot{\varphi} = -\gamma \left[2\alpha(T - T_c)\varphi + 2b\varphi^3 + \frac{\sigma^2}{\varphi^3} \right], \tag{16}$$

determining the dynamics of the order parameter at a constant stress. Introducing dimensionless variables

$$\tilde{\varphi} = \varphi / \varphi_0, \qquad \tilde{\sigma} = \sigma / \sigma_0, \qquad \tilde{t} = t / t_0, \tag{17}$$

with $\varphi_0 = (\alpha(T_c - T))^{1/2}$ being the equilibrium value of the order parameter in the absence of

deformation (see (5)), $\sigma_0 = \frac{1}{b} \left(\frac{2}{3} \alpha(T_c - T) \right)^{3/2}$ the maximum stationary elastic stress which can

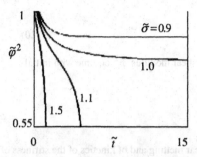

Figure 2. *Dynamics of the shear modulus (squared order parameter) according to Eqn. (18) at stresses smaller, equal and larger than the critical. Corresponding experimental data can be found in [Reiter G., Demirel A.L., Peanasky J. et al. In: Physics of Sliding Friction. Eds. B.N.J. Persson and E. Tosatti, p.119-138].*

be created in the layer (Eqn. (10)), and $t_0 = (2\alpha(T_c - T)\gamma)^{-1}$ the characteristic relaxation time of the order parameter to the equilibrium value, we can rewrite (16) in the form

$$\frac{\partial \tilde{\varphi}}{\partial \tilde{t}} = \tilde{\varphi} - \tilde{\varphi}^3 - \frac{4}{27}\frac{\tilde{\sigma}^2}{\tilde{\varphi}^3}. \qquad (18)$$

Numerical solutions of Eqn. (18) for the critical stress as well as for stresses a little bit smaller and larger as the critical value are represented in Fig.2. These kinetic curves are in an excellent correspondence with experimental data obtained in [2].

KINETICS OF THE TRANSITION TO THE SOLID STATE

The kinetics of solidification of a stress free lubrication layer is given by Eqn. 16 with $\tilde{\sigma} = 0$:

$$\frac{\partial \tilde{\varphi}}{\partial \tilde{t}} = \tilde{\varphi} - \tilde{\varphi}^3. \qquad (19)$$

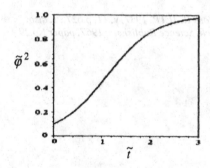

Figure 3. *Kinetics of shear modulus after unloading according to Egn. 20.*

Its solution is given in implicit form by

$$\ln\frac{\tilde{\varphi}}{\tilde{\varphi}(0)} - \frac{1}{2}\ln\frac{1-\tilde{\varphi}}{1-\tilde{\varphi}(0)} - \frac{1}{2}\ln\frac{1+\tilde{\varphi}}{1+\tilde{\varphi}(0)} = \tilde{t} . \tag{20}$$

The corresponding dependence of the dimensionless shear modulus $\tilde{\varphi}^2$ on time with initial condition $\tilde{\varphi}^2(0) = 0.1$ is represented in Fig.3.

CONCLUSIONS

We have shown that the experimental data on shear melting and of kinetics of the stiffness of thin lubricating films obtained in [2] can be simply described in the frame of the Landau's theory of phase transitions of the second order.

ACKNOWLEDGEMENTS

The financial support of the BMBF in the frame of the German-Israeli Project Cooperation "Novel Tribological Strategies from the Nano-to Meso-Scales" is gratefully acknowledged.

REFERENCES

1. *Persson B.N.J.* Sliding Friction. Physical Principles and Applications. Springer Verlag, Berlin, Heidelberg, 1998.
2. *Reiter G., Demirel A.L., Peanasky J. et al.* In: Physics of Sliding Friction. Eds. B.N.J. Persson and E. Tosatti, p.119-138.
3. *Demirel A.L., Granik S.* // J. Chem. Phys. 1998, v. 109, No. 16, pp.6889-6897.
4. *Thomson P.A., Robbins M.O. and Grast G.S.* // Israel J. Chem., 1995, v. 35, pp.93-106.
5. *Landau L.D. and Lifshitz E.M.*, Statistical Physics. Pt.1, 3rd ed. (Pergamon Press Oxford, 1980.
6. *Landau L.D. On the theory of phase transformations II. JETP. 1937, v. 11, p.627., s. also: Collected Papers of L.D. Landau. Gordon and Breach, Science Publishers, 1967, paper No.29, pt.II.*

Mat. Res. Soc. Symp. Proc. Vol. 651 © 2001 Materials Research Society

Spectrophotometric Observations of Gel-Free Reaction Front Kinetics in Confined Geometry

Sung Hyun Park[*], Stephen Parus[*], Raoul Kopelman[*] and Haim Taitelbaum[**]
[*]Department of Chemistry, University of Michigan, Ann Arbor, MI 48109, U.S.A.
[**]Department of Physics, Bar-Ilan University, Ramat-Gan 52900, Israel

ABSTRACT

We present a new experimental system to study the kinetics of the reaction front in $A + B \rightarrow C$ reaction-diffusion systems with initially-separated reactants. The set-up is composed of a CCD camera monitoring the kinetics of the front formed in the reaction-diffusion system Cu^{2+} + *tetra* $\rightarrow$ *1:1 complex* (in aqueous, gel-free solution) inside a 150 micron gap between two flat microscope slides. This is basically a two-dimensional system. The results are consistent with the theoretical predictions for the anomalous time dependence of the front's width, height, and location.

INTRODUCTION

The kinetics of the reaction front in $A + B \rightarrow C$ reaction-diffusion systems with initially-separated reactants has been studied extensively in the past decade [1-22] and has shown many exotic properties. The initial separation of the reactants is an initial condition that readily enables, in principle, experimental investigations of this system [2,5,16,18]. Indeed, some of the novel characteristics of this system, such as the non-monotonic motion of the reaction front [5], and the split front in the case of competing reactions [16], have been experimentally observed in a series of absorption measurements in a capillary where reactants diffuse and react in a gel solution.

However, some of the theoretical predictions that were recently made, such as the doubly non-monotonic motion of the reaction front [17] or the breakdown of the non-monotonic motion have not been obtained yet in experiments due to experimental difficulties, some of which related to problems with the gel solution.

In this paper, we present a new experimental system and a set of measurements of the reaction front properties in this system. The experiments are performed in a two-dimensional geometry, where the reaction and diffusion take place within a gap of 150 microns between two flat microscope slides separated by optical fibers. A sketch of the experimental set-up is shown in Figure 1.

The simple $A + B \rightarrow C$ system is assumed to be described for dimensions $d \geq 2$ by the mean-field equations for the local concentrations ρ_a, ρ_b,

$$d\rho_a / dt = D_a \nabla^2 \rho_a - k \rho_a \rho_b \qquad (1a)$$

$$d\rho_b / dt = D_b \nabla^2 \rho_b - k \rho_a \rho_b \qquad (1b)$$

where D_a and D_b are the diffusion coefficients and k is the microscopic reaction rate constant. For the initially-separated system, the initial condition reads

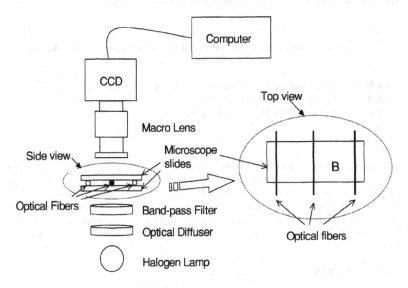

Figure 1. Schematic diagram of the experimental setup.

$$\rho_a(x, 0) = a_0 H(x), \quad \rho_b(x, 0) = b_0 [1-H(x)] \qquad (1c)$$

where a_0, b_0 are the initial densities and $H(x)$ is the Heaviside step function.

The basic quantities that describe the kinetic behavior of the reaction front are defined on the basis of the reaction term $R(x, t) = k \rho_a(x, t)\rho_b(x, t)$. These are the front width $w(t)$ which was shown by Galfi and Racz [1] to increase with time as $t^{1/6}$, an exponent which is significantly smaller than the ½ exponent associated with length scales in diffusion processes; the front height $R(x_f, t)$, which decreases with time as $t^{-2/3}$; and $x_f(t)$, the location of the reaction front center, which is defined as the position where $R(x, t)$ is maximal. This quantity has been found [5,17] to behave in a non-trivial manner, but asymptotically propagates with time as $t^{1/2}$.

In the following we describe in detail the experimental set-up and procedure and present data for the above-mentioned quantities that are in good agreement with the theoretical predictions.

EXPERIMENTAL

The experiments are performed in a thin gap between two flat microscope slides with dimensions 75 mm x 25 mm x 1 mm (optical flatness $\lambda/4$ over 75 mm x 25 mm plane). Three parallel optical fibers with a diameter of 150 microns are inserted between two microscope slides as shown in Figure 1. Two optical fibers at both ends act as spacers, while the fiber at the center of the slides acts as a boundary allowing the initial separation of two reactant solutions as well as the formation of a straight reaction front at t=0. It was not necessary to have a boundary along or seal the long edge of the slide since no evaporation effects were observed from these water-based solutions over the time of the reaction. The observed reaction is a 1:1 complex formation between copper (II) and disodium ethyl bis(5-tetrazolylazo)acetate trihydrate ("tetra") in aqueous solution, i.e., Cu^{2+} + tetra $\rightarrow$ 1:1 complex. The concentrations of the reactants are $[Cu^{2+}] = 2.0 \times 10^{-3}$ M and [tetra] $= 1.0 \times 10^{-3}$ M. Reactants are injected into each side of the thin gap with syringes and spread to fill the volume by capillary action.

A CCD camera (SpectraSourse Instruments, Westlake Village, CA, model Teleris 2 12/16) with a macro lens (Nikon AF Micro 60 mm f2.8 1:1) records the optical absorption images of the product at different times. The CCD has 512×512 pixels, and the size of the image monitored is ~ 1 cm $\times$ 1 cm, providing the spatial resolution of ~20 microns/pixel. A typical exposure time is 100 msec. An optical diffusing glass reduces spatial fluctuations of the light intensity from the halogen lamp over the illumination area. The absorption maximum of the purple-colored product is 535 nm whereas the absorption of reactants is peaked at 410 nm. An optical band-pass filter at 540 nm monitors only the product, where the reactants have negligible absorption and are not observed in the images. The entire experiment is performed at room temperature.

After reactants are injected into the gap between the slides, the optical fiber at the center of the slides is pulled out quickly to initiate the reaction. The reaction gives an immediate formation of the purple-colored product at the boundary. The first image of the product is taken immediately when the optical fiber is removed from the center of the slides (time zero). A series of images of the product are then taken with increasing time intervals from 30 seconds to 20 minutes. Product formation was typically monitored up to ~ 60 minutes.

RESULTS AND DISCUSSION

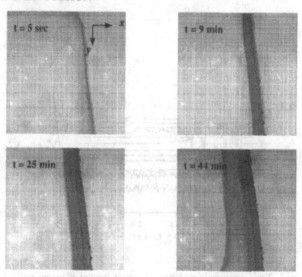

Figure 2. Typical images of accumulated reaction front at different times for Cu^{2+} + tetra $\rightarrow$ 1:1 complex

Figure 2 shows typical images of the product seen as the dark vertical band in the central one-third of the image obtained at different times. Copper is on the right side and tetra is on the left of the product in the images. Each image represents the total accumulated product formed up to that time. To obtain the mean-field properties of the dynamics of the reaction front in our experimental system, we need to integrate the product intensities along the y direction of the images, which is parallel to the reactant boundary in Figure 2. However, we notice the similarity of the product concentrations along the y direction in the image. This implies that the dynamics along the y direction is basically independent of position, and we don't have to integrate along the y-axis. This symmetry allows us to simply choose any one pixel-line along the x-axis and compare the profiles at this position within the series of images. To improve the statistics, however, we measured the parameters of the reaction front at 15 different horizontal pixel-lines along the y-axis and calculated the average values for the analyses.

According to the Beer-Lambert Law, the optical absorbance is directly proportional to the concentration of the product. By subtracting two consecutive profiles of the product and normalizing by the time interval between them, we obtain the profiles of the reaction front at different times along each pixel line, i.e. the spatial profiles of $R(x, t)$ at different times. We can measure the dynamic parameters of the reaction front from the profiles, namely, the width $w(t)$, height at the front center, $R(x_f, t)$, and the position of the reaction front, $x_f(t)$.

In Figure 3 we show a set of reaction front profiles corresponding to the production rate $R(x, t)$ of C. The profiles pertain to successive times, ranging from 15 sec up to 48 min. These profiles were analyzed in terms of their height at the approximate center, the width at half-height and center location along the x-axis.

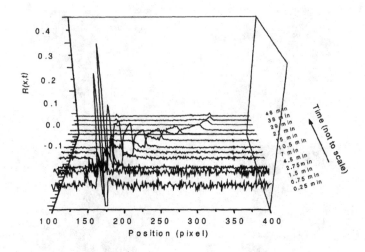

Figure 3. Typical pattern of evolution of reaction front at different times.

In Figure 4 we plot the results of the analyses for the exponents of the reaction front parameters. Each data point is the average of 15 measurements along different pixel lines. The exponents are in good agreement with the theory. Namely, the time exponent of the global rate, R(t), is −0.47 from the experiment, which is very close to the theoretical value of −0.50. The local reaction rate, $R(x_f, t)$, which is measured as height of the reaction front in experiment, shows the time exponent −0.64, and the center location of the reaction front, x_f, shows 0.60 as time exponent, both of which are reasonably well matched with the theoretical values of −0.67 and 0.50, respectively. The width of the reaction front, w, shows the time exponent 0.12, which is reasonably close to theoretical value of 0.17, within the large associated errors.

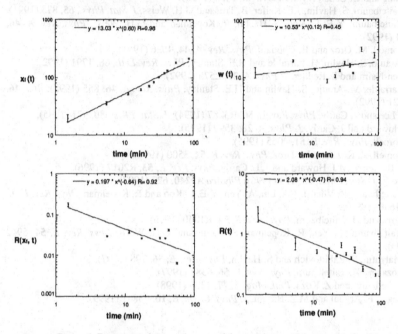

Figure 4. Experimental time exponents of the parameters of the reaction front.

CONCLUSIONS

We have presented a new experimental system for measuring reaction kinetics with initially separated reactants. The system reproduces the basic anomalous kinetics of this process. The new technique, in particular the use of a gel-free solution in which the reaction takes place, is very promising with respect to more sophisticated investigations of this system. In particular, one can attempt to control the motion of the reaction front, say by changing the initial concentrations, in order, for example, to stabilize the front motion, or to have it moving in any preferred direction, with possible non-monotonic motion. In addition, reducing the gap size may eliminate any possible convection effects.

ACKNOWLEDGEMENT

This work was supported by NSF grant DMR-9900434. H.T. thanks Israel Science Foundation for the financial support.

REFERENCES

1. L. Galfi and Z. Racz, *Phys. Rev. A*, **38**, 3151 (1988).
2. Y-E. L. Koo and R. Kopelman, *J. Stat. Phys.*, **65**, 893 (1991).
3. Z. Jiang and C. Ebner, *Phys. Rev. A*, **42**, 7483 (1990).
4. H. Taitelbaum, S. Havlin, J.E. Keifer, B. Trus and G.H. Weiss, *J. Stat. Phys.*, **65**, 873 (1991).
5. H. Taitelbaum, Y-E. L. Koo, S. Havlin, R. Kopelman and G.H. Weiss, *Phys. Rev. A*, **46**, 2151 (1992).
6. S. Cornell. M. Droz and B. Chopard, *Phys. Rev. A*, **44**, 4826 (1991).
7. M. Araujo, S. Havlin, H. Larralde and H.E. Stanley, *Phys. Rev. Lett.*, **68**, 1791 (1992).
8. E. Ben-Naim and S. Redner, *J. Phys. A*, **25**, L575 (1992).
9. H. Larralde, M. Araujo, S. Havlin and H.E. Stanley, *Phys. Rev. A*, **46**, 855 (1992); *ibid*, **46**, R6121 (1992).
10. B.P. Lee and J. Cardy, *Phys. Rev. E*, **50**, R3287 (1994); *J. Stat. Phys.*, **80**, 971 (1995).
11. M. Howard and J. Cardy, *J. Phys. A*, **28**, 3599 (1995).
12. S. Cornell, *Phys. Rev. E*, **51**, 4055 (1995).
13. S. Cornell, Z. Koza and M. Droz, *Phys. Rev. E*, **52**, 3500 (1996).
14. G.T. Barkema, M.J. Howard and J. L. Cardy, *Phys. Rev. E*, **53**, R2017 (1996).
15. Z. Koza, *J. Stat. Phys.*, 85, 179 (1996); *Physica A*, **240**, 622 (1997).
16. H. Taitelbaum, B. Vilensky, A. Lin, A. Yen, Y-E. L. Koo and R. Kopelman, *Phys. Rev. Lett.*, **77**, 1640 (1996).
17. Z. Koza and H. Taitelbaum, *Phys. Rev. E*, **54**, R1040 (1996).
18. H. Taitelbaum, A. Yen, R. Kopelman, S. Havlin and G.H. Weiss, *Phys. Rev. E*, **54**, 5942 (1996).
19. V. Malyutin, S. Rabinovich and S. Havlin, *Phys. Rev. E*, **56**, 708 (1997).
20. Z. Koza and H. Taitelbaum, *Phys. Rev. E*, **56**, 6387 (1997).
21. H. Taitelbaum and Z. Koza, *Phil. Mag. B*, **77**, 1389 (1998).
22. C. Leger, F. Argoul and M.Z. Bazant, *J. Phys. Chem. B*, **103**, 5841 (1999).

Mat. Res. Soc. Symp. Proc. Vol. 651 © 2001 Materials Research Society

A Grand Canonical Monte-Carlo Study of Argon Adsorption/Condensation in Mesoporous Silica Glasses: Application to the Characterization of Porous Materials

R.J.-M. Pellenq*, S. Rodts, P. E. Levitz
Centre de Recherche sur la Matière Divisée, CNRS et Université d'Orléans,1b rue de la Férollerie, 45071 Orléans, cedex 02, France. email: pellenq@cnrs-orleans.fr

ABSTRACT

We have studied adsorption of argon in a mesoporous silica Controlled Porous Glass (CPG) by means of Grand Canonical Monte-Carlo (GCMC) simulation. A numerical sample of the CPG adsorbent has been obtained by using an off-lattice reconstruction method recently introduced to reproduce topological and morphological properties of correlated disordered porous materials. The off-lattice functional of $(100 \, m^2/g)$-Vycor is applied to a simulation box containing silicon and oxygen atoms of cubic cristoballite with an homothetic reduction of factor 2.5 so to obtain 30Å-CPG sample. A realistic surface chemistry is then obtained by saturating all oxygen dangling bonds with hydrogen. The Ar, Kr and Xe adsorption/desorption isotherms are calculated at different temperatures. At sufficiently low temperature, they exhibit a capillary condensation transition with a finite slope by contrast to that theoretically predicted for simple pore geometries such as slits and cylinders. In the low pressure and temperature domain, we have identified different adsorption scenarios, which can be interpreted on the basis of a Zisman-type of criterium for wetting. We demonstrate that the BET surface area is strongly related to this criterion. At higher pressure, we demonstrate that the pore size distribution obtained by using the standard BJH analysis applied to both adsorption and desorption data qualitatively reproduces the main features of the chord length distribution.

INTRODUCTION

The vast majority of mesoporous materials are disordered structures. Silica controlled porous glasses constitute a large class of materials among which is Vycor, a porous silica glass widely used as a model structure to study the properties of confined fluids. In fact, there are two kinds of vycor (it is usually referred to as Vycor-7930 by the manufacturer Corning Inc.): a first type with a specific surface area around $100 \, m^2/g$ and a second at around $200 \, m^2/g$ (see Figure 1). The pores in the low-specific-surface-vycor have an average radius of about 35 Å (assuming a cylindrical geometry) [1]. Conversely, for the high-specific-surface material, the mean pore radius is around 22 Å [2]. It is known from theoretical and simulation studies on simple pore geometries (slits and cylinders) that confinement strongly influences the thermodynamics of confined fluid [3]. The effect of the matrix disorder in terms of pore morphology –the pore shape- and topology –the way the pores distribute and connect in space- on the thermodynamics of confined molecular fluid still remain to be clarified. This first raises the challenge of describing the morphology and the topology of these porous solids [4]. The aim of this work is to provide an insight in the adsorption mechanism of simple adsorbates in a disordered connected mesoporous medium such as vycor at a microscopic level.

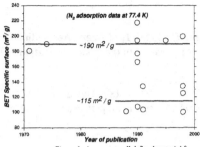

Figure 1 : is vycor a well defined material ?

SIMULATION PROCEDURES

We have used on an off-lattice reconstruction algorithm in order to numerically generate a porous structure which has the main morphological and topological properties of real low-specific-surface-area vycor in terms of pore shape: close inspection of molecular self-diffusion shows that the off-lattice reconstruction procedure gives a connectivity similar to experiment (see Figure 2). Experimental small angle diffusion data on this kind of vycor shows a correlation peak, which corresponds to a minimal (pseudo) unit-cell, size around 270 Å [5]. In order to reduce the computational cost of the GCMC adsorption/condensation simulation process, we have applied a homothetic reduction with a factor of 2.54 so that the final numerical sample is contained in a box of about 107 Å in size (see below). This transformation preserves the pore morphology but reduces the average pore size from 70 Å to roughly 30 Å (Figure 3). Note that the reconstructed minimal numerical sample hereafter named as pseudo-vycor, is still well within the mesoporous domain (>20 Å). A further effect of the homothetic reduction is to increase the specific surface area by a factor equal to the homothetic factor compare to that of the starting material (the specific surface area of all pseudo-vycor samples

Figure 2

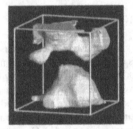

Figure 3

Figure 2 : 3D numerical reconstruction of low-specific-surface area vycor. One sees the porosity in grey through the solid matrix. The simulation box size in 100 nm and the representative elementary volume is 27^{3} nm³.

Figure 3D numerical reconstruction of pseudo-vycor. One sees the porosity in grey through the solid matrix. The representative elementary volume is 10.7^{3} nm³

is around 220 m^2/g). The atomistic description of pseudo-vycor is obtained by applying the off-lattice functional to a box containing the silicon and oxygen atoms of 15^3 unit cells of cubic cristoballite (a siliceous non-porous solid). This allows cutting out portions of the initial volume in order to create the vycor porosity. Periodic boundary conditions are applied in order to simplify the Grand Canonical Monte-Carlo (GCMC) adsorption procedure. Saturating all oxygen dangling bonds with hydrogen atoms placed in the pore voids ensures Electroneutrality. In a previous publication, we have shown that this atomistic procedure for generating silica porous glasses allows a good description of internal surface in terms of roughness [10].

In this work, we have used a PN-type potential function as reported for adsorption of rare gas in silicalite (a purely siliceous zeolite): it is based on the usual partition of the adsorption intermolecular energy which can written as the sum of a dispersion interaction term, a repulsive short range contribution and an induction term [6]. The dispersion and induction parts in the Ar/H, Kr/H and Xe/H adsorption potentials are obtained assuming that hydrogen atoms have a partial charge of $0.5e$ (q_O=-$1e$ and q_{Si}=-$2e$ respectively) and a polarizability of 0.58 Å^3; the adsorbate/H repulsive contribution (Born-Mayer term) is adjusted on the experimental low coverage isosteric heat of adsorption (13.5 kJ/mol for argon [2,7], 15 kJ/mol [8] and 17.5 kJ/mol [9]). The adsorbate-adsorbate potential energy was calculated on the basis of a Lennard-Jones function (ε=120 K, σ=3.405 Å for Ar, ε=170 K, σ=3.69 Å for Kr and ε=211 K, σ=4.10 Å for Xe). In the Grand Canonical Ensemble, the independent variables are the chemical potential (related to the temperature and the bulk pressure), the temperature and the volume [11]. The adsorption isotherm can be readily obtained from such a simulation technique by evaluating the ensemble average of the number of adsorbate molecules. Typical GCMC runs consists in 3 10^5 Monte-Carlo steps per adsorbate.

3. RESULTS AND DISCUSSION

Specific surface can be measured by calculating the first momentum of the in-pore chord length distribution [4,5] for a known value of the porosity. We have evaluated this distribution for all numerical samples by making use of the adsorption potential energy hypersurface: at each current probe position of a given chord, the energy is calculated and the interface is easily

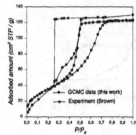

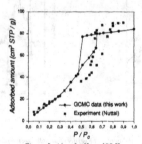

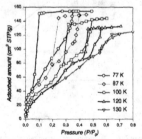

Figure 4 : Kr adsorption /desorption isotherm At 117 K.

Figure 5 : idem for Xe at 195 K

Figure 6 : idem for Ar at several temperatures

located when the adsorption energy changes of sign (negative in the pore, positive in the matrix). This allows the direct determination of the intrinsic specific surface for each porous structure taking into account atomic surface roughness. Interestingly, we found that for a given pseudo-vycor sample, chord length distributions were almost identical giving a specific surface area at $S_{sp} = 220 +/- 20$ m^2/g. Another interesting aspect is that the in-pore chord length distribution can be considered as the "true" pore size distribution.

Figures 4 and 5 present an adsorption/desorption isotherm for krypton and xenon respectively. Clearly, the simulated curves closely resemble to the their experimental counterpart: upon increasing pressure, adsorption proceeds in a continuous manner while the desorption branch is nearly vertical; the isotherms thus exhibit a hysteresis loop, in agreement with other simulation works on disordered materials. The behaviour of the hysteresis loop with increasing temperature (it shrinks, see Figure 6) can be interpreted in terms of fluid criticality by analogy with that found for unconnected cylindrical and slit-shaped pores [13]. The overall isotherm shape differs from that predicted for simple pore geometries. In particular, the shape of the adsorption branch can be considered as the signature of disordered mesoporous structure. Furthermore, the pseudo-vycor adsorption/desorption curves are shifted to the lower pressure region compared to experiment (on the high-specific-surface vycor) due to the homothetic reduction [2,7-9]. On the overall, GCMC simulations of adsorption/condensation of simple fluids in the hypothetical pseudo-vycor material as defined in this work do reproduce the main characteristic features experimentally observed. This can be rationalised on the basis of the properties of the off-lattice functional used to generate pseudo-vycor samples, which ensure that all relevant correlations will be taken into account. Futhermore the use of periodic boundary conditions does not seem to lead to unphysical behaviours when simulating adsorption/desorption processes.

At 77 K, the analysis of the adsorbed density reveals that Ar does cover the entire surface before condensation occurs by forming a continuous film. The corresponding specific surface

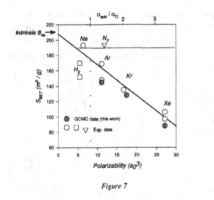

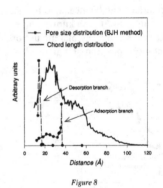

Figure 7

Figure 8

Figure 7 :specific surface area versus adsorbate polarizability α_{ads}. The vertical line indicates the polarizability value for in-silica oxygen α_o [16]. « intrinsic S_{sp}» is the geometrical value for the specific surface area as measured from chord length.

Figure 8 : comparison between the BJH pore size distribution from desorption data and the chord length distribution.

value as measured from the adsorption isotherm using the BET equation, is 157 m^2/g. By contrast, at 195 K, Xe does not "wet" the pseudo-vycor surface [11,14]: adsorption and condensation take place in the places of highest surface curvature (this corresponds to regions of space where confinement effects are maximum). This leads to an unexpected situation where regions of the pores are filled with condensate while other parts of the interface remain uncovered. The Xe specific surface is around 80 m^2/g. This difference is clearly due to the adsorption mechanism which is different between different rare gases on such a curved and rough silica surface. The values of specific surface as obtained from simulated adsorption isotherms are well below that calculated from chord-length distribtion (see above). The difference in specific surface as obtained from Ar, Kr and Xe adsorption isotherms can be rationalised on the basis of a difference of polarizability between the adsorbate and the atomic species of the substrate plus curvature effects (Zismann's type of law) [21,22] (Figure 7). It is thus clear that monolayer-based method (such as the BET approach) cannot be used for determining the specific surface especially in non-wetting situations. Note that *wetting* corresponds here as the formation of a thin adsorbate film (few adsorbate layers in thickness *ie* the so-called *statistical* monolayer capacity in the BET theory [17]). Figure 8 compares the chord length distribution of our numerical sample with the pore size distribution (PSD) as obtained from the standard BJH analysis [17] (corrected for adsorption) applied to both the adsorption and the desorption branches of the isotherm. Clearly, the BJH-PSD is quite different than the chord distribution which can be considered as an exact PSD. However, the BJH-PSD based on both adsorption and desorption data, has two main peaks of the same relative amplitude as those of the chord distribution. These peaks are shifted towards the smaller pore-size domain by roughly 1 nm in agreement with recent simulation results on similar materials[]. Interestingly enough, the adsorption branch of the isotherm leads to a maximum value for the PSD which is close to the mean value of the chord distribution ($<l_p>$=36 Å). Note that the BJH method implicitly assumes that the porous material is a set of unconnected cylinders (the BJH characteristic distances reported here correspond to pore diameters). This comparison between chord distribution and BJH results demonstrates that the BJH method does provide an oversimplified picture of disordered porous materials such as CPG's.

CONCLUSION

We have performed atomistic Grand Canonical Monte-Carlo (GCMC) simulations of adsorption of Ar, Kr and Xe in a vycor-like matrix at different temperatures. This disordered mesoporous network is obtained by using a numerical 3D off-lattice reconstruction method based on the off-lattice functional of low-specific-surface area-Vycor applied to a simulation box originally containing silicon and oxygen atoms of a non-porous silica solid. It allows creating the mesoporosity that has the morphological and the topological properties of the real vycor glass but with a homothetic decrease of the box dimensions so to produce 3nm CPG samples. The simulated isotherms show a gradual capillary condensation phenomenon: the shape of the adsorption curves differs strongly from that obtained for simple pore geometries. The BET analysis of the adsorption isotherm leads to an underestimated value of the specific surface when compared to the intrinsic value obtained from chord length analysis and depends on the contrast of polarizability between the adsorbate and the silica oxygen. The pore size distribution determined with the standard BJH method reproduces the main features of the chord distribution if both adsorption and desorption data are considered although all BJH characteristics distances are shifted toward smaller values.

ACKNOWLEGEMENTS

We thank the Institut du Développement et des Ressources en Informatique Scientifique, (CNRS, Orsay, France) for the computing grant 991153.

REFERENCES

1. P. Levitz, V. Pasquier, I. Cousin, Caracterization of Porous Solids IV, B. Mc Enanay, T. J. May, B., J. Rouquerol, K. S. W. Sing, K. K. Unger (eds.), The Royal Soc. of Chem., London, (1997), p 213.
2. M. J. Torralvo, Y. Grillet, P. L. Llewellyn, F. Rouquerol, J. Coll. Int. Sci., 206 (1998), p 527-531.
3. L. D. Gelb, K. E. Gubbins, R. Radhakrishnan, M. Sliwinska-Bartkowiak, Rep. Prog. Phys., 62 (1999), p 1573-1659.
4. P. Levitz, D. Tchoubar, J. Phys.I, 2 (1992), p 771-790.
5. P. Levitz, Adv. Coll. Int. Sci., 76-77 (1998), p 71-106.
6. R. J.-M. Pellenq, D; Nicholson, J. Phys. Chem., 98 (1994), p 13339-13349.
7. G. L. Kington, P. S. Smith, J. Faraday Trans. 60 (1964), p 705-720.
8. S. Brown, PhD thesis, University of Bristol, UK , (1963).
9. C. G. V. Burgess, D. H. Everett, S. Nuttal, Langmuir, 6 (1990), p 1734-1738.
10. R. J. M. Pellenq, S. Rodts, V. Pasquier, A. Delville, P. Levitz, Adsorption, 6 (2000), p 241-248.
11. D. Nicholson and N. G. Parsonage in "Computer simulation and the statistical mechanics of adsorption", Academic Press, 1982.
12. K. S. Page, P. A. Monson, Phys. Rev. E, 54 (1996), p 6557-6564. 23. L. Sarkisov, K. S. Page, P. A. Monson, Proceedings of the VIth Fundamental of Adsorption conference, May 1998, Giens, France, F. Meunier Ed., Elsevier, 1999, p 847.
13. R. J.-M. Pellenq, B. Rousseau, P. Levitz, Phys. Chem. Chem. Phys., in press.
14. R. J.-M. Pellenq, A. Delville, H. van Damme, P. Levitz, Stud. in Surf. Sci. and Catal., 128 (2000), p 1.
15. P. G. de Gennes, Rev. Mod. Phys., 57 (1985), p 289-305.
16. R. J.-M. Pellenq, D. Nicholson, J. Chem. Soc. Faraday Trans., (1993), p 13339-
17. F. Rouquerol, J. Rouquerol and K. Sing, in "Adsorption by Powders and Porous Solids', Academic Press, 1998.

Mat. Res. Soc. Symp. Proc. Vol. 651 © 2001 Materials Research Society

Hydrocarbon Reactions in Carbon Nanotubes: Pyrolysis

Steven J. Stuart, Brad M. Dickson, Donald W. Noid, and Bobby G. Sumpter[1]
Department of Chemistry, Clemson University, Clemson, SC 29634-0973, USA.
[1]Chemical and Analytical Science Division, Oak Ridge National Laboratory, Oak Ridge, TN 37830, USA.

ABSTRACT

Molecular dynamics simulations have been used to study the pyrolysis of eicosane ($C_{20}H_{42}$) both in the gas phase and when confined to the interior of a (7,7) carbon nanotube. A reactive bond-order potential was used to model the thermal decomposition of covalent bonds. The unimolecular dissociation is first-order in both cases. The decomposition kinetics demonstrate Arrhenius temperature dependence, with similar activation barriers in both geometries. The decomposition rate is slower by approximately 30% in the confined system. This rate decrease is observed to be a result of recombination reactions due to collisions with the nanotube wall.

INTRODUCTION

Carbon nanotubes are currently being investigated for, among many other remarkable properties, their ability to be filled to create nanoparticles and nanowires.[1-6] Materials ranging from metals[3] to salts[4] to aqueous solutions[6] have been introduced into nanotubes via capillary forces and induced to undergo chemical reactions. Hydrocarbon liquids also have surface tensions suitable for similar treatment,[2] which opens the possibility of using nanotubes as synthetic reaction vessels for organic chemistry. Because surface effects will dominate in these molecular-scale systems, chemical reactivity and product distributions are conceivably quite different than they would be in gas- or bulk-phase reactions. As part of a series of investigations aimed at studying these phenomena, we consider here the thermal decomposition of straight-chain alkanes when confined to the interior of a carbon nanotube. These decomposition reactions are compared to those observed in the gas phase.

The pyrolysis of hydrocarbons has been studied extensively in the gas phase, as well as in the confined geometries of zeolites and other porous structures. There, the emphasis has been primarily on physical dynamics, such as diffusion of the hydrocarbons through the porous lattice. When chemical effects have been considered, it is the catalytic properties of the zeolites that attract the most attention. The focus here is on the changes in reaction dynamics that occur as a result of purely physical confinement.

In particular, the thermal decomposition of eicosane ($C_{20}H_{42}$), a straight-chain alkane, is studied both in vacuum and in the interior of a (7,7) nanotube. Although differences in alkane structure and properties are certainly expected, and observed, the focus here is specifically on the effects on confinement on the decomposition reaction at elevated temperatures.

METHODS

The pyrolysis being modeled here involves the dissociation of a covalent bond, and simulation thus requires a theoretical method capable of modeling bonding changes. Statistical methods such as transition state theory are not appropriate because of incomplete information about the effects of the nanotube on the dividing surface, and the possibility of barrier recrossings due to collisions with the tube wall. While the $C_{20}H_{42}$ system is small enough to be treated by ab initio or density functional methods, these techniques are still too computationally intensive to provide long-time dynamics or averaging across multiple trajectories. Thus these simulations were performed using classical molecular dynamics and a reactive potential. The AIREBO (adaptive intermolecular reactive empirical bond-order potential) was chosen for this purpose because it is capable of accurately treating covalent bonding reactions in hydrocarbons.[7] This is a reactive, bond-order potential, based on the well-known REBO potential[8,9] with the addition of the dihedral angle and dispersion interactions that are required to treat long-chain hydrocarbons.

The focus in this study is on the geometric effects of confinement to nanoscale dimensions, rather than any specific chemical interactions with the carbon nanotube walls. For this reason, and in order to make the simulations more efficient, a cylindrically symmetric confining potential was used to represent the carbon nanotube. This potential had the form

$$V(r) = A\left(\frac{\sigma}{R-r}\right)^{2p+q} - B\left(\frac{\sigma}{R-r}\right)^{p+q} + C\left(\frac{\sigma}{R-r}\right)^{q} + D \tag{1}$$

where r is the distance of an atom from the center of the nanotube, R is the radius of the nanotube, and the remaining terms are parameters of the potential. Equation (1) represents the simplest soft-wall cylindrical confining potential that results in an adsorption minimum near the tube wall, a local maximum at the center of the tube, and an infinite interaction as a particle approaches the tube wall ($r=R$). Taking R from 0 K simulations of a (7,7) nanotube with the AIREBO potential, and selecting $p=4$ arbitrarily, the remaining parameters were chosen to fit the cylindrically averaged interaction of C and H atoms with a single-walled (7,7) nanotube using the AIREBO potential. Figure 1 illustrates the resulting potential energy surface.

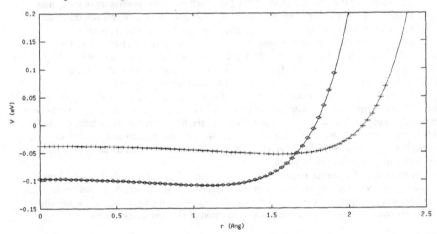

Figure 1. Cylindrically symmetric potential describing the interaction between a (7,7) nanotube wall and carbon atoms (diamonds, red) and hydrogen atoms (crosses, green). The points represent interactions obtained from the AIREBO potential and the lines represent thebest-fit potential represented by Equation (1).

The pyrolysis of the $C_{20}H_{42}$ monomer was studied by performing a series of simulations both in vacuum and in the confining nanotube potential. Initial configurations for these runs were obtained from a 1 ns trajectory in the canonical ensemble at a kinetic temperature of 2700 K. The sampled configurations were next equilibrated at temperatures of between 3100 K and 4200 K, and then simulated for 100 ps in the microcanonical ensemble. These constant-energy runs were then used to determine the thermal decomposition kinetics. Note that the simulation temperatures are necessarily higher than the typical experimental pyrolysis temperatures of between 600 and 1100 K, in order to observe sufficient decomposition reactions on the 100-ps timescales simulated here. Thus the specific decomposition rates obtained here will not be directly comparable to experimental values; our interest is primarily in the differences in reaction rate induced by nanoscale confinement and surface effects. It is assumed that these confinement effects are largely independent of temperature. Because of the high temperatures, long run durations, and the presence of fast bond vibrations, these simulations required a fairly small timestep of 0.1 fs to avoid problems with energy conservation.

RESULTS

A total of 1135 simulations were performed at temperatures between 3100 and 4200 K; approximately half in vacuum and half in the confined system. Decomposition reactions occurred in 376 of these runs, with both C—C and C—H bond dissociations being observed. Because we are interested here only in the initial thermal dissociation step, subsequent decomposition of the radical byproducts was disregarded.

For each temperature studied, the result from the series of simulations was a set of first reaction times. Combined, these data give an indication of the kinetics of the reaction. In all cases, the kinetics were consistent with first-order decay, as expected for a unimolecular decomposition. Figure 2, for example, shows the survival probability for the $C_{20}H_{42}$ monomer at 3700 K. For a first-order decomposition reaction with rate constant k, the probability of observing a dissociation event in a time interval $(t_i - \Delta t/2, t_i + \Delta t/2)$ is given by

$$P\left(t_i - \frac{\Delta t}{2} < t < t_i + \frac{\Delta t}{2}\right) = \int_{t_i-\Delta t/2}^{t_i+\Delta t/2} ke^{kt}dt = 2e^{-kt_i}\sinh(k\Delta t/2). \tag{2}$$

This distribution was used to determine a maximum-likelihood estimate for the rate constant k from the observed decay times at each temperature, with a granularity of Δt=5 ps. The maximum-likelihood fit at 3700 K is displayed along with the simulation data in Figure 2. The temperature dependence of the resulting rate constants is presented in Figure 3.

For both the gas-phase and confined systems, the straight-line fit to the data indicates that they are consistent with Arrhenius rate behavior. It is also evident from Figure 3 that the decomposition rates are consistently slower when confined to the interior of the (7,7) nanotube than when in the gas phase. This is true across the full 1100 K temperature range from 3100 K to 4200 K, with an average decrease in rate of 30% inside the nanotube, relative to the gas phase.

A weighted least squares fit to the data indicates activation barriers of 99 ± 6 kcal/mol in vacuum and 92 ± 7 kcal/mol in the nanotube. These are not distinguishable to within statistical error. Both activation energies are in reasonable agreement with experimental values of 81-85 kcal/mol for small hydrocarbons[10] and with barriers of 89-95 kcal/mol found in ab initio calculations[11]. If the 30% rate decrease in the confined system were due to an inhibitory (i.e. anti-catalytic) energetic interaction with the tube wall, the activation barriers would be different

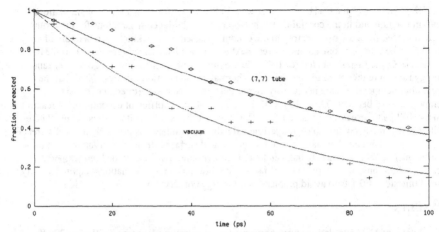

Figure 2. *Thermal decomposition of eicosane ($C_{20}H_{42}$) at 3700 K in vacuum (crosses, blue) and when confined to the interior of a (7,7) carbon nanotube (diamonds, red). The points are observed decay rates from simulation; the solid lines are exponential fits.*

in the two systems. This is not observed, and if anything the barrier may be slightly lower in the confined system. This implies that the difference in rate is due primarily to entropic effects. That is, when the molecule is confined to a smaller geometry, it is restricted to only a fraction of the phase space that was accessible to it in the gas phase. A smaller proportion of this restricted

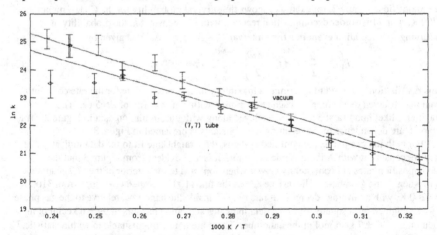

Figure 3. *Arrhenius plot for the thermal decomposition of eicosane ($C_{20}H_{42}$) in vacuum (crosses, blue) and when confined to the interior of a (7,7) carbon nanotube (diamonds, red). The straight lines are weighted linear least squares fits.*

phase space represents reactive configurations of the molecule, resulting in a decreased reaction rate.

A closer examination of the trajectories provides a microscopic explanation for this decreased reaction rate. A substantial number of the conformations that result in bond stretching to or beyond lengths that would in the gas phase result in dissociation (the classical dividing surface) are followed by collisions with the nanotube wall and subsequent deactivation or recombination. These are exactly the sort of correlated barrier recrossings that prohibit the use of transition state theory-based approaches for this problem.

CONCLUSIONS

Molecular dynamics simulations with reactive potentials have demonstrated that the thermal decomposition of eicosane ($C_{20}H_{42}$), a straight-chain alkane, proceeds at a slower rate when confined to nanoscale geometries. In particular, the decomposition is 30% slower in the interior of a (7,7) carbon nanotube than in the gas phase, at temperatures of between 3100 K and 4200 K. This change in kinetic behavior appears to be due to surface-induced recombination reactions, rather than any energetic interactions with the tube itself. The decomposition demonstrates first-order kinetics with standard Arrhenius temperature dependence for both systems.

These observations are significant in the context of current experimental investigations of filled carbon nanotubes. The decrease in reaction rates is particularly significant with respect to the current interest in nanotubes as storage and sequestration vessels. While direct control of reaction rates is significant, it will be noteworthy from a synthetic point of view primarily if modification of relative reaction rates is possible. For example, there is a possibility that pyrolysis in a nanotube could change the distribution of primary pyrolysis radicals, and thus also modify the final product distributions. The small number of reactions (and large number of distinct products) observed in this study prevents quantitative conclusions regarding this possibility, but we note that $34 \pm 3\%$ of the reactions in the nanotube resulted in C—C bond cleavage, while only $26 \pm 3\%$ of the gas-phase dissociations occurred at C—C bonds.

This study has used a rigid, nonreactive, and cylindrically symmetric confining potential to represent the (7,7) nanotube. As a consequence, it is clear that the effect on kinetics observed here is a purely geometric effect that is common to all nanoscale confined systems. Nonetheless, it would also be interesting to consider the effects of thermal vibrations and atomic-scale corrugation in the tube walls, as well as the possibility of chemical reactions with the nanotube. None of those effects were considered here. Likewise, it would be valuable to extend these studies to lower temperatures, comparable to those used experimentally. This would require simulation times on the order of microseconds or even milliseconds, however, and were impossible to achieve with current algorithms and computational hardware.

ACKNOWLEDGEMENTS

Acknowledgement is made to the Research Corporation, and to the donors of the Petroleum Research Fund, administered by the ACS, for partial support of this research. Thomas Zacharia of Oak Ridge National Laboratory is also acknowledged for a generous donation of computer resources.

REFERENCES

1. M. R. Pederson and J. Q. Broughton, *Phys. Rev. Lett.*, **69**, 2689 (1992).
2. P. M. Ajayan and S. Iijima, *Nature*, **361**, 333 (1993).
3. E. Dujardin, T. W. Ebbesen, H. Hiura, and K. Tankgaki, *Science*, **265**, 1850 (1994).
4. D. Ugarte, A. Châtelain, and W. A. de Heer, *Science*, **274**, 1897 (1996).
5. M. Terrones, N. Grobert, W. K. Hsu, Y. Q. Zhu, W. B. Hu, H. Terrones, J. P. Hare, H. W. Kroto, and D. R. M. Walton, *MRS Bull.*, **24** (8), 43 (1999).
6. S. C. Tsang, Y. K. Chen, P. J. F. Harris, and M. L. H. Green, *Nature*, **372**, 159 (1991).
7. S. J. Stuart, A. B. Tutein, and J. A. Harrison, *J. Chem. Phys.*, **112**, 6472 (2000).
8. D. W. Brenner, *Phys. Rev. B*, **42**, 9458 (1990); **46**; 1948 (1992).
9. D. W. Brenner, J. A. Harrison, C. T. White, and R. J. Colton, *Thin Solid Films*, **206**, 20 (1991).
10. I. Safarik and O. P. Strausz, *Res. Chem. Intermediat.*, **22**, 275 (1996).
11. Y. T. Xiao, J. M. Longo, G. B. Hieshima, and R. J. Hill, *Ind. Eng. Chem. Res.*, **36**, 4033 (1997).

Mat. Res. Soc. Symp. Proc. Vol. 651 © 2001 Materials Research Society

Molecular Dynamics and Relaxation Methods in the Stability Calculations for the Study of Distortions of Confined Nematic Liquid Crystals

A. Calles, R.M. Valladares and J.J. Castro[1]
Departamento de Física, Facultad de Ciencias, UNAM
Apdo. Postal 70-646, 04510 México, D.F.
[1]On sabbatical leave from Departamento de Física, CINVESTAV del IPN
Apdo. Postal 14-740, 07300 México, D.F.

ABSTRACT

We present a calculation using a relaxation method for the study of the orientational ordering of a nematic liquid crystal near the surfaces confining the system. We comment on the advantage of using this method as compared with molecular dynamics simulation. The system is simulated by a lattice model with a superposition of isotropic and anisotropic intermolecular interaction of the Maier-Saupe and induced dipole-induced dipole type force for the bulk nematic phase. For the nematic confining surface we consider a Rapini-Papoular interaction. We present simulations for negative dielectric anisotropy.

INTRODUCTION

One of the most interesting problems in liquid crystals, both for a theoretical description as well as for practical applications is the understanding of the behavior of the director near a surface. Most liquid crystal displays consist of a liquid crystal layer contained between two substrate surfaces. The surfaces impose a specific orientation of the liquid crystal director near its boundary that might considerable influence the electro-optical properties of the material. In particular we can mention the case when a nematic liquid crystal is placed between a flat surface, which induced a parallel anchoring of the director, and a grating surface treated to induced a perpendicular anchoring. It has been shown [1] that for this case, the effective anchoring becomes azimuthal at the grating surface and depends on the applied voltage when the nematic phase has negative dielectric anisotropy (that is, the director tends to align perpendicular to the applied field). This suggests the construction of new liquid crystal materials for high-quality displays that could electrically manipulate liquid crystal orientation near the surfaces, lowering the traditional operating voltages. This can be achieved making liquid crystals with weak surface anchoring [2].

Simulations of such materials with different kind of anchoring and its response to applied voltages might help for gaining physical insight on the understanding of the response of the molecular orientation near the surface. In the present work we study the behavior of the director of a nematic liquid crystal near a surface with different boundary conditions that simulates particular anchoring situations, with and without external electric fields. This is made by the use of a relaxation method and visualization algorithms that allow to follow the distribution of the director towards equilibrium. We also present a comment on molecular dynamics simulation and the advantages of using a relaxation method for this particular system. Studies of the behavior of a nematic liquid crystal near a surface using different assumptions have been reported [3-5].

THEORY

We consider a simple-hexagonal lattice model of a nematic liquid crystal between two parallel plates, where the molecules are fixed at each lattice site. The model is limited to planar deformations and the molecules are allowed to rotate in parallel planes that are perpendicular to both walls. The hexagons of the simple-hexagonal lattice lay in a plane perpendicular to the surfaces and introduces no bulk easy axis [3]. The molecules are described by molecular directors $n(R)$, where R determines the lattice site. A pair of molecules with directors $n = n(R)$ and $n' = n(R')$ separated by a distance $r = R'-R$ are interacting through the potential

$$V_B (n,n',r) = -\frac{C}{r^6} \left[n \cdot n' - 3 \, \frac{\varepsilon}{r^2} (n \cdot r)(n' \cdot r) \right]^2 \qquad (1)$$

where $C > 0$ is the interaction strength and ε is the anisotropy parameter which can vary between 0 and 1. For $\varepsilon = 0$ we obtain the isotropic Maier-Saupe interaction, whereas for $\varepsilon = 1$ we obtain the induced dipole-induced dipole interaction [6].

For the nematic-substrate interaction we assume the Rapini-Papoular form [7]

$$V_S (n,\Pi,r) = -\frac{D}{r^6} \left[n \cdot \Pi \right]^2 \qquad (2)$$

where $D > 0$ is the interaction strength and Π is the substrate-induced easy axis.

In the model we take into account all interactions between nematic molecules in the bulk and the interactions between the nematic molecules and the molecules that conform the substrate. The director of the nematic molecules is parametrized by the angle ϑ between the director and the surface normal. For the purpose of simulating cases of interest for possible applications, we consider situations with different anchoring on each surface. Strong and weak anchoring is simulated either by taking different interaction strength on each surface or by taking different number of molecular layers for each substrate.

The external electric field is assumed uniform within the nematic and we study the response of the director near the surface as a function of the applied voltage and the anchoring strength.

The total energy of the system V, is composed of the bulk and the surface contributions as well as the energy coming from the external electric field

$$V = V_B + V_S + V_E \qquad (3)$$

We look for the stabilization of the system trough the use of a relaxation method that allows us to find the minimum of the total energy as a function of the distribution of the director $n = n(R)$, that is:

$$\frac{\partial V}{\partial n(R)} = 0 \qquad (4)$$

The model we use in this work is in principle similar to that found in Reference [3]. However the main contribution of this work has been a generalization that permits dealing with more general boundary conditions and simulates in a more proper way surface characteristics of materials that are in contact with nematic liquid crystals.

RESULTS

In the present work we will restrict to the discussion of one particular case for the anchoring conditions at each surface. We chose parallel anchoring of the director at one surface and perpendicular at the other. The examples presented correspond to a nematic phase with negative dielectric anisotropy. This configuration simulates conditions similar to the ones proposed for liquid crystal materials that could electrically manipulate molecular orientation at the surface [1,2]. All the results we present have been obtained trough the application of a relaxation method for the total energy, looking for the director distribution that stabilized the system.

We carried out molecular dynamics simulations using the Verlet algorithm for the same system at the Supercomputing Facility at UNAM. We found strong instabilities and hence long computational times.

In figure 1 we show the distribution at zero applied external electric field. The anchoring at each surface is indicated by a line at the top of each surface. This state is obtained after relaxation of the system after initiating from a random director distribution. Figure 2 shows the visualization of the director distribution as in figure 1, but now at a non zero external electric field. The electric field is always taken in the z direction. This figure is shown just as an example of how the nematic molecules would reorient in presence of an external field.

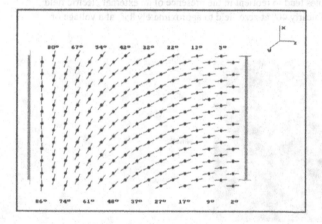

Figure 1. *Director distribution for each layer at zero electric field with left anchoring at 90°* *and right anchoring at 0°.*

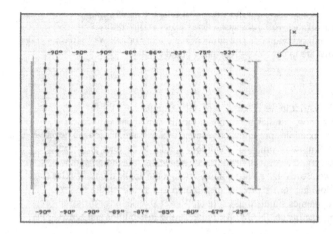

Figure 2. *Director distribution for each layer at non zero electric field with left anchoring at 90° and right anchoring at 0°.*

Figure 3 shows the director orientation as a function of the applied voltage. The angle is measured from the normal to the surface. The units for the applied voltage are arbitrary. In figure 4 we plot the order parameter as a function of the applied voltage. Last two figures clearly show how the nematic molecules tend to reorient in the presence of an external electric field going from a director angle of nearly 40° at zero field to approximately 85° at a voltage of 1 (a.u.).

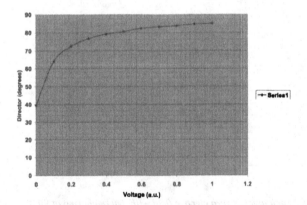

Figure 3. *Bulk director orientation as a function of the applied voltage (arbitrary units).*

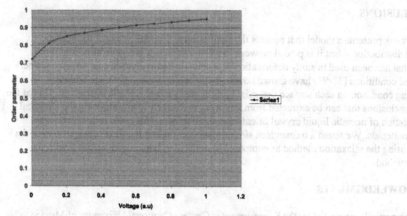

Figure 4. *Order parameter as a function of the applied voltage (arbitrary units).*

In figure 5 we show the director angle at each lattice site between the two surfaces (arbitrary divided in 17 layers) as a function of the applied voltage. From this figure we can see how at zero voltage (this corresponds to figure 1) the molecules near the left hand surface are oriented at nearly 90° (parallel to the surface), and at the right hand surface they are oriented close to 0° (perpendicular to the surface). For non zero external voltage this figure clearly shows the strong dependence of the liquid crystal orientation, near the surface of perpendicular anchoring, to the applied voltage.

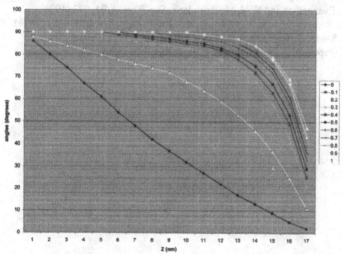

Figure 5. *Director angle (degrees) for each layer as a function of the applied voltage (a.u.).*

CONCLUSIONS

This work presents a model that permits the simulation and visualization of the nematic director distribution when it is placed between two flat surfaces. It is a generalization of the model that has been used to study deformation of a nematic liquid crystals near a surface under restricted conditions [3]. We have shown how with this model we can simulate arbitrary anchoring condition on each surface that might mimic real surface nematic interaction. One of the main conclusions that can be extracted from this work is the facility in simulating the electrical manipulation of nematic liquid crystal orientation near surfaces, that could mimic real liquid crystal materials. We found a tremendous simplification in finding the final configuration of the system using the relaxation method as compared to molecular dynamics calculation with the Verlet method.

ACKNOWLEDGMENTS

Research partially supported by the Supercomputer Center at National University of Mexico under contract No. UNAM-CRAY-I-941008.

REFERENCES

1. G. P. Bryan-Brown, C. V. Brown, I. C. Sage and V. C. Hui, *Nature*, **392**, 365 (1998).
2. G. P. Bryan-Brown, E. L. Wood and I. C. Sage, *Nature*, **399**, 338 (1999).
3. G. Skačej, V. M. Pergamenshchik, A. L. Alexe-Ionescu, G. Barbero and S. Žumer, *Phys. Rev.* **E56**, 571 (1997).
4. G. Barbero, L.R. Evangelista and N. V. Madhusudana, *Eur. Phys. J.* **B1**, 327 (1998).
5. M. Rajteri, G. Barbero, P. Galatola, C. Oldano and S. Faetti, *Phys. Rev.* **E53**, 6093 (1996).
6. W. Maier and A. Saupe, *Z. Naturforsch.* **A14**, 882 (1959); **A15**, 287 (1960).
7. A. Rapini and M. Papoular, *J. Phys.* (France), Colloq. **30**, C4-54 (1969).

Mat. Res. Soc. Symp. Proc. Vol. 651 © 2001 Materials Research Society

THE INFLUENCE OF BOUNDARY CONDITIONS AND SURFACE LAYER THICKNESS ON DIELECTRIC RELAXATION OF LIQUID CRYSTALS CONFINED IN CYLINDRICAL PORES

Z. NAZARIO[a], G. P. SINHA[b] and F.M. ALIEV[a,c]
[a] Dept. of Physics, University of Puerto Rico, San Juan, PR 00931, USA
[b] Dept. of Physics, Case Western Reserve University, Cleveland, OH 44106, USA
[c] Yokoyama Nano-structured Liquid Crystal Project, TRC, Tsukuba, Ibaraki, 300-2635, Japan

ABSTRACT

Dielectric spectroscopy was applied to investigate the dynamic properties of liquid crystal octylcyanobiphenyl (8CB) confined in 2000 Å cylindrical pores of Anopore membranes with homeotropic and axial (planar) boundary conditions on the pore walls. Homeotropic boundary conditions allow the investigation of the librational mode in 8CB by dielectric spectroscopy. We found that the dynamics of the librational mode is totally different from the behavior observed in investigations of relaxation due to reorientation of molecules around their short axis. The interpretation of the temperature dependence of relaxation times and of the dielectric strength of the librational mode needs the involvement of the temperature dependence of orientational order parameter. For samples with axial boundary conditions, layers of LCs with different thickness were obtained on the pore walls as a result of controlled impregnation of porous matrices with 8CB from solutions of different liquid crystal concentration. The process due to rotation of molecules around their short axis with single relaxation time observed for bulk 8CB is replaced by a process with a distribution of relaxation times in thin layers. This relaxation process broadens with decreasing layer thickness.

INTRODUCTION

Alkylcyanobiphenyls are liquid crystals, which have been thoroughly investigated in the past. These materials are stable and their molecules have a large dipole moment (~5 D) oriented along the molecular long axis. The dielectric properties of the bulk alkylcyanobiphenyls have been studied extensively [1-7] and have been quite clearly understood. For a geometry in which the electric field E is parallel to the director n i.e. E∥n, the Debye type process due to the restricted rotation of the molecules about their short axis exists. The characteristic frequency of this process is ~5 MHz and the temperature dependence of the corresponding relaxation times obeys the empirical Arrhenius equation. For the geometry in which the electric field E is perpendicular to the director n, i.e. E⊥n, the most prominent relaxation process has a characteristic frequency at about 70 MHz. In dielectric investigation of 5CB confined in cylindrical porous membrane [8] and random porous matrices [9] this process has been attributed to the librational motion of the molecules. The available information about this mode, to our best knowledge, is scarce.

To obtain a quantitative description of the librational mode we have applied dielectric spectroscopy and investigated the relaxation properties of 8CB confined in 2000 Å cylindrical pores of Anopore membranes treated with lecithin. Lecithin treatment provides homeotropic boundary conditions for confined 8CB. Since the pore axis is parallel to the probing electric field and the molecular dipole moment is oriented perpendicularly to its direction, such a configuration makes it possible to investigate the dynamics of librational (tumbling) mode by dielectric method. These results are compared with the relaxation due to rotation of 8CB molecules around their short axis in pores of the same size. In addition, using dielectric spectroscopy we have investigated the influence of layer thickness on the dynamics of reorientational motion of 8CB molecules - partially filling cylindrical pores.

EXPERIMENT

We used Anopore membranes with cylindrical parallel pores 2000 Å in diameter as matrices. For the preparation of the sample with homeotropic boundary conditions, the porous membrane was dipped inside a 3% mass concentration solution of lecithin in hexane for one hour. The matrix was dried until complete evaporation of hexane from the pores occurred. After that the matrix treated with lecithin was impregnated with 8CB from isotropic phase.

Layers of different thickness are formed [10] on the pore walls as a result of controlled impregnation of porous matrices with 8CB from solutions of different liquid crystal concentration. The thickness of LC layers formed on pore walls after evaporation of solvent is determined by the concentration of 8CB in the solution. In these experiments the probing electric field was parallel to the axis of the cylindrical pores.

8CB has a smectic phase in the temperature range of 21.1°C - 33.5°C in addition to the nematic range of 33.5°C - 40.8°C.

Dielectric measurements were performed using the broadband dielectric spectrometer based on SI 1260, Novocontrol Broad Band Dielectric Converter with active sample cell and HP4291A. We focus on relaxation observed at frequencies above 1MHz that was investigated using the high frequency part of the spectrometer. For the quantitative analysis of the dielectric spectra, the Havriliak-Negami function [11] has been used:

$$\varepsilon^* = \varepsilon_\infty + \frac{\Delta\varepsilon}{[1+(i2\pi f\tau)^{1-\alpha}]^\beta}$$

where ε_∞ is the high-frequency limit of the real part of the dielectric permittivity, $\Delta\varepsilon$ - the dielectric strength and τ - the mean relaxation time. The exponents α and β describe the distribution of relaxation times.

RESULTS AND DISCUSSION

Figure 1 illustrates the difference in relaxation observed in bulk 8CB and 8CB confined in lecithin treated matrices (homeotropic orientation).

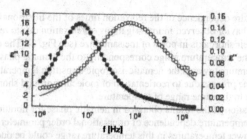

Figure 1: *Dielectric spectra (at T=310K) of: 1 - bulk 8CB (left axis) - relaxation due to the rotation of molecules around their short axis; 2 - 8CB confined in lecithin treated Anopore membrane (right axis) - librational mode. Symbols: experiment, lines: fitting.*

The relaxation in bulk 8CB is due to reorientation of molecules around their short axis. It is clear that the relaxation process in lecithin treated matrix is observed at much higher frequencies than in bulk sample and therefore cannot be assigned to the same mechanism. Since molecules of 8CB do not have a component of dipole moment perpendicular to the long axis, the only explanation of the observed high frequency relaxation in treated sample is to associate it with librational motion of molecules. This mode should be observed in 8CB if the probing electric field is perpendicular to the director as it takes place in the treated sample.

The lines in Fig. 1 represent the use of fitting analysis according to relation (1). For the bulk LC, the fitting line corresponds to Debye relaxation function ($\alpha=0$, $\beta=1$). The experimental results for liquid crystal in lecithin treated matrix could be fitted with a single relaxation process. The process for the librational mode observed in the lecithin treated porous matrix is however of non-Debye type as the parameter $\alpha = 0.25$. The spectra are wider and the asymmetry in the spectra are greater in the homeotropic alignment than in the axial orientation.

The temperature dependence of the relaxation times obtained for 8CB confined in lecithin treated cylindrical pores is illustrated in Fig. 2.

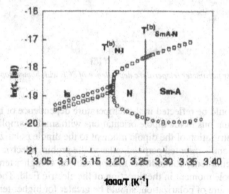

Figure 2: *Temperature dependence of relaxation times of 8CB confined in 2000 Å cylindrical pores: open circles - librational mode, open squares - mode due to reorientation of molecules around their short axis.*

The temperature dependence of the relaxation times of the librational mode is totally different from the behavior observed in investigations of relaxation due to reorientation of molecules around their short axis in pores of the same size (see Fig. 2). The relaxation time of librational mode in the temperature range corresponding to the nematic phase increases upon increasing the temperature towards the nematic-isotropic transition temperature. In contrast, the relaxation times of the process due to reorientation of molecules around short axis decrease upon temperature increase in the same range of temperature.

The interpretation of the results obtained in nematic phase for librational mode needs the involvement of the temperature dependence of orientational order parameter. The decrease of relaxation times at lower temperatures in this temperature range could be due to acceleration of the process with increasing of order parameter. In the sample with homeotropic boundary conditions, in the case of perfect order ($S=1$) and taking into account that the dipole moment of the molecule is parallel to its long axis, the projection of the dipole moment on the direction of the electric field that is along pore axis is minimal because the fluctuations of molecular orientations with respect to radial direction are very small. These fluctuations of the dipole moment orientation (or molecular long axis) correspond to the librational motion of the molecule and the amplitude of the fluctuations determines the relaxation rate of dipole in the viscous media. At higher temperatures these deviations (fluctuations) are of greater amplitude (the order parameter is smaller) and this requires longer time to complete one librational cycle. As a result the relaxation rate is smaller for fluctuations of greater amplitude and *vice versa*. Such behavior has resulted in the particular temperature dependence of relaxation times observed for the librational mode in the nematic phase temperature range. In smectic phase the temperature dependence of relaxation times is mainly determined by the variations of viscosity.

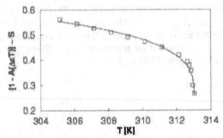

Figure 3: *Orientational order parameter temperature dependence of 8CB with homeotropically aligned molecules in cylindrical pores.*

This scenario should be reflected in the temperature dependence of the dielectric strength of the process. For fluctuations of the dipole orientations with smaller amplitude i.e. with greater order parameter, the contribution of the dipole moment to the dipole polarization is minimal because the dipole moment is oriented perpendicular to the probing electric field. The increase in the amplitude of the fluctuations, due to lower order parameter at higher temperatures, increases the projection of the dipole moment in the direction of the electric field. Therefore the dielectric strength, which is a measure of polarization, should be greater for higher temperatures than for lower ones. For LCs with molecules having dipole moment μ parallel to long axis and homeotropically aligned in cylindrical pores, the dielectric strength $\Delta\varepsilon = \varepsilon_s - \varepsilon_\infty \propto <\mu^2>/3kT$ can be obtained from Maier and Meier equations[12] as

$$\Delta\varepsilon \sim \mu^2(1-S)/kT$$

Therefore the product $\Delta\varepsilon \cdot T$ is a measure of $S(T)$. Using experimentally obtained dielectric strength for each temperature, the temperature dependence of $S(T)$ multiplied with an arbitrary constant A for 8CB is presented in Fig. 3.

In the bulk LCs the temperature dependence of order parameter exhibits[13] the mean-field theory critical exponent expected near a second-order transition at a temperature T^* preempted by a first order phase transition at T_{NI} as:

$$S = S_0 + a(T^* - T)^{\beta'}$$

where β' is a critical exponent of order parameter that in the Landau theory is equal to 0.5. The parameters obtained from fitting the data with this equation were $T^* = 313.05$ K and $\beta' = 0.25$ for 8CB. The fitting line with the above parameters is presented in Fig. 3. The obtained values of T^* and β' should not be considered as parameters quantitatively describing the temperature dependence of the order parameter of 8CB confined in cylindrical pores with radial orientation. The above consideration is just a qualitative self-consistent picture of dipole librational reorientations in the sample under investigation.

We have additionally investigated the influence of the thickness of the layer of liquid crystal formed on the pore walls as a result of controlled impregnation of porous matrices with 8CB from solutions of different liquid crystal concentration.

Figure 4 illustrates the difference between relaxation due to molecular reorientations around their short axis in bulk liquid crystal and liquid crystal partially filling cylindrical pores for three samples: bulk 8CB, matrix partially filled with 8CB from 50% concentrated solution of 8CB in hexane, and 13% concentrated solution.

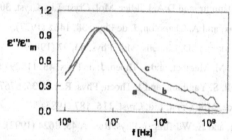

Figure 4: Dielectric spectra at T = 306 K for (a) bulk 8CB, matrix partially filled with 8CB from 50% concentrated solution of 8CB in hexane, and (c) - from 13% concentrated solution.

The broadening of dielectric spectra is observed in confined LC. This broadening increases with decrease in filling concentration, i.e. with decrease in liquid crystal layer thickness.

The characteristic frequencies corresponding to the maximum in the spectrum of the imaginary part of the dielectric permittivity are shifted towards high frequencies for samples with partially filled pores. This could be related to the facts that nematic-isotropic phase transition temperatures in samples with partially filled pores are shifted towards lower temperatures and the mass density of LC in these samples is lower than in the bulk. The results of detailed investigations of these samples will be published separately. Broadening of the spectra seems to be a general dynamical property of thin confined fluid films [14] and has been observed even for simple liquids [15].

CONCLUSION

The relaxation of librational mode in lecithin-treated cylindrical pores is observed in dielectric spectroscopy experiment. This is a proof of hometropic boundary conditions in this sample. The dynamics of the librational mode is totally different from the behavior observed in investigations of relaxation due to reorientation of molecules around their short axis. The interpretation of the temperature dependence of relaxation time needs the involvement of the temperature dependence of orientational order parameter and suggests that the better orientational order the faster the process due to libration of molecules. Dielectric spectroscopy experiments show that the dynamics of molecular mode is influenced by thickness of LC layer formed on the pore wall surface. The main relaxation process broadens with decreasing thickness of the layers and the characteristic frequencies are shifted towards the higher frequencies for thinner layers.

ACKNOWLEDGEMENTS

This work was supported by Naval Research Office grant N00014-99-1-0558. F.A. acknowledges support from Japan Science and Technology Corporation.

REFERENCES

1. P.G. Cummins, D.A. Danmur, and D.A. Laidler, Mol. Cryst. Liq. Cryst. **30**, 109 (1975).

2. D. Lippens, J.P. Parneix, and A. Chapoton, J. de Phys. **38**, 1465 (1977).

3. J.M. Wacrenier, C. Druon, and D. Lippens, Mol. Phys. **43**, 97 (1981).

4. T.K. Bose, R. Chahine, M. Merabet, and J. Thoen, J. de Phys. **45**, 11329 (1984).

5. T.K. Bose, B. Campbell, S. Yagihara, and J. Thoen, Phys. Rev. A **36**, 5767 (1987).

6. A. Buka and A. H. Price, Mol. Cryst. Liq. Cryst. **116**, 187 (1985).

7. H. -G. Kreul, S. Urban, and A. Würflinger, Phys. Rev. A **45**, 8624 (1992).

8. S.R. Rozanski, R. Stanarius, H. Groothues, and F. Kremer, Liq. Cryst. **20**, 59 (1996).

9. G.P. Sinha and F.M. Aliev, Phys. Rev. E **58**, 2001 (1998).

10. B. Zalar, S. Zumer, and D. Finotello, Phys. Rev. Let. **84**, 4866 (2000).

11. S. Havriliak and S. Negami, Polymer **8**, 101 (1967).

12. W. Maier and G. Meier, Z. Naturforsch. **16a**, 262 (1961).

13. P.G. de Gennes and J. Prost, *The Physics of Liquid Crystals*, second ed., (Clarendon Press, Oxford 1993).

14. P.A. Thompson, G.S. Grest, and M.O. Robbins, PRL, **68**, 3448 (1992).

15. A.L. Demirel and S. Granik, Phys. Rev. Lett., **77**, 2261 (1996).

Mat. Res. Soc. Symp. Proc. Vol. 651 © 2001 Materials Research Society

Mechanism for Ferromagnetic Resonance Line Width Broadening in Nickel-Zinc Ferrite

Hee Bum Hong, Tae Young Byun, Soon Cheon Byeon and Kug Sun Hong
School of Materials Science and Engineering, Seoul National University,
Seoul, 151-742, KOREA

ABSTRACT

The systematic variation in line width of ferromagnetic resonance with the Fe content was observed at X band (9.78GHz) in $(Ni_{0.5}Zn_{0.5})_{1-x}Fe_{2+x}O_4$ ($-0.2 \leq x \leq 0.2$). The line width of the stoichiometric composition ($x = 0$) showed minimum value, 50 Oe. In contrast, the line width of the non-stoichiometric compositions sharply increased to 210 Oe with increasing non-stoichiometry (x). The mechanism for this line width broadening was investigated using thermoelectric power and electrical resistivity, since the contribution of anisotropy and porosity to the line width was negligible in this compositional region. In Fe excess region, Fe^{2+} ion concentration increased with increasing Fe content, resulting in line width broadening due to relaxation. But, it was suggested that Ni^{3+} and Fe^{2+} ions coexist in Fe deficient region. Therefore the increase of line width in nickel-zinc ferrites originated from the Fe^{2+}/Fe^{3+} magnetic relaxation in Fe excess region, and the Fe^{2+}/Fe^{3+}, Ni^{2+}/Ni^{3+} magnetic relaxation in Fe deficient region.

INTRODUCTION

Ferrites are extensively used in microwave devices, communication systems, magnetic recording systems, etc [1]. Magnetic semiconductor ferrites have a wide spread role in high frequency circuit components such as transformers, noise filters, and magnetic recording heads because both of their electrical resistivities and initial permeabilities are much higher than those of metals at high frequencies [2]. Recently, applications are being extended to include components being used to shield the electromagnetic interference (EMI) [1].

Nickel-zinc ferrites are being used as high frequency components because of its high electrical resistivity. One criterion to be useful at high frequency is low loss. Therefore there are many researches on reduction of loss at high frequencies especially in the viewpoint of fabrication process. But there is little report on the mechanism for this loss. In this study a systematic investigation on the loss mechanism with iron concentration in Nickel-zinc ferrites is performed.

EXPERIMENT

Nickel-zinc ferrites of the composition $(Ni_{0.5}Zn_{0.5})_{1-x}Fe_{2+x}O_4$ ($x=-0.2$, -0.1, -0.05, -0.025, 0, 0.025, 0.05, 0.1, 0.2) were prepared using the conventional mixed oxide method. Ferric oxide (α-Fe_2O_3, 99.99% high purity chemical, Japan), manganese oxide (Mn_3O_4 99.9%, MMC, South Africa), and zinc oxide (ZnO, 99.9%, Jung-Dong Chemical, Japan) powders were mixed for 12 hours. The mixed slurry was dried and then calcined at 900°C for 2 hours in air. The powder was granulated after milling the calcined powder for 24 hours. The powder was uniaxially pressed at a pressure of 1000 kg/cm^2 to form a bar-shape sample. After binder burnout at 600°C for 1 hour, the specimens were sintered at 1350°C for 5 hours in a controlled atmosphere. To maintain equilibrium in ferrous ion concentration during sintering the oxygen partial pressure was controlled using the following equation suggested by Morineau [3], where a is the atmospheric parameter which represents an oxidation potential.

$$\log P_{O2} = a - \frac{18950}{T(K)} \tag{1}$$

In the present study, specimens were prepared with a constant atmospheric parameter, a=11.5. The cooling rate was 100°C/h. Ferromagnetic resonance (FMR) spectra were recorded at 300K using an X-band (9.78GHz) spectrometer (Bruker, Model ER 301). Samples were cut and polished to a sphere with a diameter of 600μm and were placed at the center of a cavity oscillating in the rectangular TE$_{102}$ mode. A conventional field modulation was used. Saturation magnetization (Ms) was measured using a vibrating sample magnetometer (VSM, Digital Instruments, model DMS 880, USA) under a field of 10 kOe. Electrical resistivity and permeability was measured at room temperature using impedance spectroscopy. Permeability was measured in the frequency range 100 Hz to 100 MHz using an impedance/gain phase analyzer (Hewlett Packard, Model HP 4194A). Thermoelectric power was measured at 400°C using dc method. Porosity was calculated from bulk density measured by Archimedes' method.

RESULTS AND DISCUSSION

Figure 1(a) shows FMR spectra with the concentration of iron in nickel-zinc ferrite. These spectra were recorded as a first derivative of the absorption. As the composition deviates from the stoichiometric composition (x=0), the resonance line broadened and the resonance field decreased. The line width is a measure of loss. So the loss increases with the compositional deviation from the stoichiometric composition. Figure 1(b) shows a replot of the line width in figure 1(a) with the concentration of iron. At the stoichiometric composition the line width is minimum (50 Oe) and the line width increase with the deviation from the stoichiometric composition. At the compositions of x=0.2 and x=-0.2 the line width is over 200 Oe. X-ray powder diffraction (XRD, not shown here) showed spinel single phase in all the compositions investigated. Secondary phase is not detected in XRD pattern or in a scanning electron microscopy (SEM). Therefore the line width broadening cannot be attributed to the second phase.

The line width is ordinarily attributed to anisotropy, porosity, saturation magnetization, eddy current, and unknown factors like relaxation. And it is presented by the sum of the above factors. The line width of the FMR of polycrystalline ferrites is a sum of various contributions. According to Scholmann's report the line width of single-phase ferrite (ΔH) can be given as follows [4].

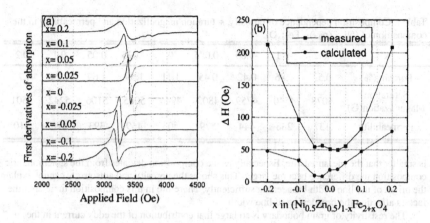

Figure 1. (a) Iron concentration dependent ferromagnetic resonance spectra and (b) comparison between measured line width and calculated line width considering porosity and anisotropy

$$\Delta H = \Delta H_{anisotropy} + \Delta H_{porosity} + \Delta H_{eddy} + \Delta H_{others}$$

$$= \frac{8\pi\sqrt{3}}{21} \frac{H_A^2}{4\pi M_s} + \frac{8}{\sqrt{3}\pi} 4\pi M_s + \frac{P_{eff}}{1+P_{eff}} + \Delta H_{eddy} + \Delta H_{others} \qquad (2)$$

where H_A is anisotropy field, $4\pi M_s$ saturation magnetization, and P_{eff} is effective porosity.

Based on the above equation the contribution of each factor to the line broadening is examined. In order to calculate the contribution of porosity, the porosity is measured by the difference between the theoretical density and the measured density. The measured porosity is listed in Table I. Also in order to calculate the contribution of the anisotropy, the anisotropy field in equation (2) should be known. The cubic anisotropy field ($H_A = 4K_1/3M_s$) [4] was calculated using anisotropy constant. Magnetic anisotropy constant K_1 for this nickel-zinc ferrite is reported only at the stoichiometric composition (-2.8×10^4 erg/cm^3). Therefore in this study the magnetic anisotropy constant is calculated using the equation suggested by Globus [5]. Permeability is constant up to a certain frequency and decays at higher frequency. The anisotropy constant can be calculated using this constant permeability. The measured permeability and the calculated anisotropy constant are also listed in Table 1.

Line width due to porosity and anisotropy

Using Table I and the equation 1 the contribution to the line width by porosity and anisotropy is calculated and plotted in Figure 1(b). The line width due to porosity and anisotropy

Table I. Composition dependence of porosity, saturation magnetization and permeability in the composition of $(Ni_{0.5}Zn_{0.5})_{1-x}Fe_{2+x}O_4$.

x	-0.2	-0.1	-0.05	-0.025	0	0.025	0.05	0.1	0.2
Porosity(%)	0.53	1.86	0.42	0.45	1.11	1.59	1.69	2.24	4.73
Saturation Magnetization(G)	4078	4540	4759	4803	4946	5085	5106	5242	5501
Permeability	137	288	444	579	503	469	393	426	310

is smaller that the measured one. Especially as the composition deviates from the stoichiometric composition the difference become larger. This shows the porosity and anisotropy cannot explain the origin of the line width broadening sufficiently. Therefore in this composition there are other factors affecting the broadening of line width.

The resistivity of grain boundary is so large that contribution of the eddy current in the broadening of the line width is negligible in manganese-zinc ferrites [1]. The resistivity of Ni-Zn ferrites is much higher than that of manganese-zinc ferrite, so the contribution of the eddy current can be excluded in this composition.

Electrical resistivity and thermoelectric power

Figure 2(a) shows the dependence of resistivity on iron concentration. The resistivity decreases significantly with the increase of iron concentration. The conduction in ferrite occurs via polaron hopping between Fe^{2+} and Fe^{3+} in octahedral site [6]. Therefore resistivity depends on the concentration of charge carriers, especially Fe^{2+}. So the decrease of resistivity with the iron concentration is mainly due to the increase of Fe^{2+} concentration. But in the iron-deficient composition, from x=-0.2 to x=-0.05, the resistivity increase in spite of the increase of the iron concentration. So Fe^{2+} concentration variation can not explain the resistivity in this region.

The resistivity is reported to decreases monotonically as the composition varies from the iron-deficient to iron-excess in manganese-zinc ferrites [7]. This is attributed to the ionic distribution in spinel structure. $MnFe_2O_4$ has almost normal spinel structure. At the stoichiometric composition 90% of Mn ions are in the tetrahedral sites (A site) and 10% of Mn ions are in the octahedral sites (B site) in $MnFe_2O_4$. Therefore manganese concentration in octahedral site is too low to contribute to the conduction in $MnFe_2O_4$. So the conduction is due to Fe^{2+} ions in $MnFe_2O_4$ and this makes the sample n-type semiconductor. Therefore resistivity decreases with Fe concentration due to increase in the concentration of Fe^{2+} ions. But this is contrary to our result in iron-deficient region in Figure 2(a).

However Jonker et al. reports p-n transition when the composition varies from the iron-deficient region to the iron-excess region in $CoFe_2O_4$ [8]. Iron-excess Co-ferrite shows n-type conduction and resistivity decreases with iron content in iron-excess region. Cobalt-excess Co-ferrite shows p-type conduction and resistivity decreases as iron content decreases in iron-deficient region. This also comes from the ionic distribution in the structure of $CoFe_2O_4$. Co-ferrite has an inverse spinel structure. In the stoichiometric composition, half of Co ions are in B site and the others are in A site. In iron-deficient (or cobalt-excess) composition, the excess Co ions exist in the state of Co^{3+} and the concentration of Fe^{2+} is small, resulting in the main

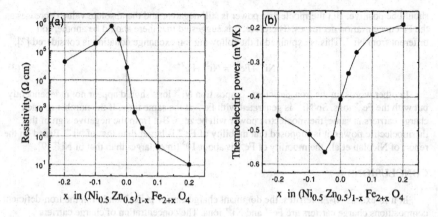

Figure 2. (a) Resistivity and (b) thermoelectric power as a function of iron concentration.

conduction via Co^{3+} and Co^{2+} ions. So the sample shows p-type conduction. It is noteworthy to mention that the decrease of iron concentration does not increase resistivity because Co^{3+} ions become dominant charge carrier and reduce the resistivity in iron-deficient region.

Conduction in Ni-Zn ferrite is a compromise between the conduction in Mn-ferrites and Co-ferrites. In Ni-Zn ferrites, Zn ions do not affect the conduction because they occupy A site. So the ions affecting the conduction are Ni and Fe ions. Ni-Zn ferrite has inverse spinel structure, so in the stoichiometric composition $(Ni_{0.5}Zn_{0.5}Fe_2O_4)$ 0.5mol Fe ions and 0.5 mol Zn ions are in A site and 1.5mol Fe ions and 0.5mol Ni ions are in B site. As the iron concentration decreases below the stoichiometric composition, excess Ni ions are transformed to Ni^{3+} ions like in Co-ferrites, resulting in the decrease of the resistivity. But reduction of the Fe^{2+} ion concentration in the iron-deficient region will increase the resistivity. So there are two competing ions in the conduction in Ni-Zn ferrite. In order to discuss in detail thermoelectric powers are measured.

Thermoelectric power shows the type and concentration of charge carriers. All the compositions investigated show n-type conduction and shown in Figure 2(b). In iron-excess composition, the absolute value of thermoelectric power decreases as the concentration of Fe increases. This means that the concentration of charge carrier increases with Fe concentration in iron-excess composition. This agrees with the decrease of resistivity with Fe concentration. The negative sign of the thermoelectric power shows the main charge carrier is Fe^{2+}. Therefore in Ni-Zn ferrites, the conduction in iron-excess compositions is due to polaron hopping between Fe^{2+} and Fe^{3+} ions and the resistivity is dependent on the iron concentration. In the iron-deficient composition the absolute value of the thermoelectric power shows maximum. The absolute value of thermoelectric power increases as the Fe concentration increases in the composition range of $-0.05 \leq x \leq 0$. This means the dominant charge carrier is Fe^{2+} ion in this region. At x=-0.05 the absolute value of the thermoelectric power is maximum and the thermoelectric power decreases as the iron concentration decreases. This suggests that opposite charge carrier is becoming dominant. If Ni^{3+} ion is dominant charge carrier like in Jonker's reports, thermoelectric power should be positive. But thermoelectric power is still negative and the absolute value decreases as the Fe concentration decreases. Therefore, it is suggested that there should be another ion different from Ni^{3+}. This is explained if the following ion exchange equation is considered [2].

$$Ni^{2+}+Fe^{3+} = Ni^{3+}+Fe^{2+} \tag{3}$$

In other words, in the iron-deficient composition Ni^{3+} ions should appear not independently but with the Fe^{2+} ions. So Ni^{3+} is generated with Fe^{2+} at the same time. If the number of each charge carriers is same, thermoelectric power will be zero. But from the negative sign of the thermoelectric power, it is supposed that mobility of Fe^{2+} is larger than that of Ni^{3+}. Based on the report of Nicolau et al., the mobility of Fe^{2+} is about 10^5 times larger than that of Ni^{3+} [9].

CONCLUSIONS

In iron-excess Ni-Zn ferrites, the dominant charge carrier is Fe^{2+} ion but in the iron-deficient compositions charge carriers are Fe^{2+} and Ni^{3+} ions. The concentration of charge carriers increases with the deviation of composition from the stoichiometric composition. In the iron-deficient compositions the absolute value of thermoelectric power looks large not by the low concentration of charge carriers but by the compensation of the charge carriers, Fe^{2+} and Ni^{3+}.

In the microwave frequency, loss can be increased by these mobile ions. According to Galt and Clogston, the relaxation between different states of the same kinds of ion will give rise to the loss. Since the magnetization is coupled to the lattice by the magnetostriction, motion of the magnetization produces varying stresses in the lattice. These variations in stress cause the arrangement of the divalent and trivalent ions to vary. But it takes some finite time in this rearrangement and thus phase lag results in the line broadening [1, 10-12].

In Ni-Zn ferrites, there is Fe^{2+}/Fe^{3+} polaron in all compositions and there is Ni^{3+}/Fe^{2+} coupling in the iron-deficient composition. Therefore these divalent and trivalent ions cause relaxation and increase loss. As a result, the significant increase of line width in the non-stoichiometric composition is mainly due to the coexistence of these divalent and trivalent ions. As the concentration of these ions increases, the line width increases.

REFERENCES

1. S. C. Byeon, K. S. Hong, J. G. Park, and W. N. Kang, J. Appl. Phys., **81**, 7835 (1997).
2. A. I. El Shora, M. A. El Hiti, M. K. El Nimr, M. A. Ahmed, and A. M. El Hasab, J. Mag. Mag. Mater., **204**, 20 (1999).
3. R. Morineau and M. Paulus, Phys. Status Solidi A, **20**, 373 (1973).
4. E. Schlomann, J. Phys. Chem. Solids, **6**, 242 (1958).
5. A. Globus, P. Duplex and M. Guyot, IEEE Trans. Magn., **7**, 617 (1971).
6. B. Gillot, Phys. Status Solidi A, **69**, 719 (1982).
7. F. K. Lotgering, J. Phys. Chem. Solids, **25**, 95 (1964).
8. G. H. Jonker, J. Phys. Chem. Solids, **9**, 165 (1959).
9. P. Nicolau, I. Bunget, M. Rosenberg, I. Belciu, IBM J. Res. Develop., **14**, 248 (1970).
10. J. K. Galt, Bell Syst. Tech. J., **33**, 1023 (1954).
11. J. K. Galt, W. A. Yager, and F.R. Merritt, Phys. Rev., **93**, 1119 (1954).
12. A. M. Clogston, Bell Syst. Tech. J., **34**, 739 (1955).

Mat. Res. Soc. Symp. Proc. Vol. 651 © 2001 Materials Research Society

Studies of Temperature-Dependent Excimer-Monomer Conversion in Dendrimeric Antenna Supermolecules by Fluorescence Spectroscopy

Youfu Cao[a], Jeffrey S. Moore[b] and Raoul Kopelman[a]*

[a] Department of Chemistry, University of Michigan, Ann Arbor, MI 48109

[b] Department of Chemistry, University of Illinois, Urbana, IL 61801

* Correspondence Author: kopelman@umich.edu

Abstract: Phenylacetylene (PA) dendrimer labeled with perylene (See Fig 1) is discovered to exhibit temperature-dependent emission spectra in certain organic solvents over the temperature range of 20-65°C. The monomer signal is increasing rapidly when temperature increases, while the excimer signal decreases slowly. Models of excimer formation and weakly associated pairs (M+M) dissociation dynamics are included, and the equilibrium constants at different temperatures are calculated. This behavior suggests potential applications in fluorescence-based thermometry.

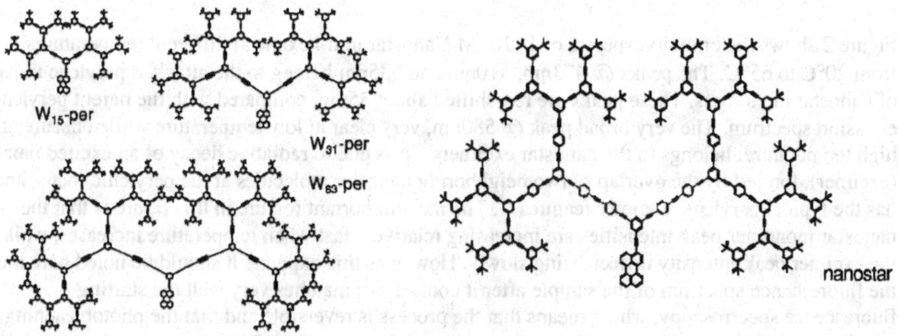

Fig 1: The molecular structures of perylene-substituted phenylacetylene (PA) dendrimers (W_{15}-per, W_{31}-per, W_{63}-per and nanostar), the colors (blue, green, orange, red) represent the emission energy of the corresponding subunits.

Introduction

Phenylacetylene (PA) dendrimers [1] and their perylene-substituted derivatives, both compact and extended series, have been investigated both experimentally[2, 3] and theoretically[4, 5], focusing on their photophysical and photochemical properties. In those perylene-substituted derivatives, perylene molecules are attached to the locus, functioning as energy traps. These energy traps help understand the PA dendrimers. Due to the high efficiency of energy transfer in these molecules and their Cayley tree (or Bethe tree) structures and large molecular sizes, they have potential applications in drug delivery systems[6, 7], optical nano-probes[8, 9], near field microscopy (single molecule light sources) [10, 11] and photoelectric devices. Synthetic chemists are building more supermolecules with similar structures[12], trying to bring them into industrial and scientific applications. In this paper we report quite-unexpected temperature-dependent emission spectra in these big molecules, over the temperature range of 20–65°C, for the "Nanostar", one typical PA dendrimer composed of 5 generations.

Experiments

In the experiment, a 1 $\times$ 10^{-6}M Nanostar solution in spectrophotometric grade n-hexane (Purchased from Fisher, used as received) is bubbled with N_2 to avoid excited state quenching by oxygen. The sample is then filled into an airtight quartz cell, preventing solvent evaporation and oxygen contacts, keeping the concentration constant. Steady state fluorescence spectra are taken on a JOBIN YVON-SPEX Instruments S.A. FluoroMax-2 fluorometer with temperature control. The excitation wavelength is 310nm, the absorption peak of the outermost linear PA chains of 2 units [13]. The temperature of the sample cell was increased by 5°C, beginning from 20°C. The fluorescence spectra are taken after 30min at each temperature, to allow enough time for the sample to reach thermal equilibrium.

Figure 2 shows the emission spectra of 1 $\times$ 10^{-6}M Nanostar in n-hexane at different temperatures, from 20°C to 65°C. The peaks @ 473nm, 510nm and 535nm belong to the attached perylene traps of nanostar monomers. These peaks are red-shifted about 35nm, compared with the parent perylene emission spectrum. The very broad peak @ 580nm, very clear at low temperature while obscure at high temperature, belongs to the nanostar excimers[1]. It is due to radiative decay of an excited pair (excimer), formed by the overlap of two neighboring nanostar molecules at the perylenic locus, and has the typical perylenic excimer features [14]. The most important feature in this figure is that the nanostar monomer peak intensities are increasing relatively fast when temperature increases, while the excimer peak intensity is decreasing slowly. How does this happen? It should be noted here that the fluorescence spectrum of the sample after it cools down matches very well the starting fluorescence spectroscopy, which means that the process is reversible and that the photobleaching of nanostar is minimal. We also did the same experiments on 1×10^{-6}M perylene in n-hexane, where we did see some photobleachings but we didn't see the dramatic temperature behavior of the nanostar. We also note that due to the back transfer to the antenna in these PA dendrimer derivatives, they usually have excellent photostability [13]. Further aspects of their photostability and photobleaching are still being investigated.

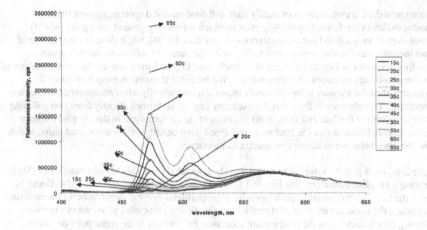

Fig 2. The fluorescence spectra of 1×10^{-6}M nanostar in n-hexane at different temperatures (15-65°C).

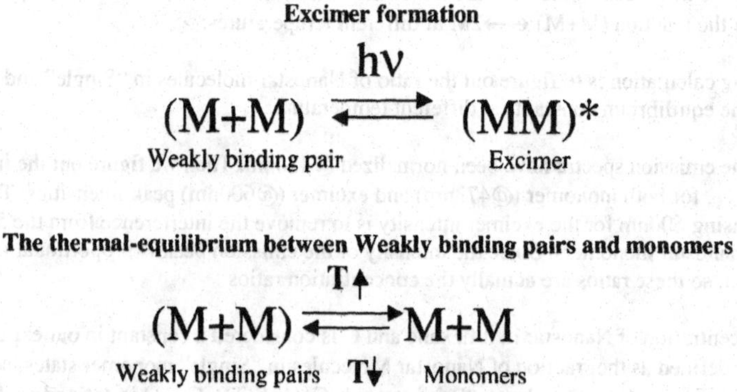

Excimer formation

$$hv$$
$$(M+M) \underset{}{\overset{}{\rightleftharpoons}} (MM)^*$$

Weakly binding pair   Excimer

The thermal-equilibrium between Weakly binding pairs and monomers

$$T\uparrow$$
$$(M+M) \rightleftharpoons M+M$$

Weakly binding pairs $T\downarrow$ Monomers

Fig 3. Schematic descriptions for excimer formation and the thermal-equilibrium between weakly binding pairs and monomers.

Discussion

In a recently published paper by Swallen et al. [14], the dependence of excimer formation in PA dendrimers on concentration, solvents and excitation wavelengths was discussed in detail. Their experimental data, a combination of steady state and time-resolved spectra, support that the nanostar excimers are formed by weakly associated noninteracting ground state pairs (M+M) present even in very diluted nanostar solutions (lower than 10^{-5}M), while there is no evidence of dimer formation in the absorption spectra. Figure 3 schematically shows the excimer spectra. Actually the spectra in Figure 2 support this conclusion. Figure 2 indicates that the concentration of nanostar monomers increases with temperature. We note that if excimers were formed by 2 monomers, then the excimer intensity would increase quardratically when monomer concentration is increased. Therefore we believe that the excimers can not be formed directly from two colliding monomers. Adding the fact that there is no evidence of dimer formation in the sample, the only reasonable conclusion is that the excimers are formed from certain weakly associated pairs, which show the same behavior in absorption spectra as monomers.

Figure 2 suggests that the ratio between the weakly associated pairs (M+M) and monomers (M) is changing with the temperature. See Figure 3 for the thermal-equilibrium between (M+M) and M. Due to the fact that the emission peak intensity is proportional to the concentration of the emission sources and the excimers are formed directly from the weakly associated pairs, we can draw the following conclusion: when the temperature increases, the amount of the weak pairs decreases, while that of monomers increases. Therefore, when the temperature increases, the excimer intensity decreases while the monomer intensities increase. But it is also notable that the increase in monomer intensities is much bigger than the decrease in excimer intensity. What will make this happen? Only if the weak pairs initially dominate in the nanostar solution. Same experiments were performed on perylene solutions in n-hexane with the same concentration, but no similar data obtained. Now let's do some calculations:

Based on Figure 2, we can figure out the ratio of the weakly associated pairs and the equilibrium constants for the reaction (M+M) $\longleftrightarrow$ 2M at different temperatures:

The following calculation is to figure out the ratio of Nanostar molecules in "Single" and "Pair" states, and the equilibrium constants at different temperature.

First all of the emission spectra have been normalized @420nm. Then we figure out the intensity ratio $I_{65°C}/I_{20°C}$, for both monomer (@473nm) and excimer (@600nm) peak intensities. The reason that we are using 600nm for the excimer intensity is to remove the interference from the 535nm peak of the nanostar monomers. Since the intensity of the emission peak is proportional to the concentration, so these ratios are actually the concentration ratios.

C is the concentration of Nanostar in n-hexane and C is considered a constant in our experiment. At T°C, β(T) is defined as the fraction of Nanostar Molecules in "Single" monomer states, so the concentration of nanostar molecules in "Pair" states is C*(1−β(T)). C_s(T) is defined as the concentration of nanostar molecules in "Single" states, C_p(T) is defined as the concentration of nanostar molecules in "Pair" states, and R_s(T) and R_p(T) are respectively defined as

$R_s(T) = C_s(T) / C_s(20 \ C)$
$R_P(T) = C_p(T) / C_p(20 \ C)$

Based on the conservation principle, we have

$R_S(T) * C * \beta(20\ C) + R_p(T) * C * (1-\beta(20\ C)) = C$

Then $\beta(20\ C) = \dfrac{1 - R_P(T)}{R_S(T) - R_P(T)}$

$\beta(T) = \beta(20\ C) * R_S(T)$

Equilibrium constant K (T) is defined as

$K(T) = [monomers]^2 / [weak\ pairs] = (C*\beta(T))^2 / [C*(1-\beta(T))/2]$
$= 2*C*\beta(T)^2 / (1-\beta(T))$

Table 1 lists the ratio of weak pairs and the equilibrium constants at different temperatures.

T/°C	$R_S(T)$ @473nm	$R_P(T)$ @600nm	$\beta(T)$	K (T) / (mol/L)
20	1	1	0.01	2.02×10^{-10}
30	2.0	0.99	0.02	8.16×10^{-10}
40	4.9	0.96	0.05	5.26×10^{-9}
50	11	0.91	0.11	2.72×10^{-8}
60	26	0.77	0.26	1.83×10^{-7}
65	36	0.68	0.36	4.05×10^{-7}

Conclusions

In conclusion, the weakly associated pairs dominate in the solution at low temperatures, but when the temperature increases, these pairs dissociate into monomers. So what we see in the emission spectra is that the monomer intensities increase rapidly when the temperature increases, while the excimer intensity decreases slowly. Of particular interest is that around 64% nanostar molecules are in weak pair forms even when the temperature (65°C) is very close to the boiling point of n-hexane (69°C).

We expect the similar temperature behaviors in all extended perylene-substituted PA dendrimer derivatives. Further temperature-dependence investigations about these two different PA dendrimer derivatives are underway.

Recently fluorescence-based-temperature sensing is receiving increasing interest, due to its unique applications in monitoring the temperature within micro-sized domains or hostile environments[15, 16]. A big limitation of these temperature sensors is the inherent fluctuations in the emission sources, like photobleaching and quenching, the ubiquitous problem in developing fluorescence-based optical sensors. Most of the fluorescence-based-optical thermometers rely on the changes in the absolute fluorescence emission peak intensity, and actually in many cases, it involves the conversion between dimers/excimers and monomers when the temperature changes. Our experimental data suggest that the Nanostar and other PA derivatives are good candidates for developing fluorescence-based optical temperature sensors, especially when we consider their high emission intensity and excellent photostability.

Acknowledgements

We acknowledge the support for this project from the National Science Foundation, Division of Materials Sciences, grant DMR-9900434.
Also our thanks go to Murphy Brasuel for his help in the experiments, and to Dr. Eric Monson and SungHyun Park for their suggestions and discussions.

References:

1. Xu, Z.; Moore, J. S. *Acta Polym* 1994, 45, 83.
2. Swallen, S. F; Shortreed, M. R.; Shi, Z. Y.; Tan, W., Xu, Z.; Devadoss C.; Moore, J.S.; Kopelman, R. (Cairo) Dendrimeric Antenna Supermolecules with Multistep Directed Energy Transfer in *Science and Technology of Polymers and Advanced Materials*; Prasad, P. N.; Plenum Press: New York, 1998
3. Shortreed, M. R.; Swallen, S. F.; Shi, Z. -Y.; Tan, W.; Xu, Z.; Devadoss, C.; Moore, J. S.; Kopelman, R. *J. Phys. Chem.* B. 1997, 101, 6318-22.
4. Bar-Haim, A.; Klafter, J; Kopelman, R. *J. Am. Chem. Soc.* 1997, 26, 6197
5. Tretiak, S.; Cherniak, V.; Mukamel, S. *J. Phys. Chem. B* 1998, 102, 3310-15
6. Junge, B. M.; McGrath, D. V. *Chem Commun.* 1997, 9, 857
7. Tomalia, D. A.; Naylor, A. M.; Goddard, W. A. *Angew. Chem., Int. ED Engl* 1990, 29, 138.
8. Kopelman, R.; tan, W. *Appl. Spectrosc. Rev.* 1994, 29, 39
9. Tan, W.; Kopelman, R. In *Fluorescence Imaging Spectroscopy and Microscopy*; Wang, X. F., Herman, B., Eds.; Wiley: New York, 1996; pp407-475.
10. Kopelman, R. & Tan, W. *Science* 262, 1382-1384 (1993)
11. Michaells J.; Hettich, C.; Mlynek, J. & Sandoghdar V., *Nature*, 405, 325-327 (2000)
12. Aida, T.; Sata, T.; Jiang, D.L., *J. Am. Chem. Soc.* 121, 10658, 1999
13. Swallen, S.F.; Kopelman, R.; Moore, J.S.; Devadoss C., Journal of Molecular Structure, 485-486 (1999) 585-597
14. Swallen, S.F.; Xu, Z.G.; Moore, J. S.; Kopelman, R, *J. Phys. Chem B*, Vol 104, No.16, 2000
15. Lou, J. F.; Hatton, T. A.; Laibinis, P. E., *Anal. Chem.*, 1997, 69, 1262-1264
16, Lou, J. F.; Finegan, T. M.; Mohsen, P.; Hatton, T. A. & Laibinis, P. E. *Reviews in Analytical Chemistry, Vol. 18, No. 4, 1999*

Mat. Res. Soc. Symp. Proc. Vol. 651 © 2001 Materials Research Society

Ordered Morphologies of Confined Diblock Copolymers

Yoav Tsori and David Andelman
School of Physics and Astronomy
Raymond and Beverly Sackler Faculty of Exact Sciences
Tel Aviv University, 69978 Ramat Aviv, Israel

ABSTRACT

We investigate the ordered morphologies occurring in thin-films diblock copolymer. For temperatures above the order-disorder transition and for an arbitrary two-dimensional surface pattern, we use a Ginzburg-Landau expansion of the free energy to obtain a linear response description of the copolymer melt. The ordering in the directions perpendicular and parallel to the surface are coupled. Three dimensional structures existing when a melt is confined between two surfaces are examined. Below the order-disorder transition we find tilted lamellar phases in the presence of striped surface fields.

INTRODUCTION

The self-assembly of block copolymers (BCP) has been the subject of numerous studies [1]-[12]. These macromolecules are made up of chemically distinct subunits, or blocks, linked together with a covalent bond. This bonding inhibits macrophase separation, and leads to formation of mesophases with typical size ranging from nanometers up to microns. For di-block copolymers, which are made up of two partially incompatible blocks (A and B), the phase diagram is well understood [1]-[6], and exhibits disordered, lamellar, hexagonal and cubic micro-phases. The phase behavior is governed by three parameters: the chain length $N = N_A + N_B$, the fraction $f = N_A/N$ of the A block, and the Flory parameter χ, being inversely proportional to the temperature.

The presence of a confining surface leads to various interesting phenomena [7]-[17]. The surface limits the number of accessible chain configurations and thus may lead to chain frustration. Transitions between perpendicular and parallel lamellar phases with respect to the confining surfaces have been observed in thin films [18]-[22]. In addition, the surface may be chemically active, preferring adsorption of one of the two blocks, and usually stabilizing the formation of lamellae parallel to the surface [12, 19, 21].

More complicated behavior occurs when the surface is chemically heterogeneous; namely, one surface region prefers one block while other regions prefer the second block. Compared to bulk systems, new energy and length scales enter the problem adding to its complexity. Thin-film BCP in presence of chemically patterned surfaces is of importance in many applications, such as dielectric mirrors and waveguides [23], anti-reflection coating for optical surfaces [24] and fabrication of nanolithographic templates [25].

THE MODEL

The copolymer order parameter ϕ is defined as $\phi(\mathbf{r}) \equiv \phi_A(\mathbf{r}) - f$, the deviation of the local A monomer concentration from its average f. Consider a symmetric ($f = 1/2$) BCP melt in its disordered phase (above the bulk ODT temperature) and confined by one or two flat, chemically patterned surfaces. The free energy (in units of the thermal energy $k_B T$) can be written as [3, 4, 5, 26, 27]:

$$F = \int \left\{ \frac{1}{2}\tau\phi^2 + \frac{1}{2}h \left[\left(\nabla^2 + q_0^2\right)\phi\right]^2 + \frac{u}{4!}\phi^4 - \mu\phi\right\} d^3\mathbf{r} \qquad (1)$$

The polymer radius of gyration R_g sets the periodicity scale $d_0 = 2\pi/q_0$ via the relation $q_0 \simeq 1.95/R_g$. The chemical potential is μ and the other two parameters are $h = 1.5\rho c^2 R_g^2/q_0^2$ and $\tau = 2\rho N \left(\chi_c - \chi\right)$. The Flory parameter χ measures the distance from the ODT point ($\tau = 0$), having the value $\chi_c \simeq 10.49/N$. For positive $\tau \sim \chi_c - \chi$, the system is in the disordered phase having $\phi = 0$, while for $\tau < 0$ (and $f = 1/2$) the lamellar phase has the lowest energy. Finally, $\rho = 1/Na^3$ is the chain density per unit volume, and c and u/ρ are constants of order unity [4]. This Ginzburg-Landau expansion of the free energy in powers of ϕ and its derivatives can be justified near the critical point, where $\tau \ll hq_0^4$ and ordering is weak. The lamellar phase can approximately be described by a single q-mode there: $\phi = \phi_q \cos(\mathbf{q_0} \cdot \mathbf{r})$. We note that similar types of energy functionals have been used to describe bulk and surface phenomena [28, 29, 30], amphiphilic systems [31, 32], Langmuir films [33] and magnetic (garnet) films [34].

We model the chemically heterogeneous surfaces by a short-range surface interaction, where the BCP concentration at the surface is coupled linearly to a surface field $\sigma(\mathbf{r_s})$:

$$F_s = \int \sigma(\mathbf{r_s})\phi(\mathbf{r_s})d^2\mathbf{r_s} \qquad (2)$$

The vector $\mathbf{r} = \mathbf{r_s}$ defines the position of the confining surfaces. Preferential adsorption of the A block is modeled by a $\sigma < 0$ surface field, and a constant σ results in a parallel-oriented lamellar layers (a perpendicular orientation of the chains). Without any special treatment, the surface tends to prefer one of the blocks, but by using random copolymers [16, 17] one can reduce this affinity or even cancel it altogether. A surface with spatially modulated pattern, $\sigma(\mathbf{r_s}) \neq 0$, induces preferential adsorption of A and B blocks to different regions of the surface.

RESULTS AND DISCUSSION

We consider first a system of polymer melt in contact with a single flat surface. The system is assumed to be above the ODT temperature, in the disordered bulk phase. The results are then extended to two confining surfaces and to BCP systems below the ODT.

Above ODT

Above the ODT the bulk phase is disordered, and the free energy, Eq. (1), is convex to second order in the order parameter ϕ. Thus, the ϕ^4 term can be neglected. The melt is confined to the semi-infinite space $y > 0$, bounded by the $\mathbf{r_s} = (x, y = 0, z)$ surface. The chemical pattern $\sigma(\mathbf{r_s}) = \sigma(x, z)$ can be decomposed in terms of its inplane q-modes $\sigma(x, z) = \sum_{\mathbf{q}} \sigma_{\mathbf{q}} \exp\left(i\left(q_x x + q_z z\right)\right)$, where $\mathbf{q} \equiv (q_x, q_z)$, and $\sigma_{\mathbf{q}}$ is the mode amplitude. Similarly, the order parameter is $\phi(x, y, z) = \sum_{\mathbf{q}} \phi_{\mathbf{q}}(y) \exp\left(i\left(q_x x + q_z z\right)\right)$. Substituting ϕ in Eq. (1), and applying a variational principle with respect to $\phi_{\mathbf{q}}$, results in a linear fourth-order differential equation [35, 36]:

$$\left(\tau/h + \left(q^2 - q_0^2\right)^2\right)\phi_{\mathbf{q}}(y) + 2(q_0^2 - q^2)\phi_{\mathbf{q}}''(y) + \phi_{\mathbf{q}}''''(y) = 0 \tag{3}$$

In the semi-infinite geometry, $y > 0$, the solution to Eq. (3) has an exponential form $\phi_{\mathbf{q}}(y) = A_{\mathbf{q}} \exp(-k_{\mathbf{q}} y) + B_{\mathbf{q}} \exp(-k_{\mathbf{q}}^* y)$, where $k_{\mathbf{q}}$ is given by

$$
\begin{aligned}
k_{\mathbf{q}}^2 &= q^2 - q_0^2 + i\sqrt{\tau/h} \\
&= q^2 - q_0^2 + i\alpha\left(N\chi_c - N\chi\right)^{1/2}
\end{aligned}
\tag{4}
$$

with $\alpha \simeq 0.59 q_0^2$. The values of $\xi_q = 1/\mathrm{Re}(k_{\mathbf{q}})$ and $\lambda_q = 1/\mathrm{Im}(k_{\mathbf{q}})$ correspond to the exponential decay and oscillation lengths of the surface q-modes, respectively. For fixed χ, ξ_q decreases and λ_q increases with increasing q [20, 35, 36]. Close to the ODT (but within the range of validity of the model), and for q-modes such that $q > q_0$ we find finite ξ_q and $\lambda_q \sim (\chi_c - \chi)^{-1/2}$. The propagation of the surface imprint (pattern) of q-modes with $q < q_0$ into the bulk can persist to long distances, in contrast to surface patterns with $q > q_0$ which decays off close to the surface. This is seen by noting that q-modes in the band $0 < q < q_0$ are equally "active", i.e., these modes decay to zero very slowly in the vicinity of the ODT as $y \to \infty$: $\xi_q \sim (\chi_c - \chi)^{-1/2}$ and λ_q is finite.

The boundary conditions for $\phi_{\mathbf{q}}$ at $y = 0$ are

$$
\begin{aligned}
\phi_{\mathbf{q}}''(0) + \left(q_0^2 - q^2\right)\phi_{\mathbf{q}}(0) &= 0 \\
\sigma_{\mathbf{q}}/h + \left(q_0^2 - q^2\right)\phi_{\mathbf{q}}'(0) + \phi_{\mathbf{q}}'''(0) &= 0
\end{aligned}
\tag{5}
$$

The amplitude A_q is found to be $A_q = -\sigma_q \left(2\mathrm{Im}(k_q)\sqrt{\tau h}\right)^{-1}$. Thus, the copolymer response diverges upon approaching the critical point as $(N\chi_c - N\chi)^{-1/2}$.

Although our analysis allows for arbitrary surface patterns, let us consider first the simple case of a surface attractive for the A block, but which has a single localized stripe, or surface "disturbance", preferentially attracting the B block (see Fig. 1 (a)). Far from this stripe A-blocks are adsorbed on the surface and the copolymer concentration profile has decaying lamellar order as was explained above. However, a non-trivial BCP morphology appears close to the stripe, reflecting the adsorption of B monomers. This behavior is indeed seen in Fig. 1 (b), where the grey scale is such that A-rich regions are white, while B-rich are black. A somewhat different situation is shown in Fig. 2 (a), where the black stripe is still preferential to the B monomers, but the rest of the surface (in grey) is neutral, $\sigma(x, z) = 0$. In this case no lamellar ordering parallel to the surface is expected. In (b) curved lamellae appear around the surface disturbance, optimizing interfacial energy. These curved lamellae are exponentially damped both in directions parallel and perpendicular to the surface. Clearly, the ordering in the perpendicular and lateral directions are coupled.

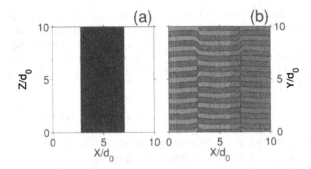

Figure 1. *A BCP melt confined by one patterned surface having a central stripe of width $5d_0$. In (a) the surface located at $y = 0$ has a localized "surface disturbance" of $\sigma = 1/2$ inside the black stripe, favoring the adsorption of the B monomers, and $\sigma = -1/2$ outside of the stripe (favoring A monomers). The morphology in the x-y plane is shown in (b), where farther away from the stripe (large $y \gg d_0$) the lamellae have a decaying order, while close to the surface ($y \lesssim d_0$) the lamellae are distorted to optimize the interfacial energy. The system is above the ODT with $\chi N = 10$. Lengths are scaled by d_0, the lamellar periodicity. B-rich regions are mapped to black shades and A-rich to white. In all plots we set the monomer length as $a = 1$, and choose in Eq. (1) $c = u/\rho = 1$, $R_g^2 = \frac{1}{6}Na^2$ and $N = 1000$ to give $\alpha \simeq 0.59q_0^2$.*

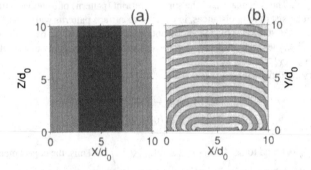

Figure 2. *Same as in Fig. 1, but here the surface area outside of the stripe is neutral to polymer adsorption, $\sigma = 0$. The B monomers are found close to the stripe, inducing a curved lamellar structure that decays away from the surface.*

A different scenario is presented in Fig. 3, where the surface has a sinusoidal variation in its affinity for the monomer type: $\sigma(x) = \sigma_0 + \sigma_q \cos(qx)$, with periodicity $d = 2\pi/q$ and average attraction $\langle \sigma \rangle = \sigma_0 > 0$ for the B monomers. The modulated surface field of amplitude σ_q induces lateral order, with A- and B-rich regions near the surface. Near the top of the figure ($y \gg d_0$), only lamellar-like ordering is seen, because the σ_q ($q > 0$) term decays faster than the σ_0 ($q = 0$) term.

Using this formulation, any two-dimensional chemical pattern $\sigma(x, z)$ can be modeled. For surface feature of size larger than d_0, the characteristic copolymer length, the chemical surface pattern can propagate via the BCP melt into the bulk. To demonstrate this we take in Fig. 4 (a)

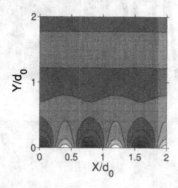

Figure 3. *Copolymer morphology for a melt confined by one surface at $y = 0$ having a chemical affinity of the form $\sigma(x) = \sigma_0 + \sigma_q \cos(q_x x)$, with $q_x = \frac{5}{4}q_0$ and $\sigma_0 = \sigma_q = 0.1$. The lamellar-like order due to the σ_0 term persists farther away from the wall than the lateral order due to the σ_q term. The Flory parameter is $\chi N = 10.4$.*

a surface pattern in the shape of the letters 'MRS'. Inside the letters $\sigma > 0$ (B monomers are attracted), while for the rest of the surface $\sigma = 0$ (neutral). The resulting patterns in the x-z planes are shown for $y/d_0 = 0.5$, 2 and 3.5 in (b), (c) and (d), respectively. An overall blurring of the image is seen as the distance from the surface is increased. The fine details (e.g. sharp corners of the letters) disappear first as a consequence of the fast decay of high surface q-modes. The second point to notice is the A↔B interchange of monomers that occurs for surfaces separated roughly by a distance of $(n + \frac{1}{2})d_0$, for integer n. This monomer interchange mimics the formation of lamellae in the bulk, although the temperature here is above the ODT. In addition, the appearance of lateral order is clearly seen, as lamellae form parallel to the edges of the letters.

We briefly mention the case where a BCP melt is confined between two flat parallel surfaces located at $y = \pm L$. The calculation of the response functions $\{\phi_q\}$ can easily be generalized to handle two confining surfaces by including the appropriate boundary conditions in Eqs. (5). If the distance $2L$ is very large, the copolymer orderings induced by the two surfaces are not coupled, and the middle of the film ($|y| \ll L$) is disordered, $\phi \approx 0$. Decreasing the film thickness to a distance comparable to the copolymer correlation length results in an overlap of the two surface fields.

Complex three dimensional morphologies can also be achieved by using only one dimensional surface patterns, if the two patterns are rotated with respect to each other. Such an example is shown in Fig. 5, for two surfaces at $y = \pm L = \pm d_0$ with perpendicularly oriented stripes. The surface patterns are shown in (a) and (c), while the mid-plane checkerboard morphology ($y = 0$) is shown in (b). In the next subsection we will show results for the more complicated situation of BCP melt below the ODT temperature, where the bulk phase is lamellar.

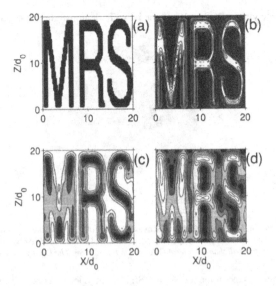

Figure 4. *Propagation of surface order into the bulk BCP melt. (a) The surface at $y = 0$ is taken to have a pattern in the shape of the letters 'MRS'. Inside the letters ($\sigma = 1$) the field attracts the B monomers, while the rest of the surface is neutral ($\sigma = 0$). In (b), (c) and (d) the morphology is calculated for x-z planes located at increasing distances from the surface, $y/d_0 = 0.5$, 2 and 3.5, respectively. Note that in (b) and (d) black (white) shades can be mapped into white (black) shades in (c), because their y-spacing is a half-integer number of lamellae. The Flory parameter is $\chi N = 9.5$.*

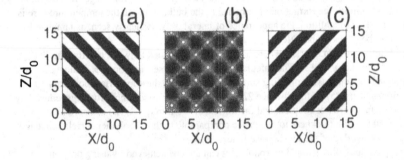

Figure 5. *Creation of three dimensional structure in a thin BCP film as a result of two one-dimensional surface patterns. (a) and (c) are stripe surfaces located at $y = -L = -d_0$ and $y = L = d_0$, respectively. The mid-plane ($y = 0$) ordering is shown in (b) and is created by a superposition of the two surface fields. $\chi N = 9$.*

Below ODT

The prevailing bulk phase below the ODT has an inherent lamellar ordering. This order interferes strongly with the surface induced order, and it is not a simple task to obtain order-parameter expressions as a function of arbitrary surface pattern. Mathematically, the difficulty lies in the fact that the ϕ^4 term in the free energy cannot be neglected. As in the case above the ODT, we expand the free energy to second order around the bulk phase, only this time the bulk phase is lamellar. As we will see, this approach has also been used to describe defects in bulk phases such as chevron and omega-shaped tilt grain boundaries [26, 27]. We consider first a BCP melt confined by one stripe surface whose one dimensional pattern is of the form:

$$\sigma(x, z) = \sigma_q \cos(q_x x) \tag{6}$$

The surface periodicity $d_x = 2\pi/q_x$ is assumed to be larger than the bulk lamellar spacing $d_0 = 2\pi/q_0$, $d_x > d_0$ throughout the analysis. The system reduces its interfacial energy by trying to follow the surface modulations, and an overall tilt of the lamellae follows regardless of the fine details of chain conformations near the surface. The tilt angle with respect to the surface is defined as $\theta = \arcsin(d_0/d_x)$ [20]. Consequently, we use the single q-mode approximation to describe the bulk phase ϕ_b

$$\phi_b(x, y) = -\phi_q \cos(q_x x + q_y y) \tag{7}$$

where $q_y \equiv q_0 \sin(\theta)$ and $q_x = q_0 \cos(\theta)$. All surface effects are contained in the correction to the order parameter: $\delta\phi(\mathbf{r}) = \phi(\mathbf{r}) - \phi_b(\mathbf{r})$. We choose the in-phase one harmonic form for $\delta\phi$

$$\delta\phi(x, y) = g(y) \cos(q_x x) \tag{8}$$

with a y-dependent amplitude $g(y)$. This correction $\delta\phi$ is expected to vanish far from the surface, $\lim_{y\to\infty} \delta\phi = 0$, recovering the bulk phase. The free energy in Eq. (1) is expanded to second order in the small correction $\delta\phi$, $\phi_b \gg \delta\phi$. We integrate out the x dependence and retain only the y dependence. Then, use of a variational principle with respect to the function $g(y)$ yields a linear differential equation:

$$[A - C \cos(2q_y y)] g + Bg'' + g'''' = 0 \tag{9}$$

where A, B, and C are parameters depending on the tilt angle θ as well as on the temperature [26]. This equation is similar to the Mathieu equation describing an electron under the influence of a periodic one-dimensional potential, and has a solution in the form of a Bloch wave function.

The main effect of the surface stripes on the melt is to introduce a tilt of the lamellae. This effect is seen in Fig. 6, where the surface periodicity d_x (in units of d_0) is chosen to be 1 in (a), and induces a perpendicular ordering. Larger ratios of d_x/d_0 cause a tilt ordering. In (b) $d_x/d_0 = 3/2$ and in (c) $d_x/d_0 = 3$, and the deviation from a perfect lamellar shape can be seen near the surface.

Tilt Grain Boundaries

Our approach can be used to describe defects in bulk systems as well. A tilt grain boundary forms when two lamellar bulk grains meet with a tilt angle φ between the lamellae normals. For symmetric tilt grain boundaries, the plane of symmetry is analogous to the patterned

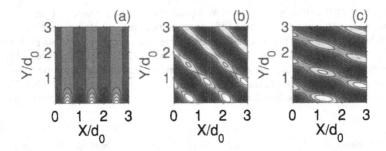

Figure 6. *BCP melt confined by one stripe surface of periodicity $d_x > d_0$ below the ODT. The surface stripes are described by $\sigma(x) = \sigma_q \cos(q_x x)$. Tilted lamellae with respect to the surface at $y = 0$ are formed and adjust to the surface imposed periodicity. Far from the surface the lamellae relax to their undistorted d_0 spacing. The mismatch ratio d_x/d_0 is 1 in (a) yielding no tilt, $3/2$ in (b) and 3 in (c). The Flory parameter is taken as $\chi N = 10.6$ and $\sigma_q = 0.08$.*

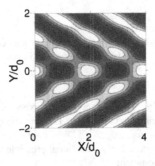

Figure 7. *Symmetric tilt grain boundary between two bulk lamellar grains. The angle between lamellae is determined by external boundary conditions at $y \to \infty$ and is chosen here to be $\varphi = 120°$. The morphology is invariant in the z direction and $\chi N = 11$.*

surface in a thin BCP film, and similar form of a correction field $\delta\phi$ can be used, see Fig. 7. The upper half plane $y > 0$ thus corresponds to Fig. 6 (b) and (c).

The correction is small in the so-called chevron region (small tilt angle φ), but becomes important for larger tilt angles [26] where the grain boundary has the form of the letter Omega. Close to the ODT point the modulations at the interface become long range and persist into the bulk up to large distances. One of the differences between tilt grain boundaries and tilt induced by surface field is that in the former case there are no real surface fields. The tilt angle between adjacent grains is determined by constraints imposed far from the $y = 0$ interface.

CONCLUSIONS

An analytical expansion of the free energy, Eq. (1), can be used to obtain order parameter expressions for a BCP melt confined by one or two flat surfaces whose pattern has arbitrary (two dimensional) shape. The thin film morphology can also have a complex three dimensional form even if the two surfaces have one dimensional patterns with different orientations with respect to each other, see Fig. 5. A BCP melt close to and above the ODT point shows decaying oscillations of the local concentration towards the bulk disorder value. Our analysis shows that close to the ODT large surface spatial features can be transferred into the BCP bulk far from the surface, while small surface details are greatly damped.

Below the ODT, confinement of the melt by patterned surfaces with periodicity $d_x > d_0$ leads to a formation of a surface layer characterized by lateral periodicity of d_x, and the lamellae relax to their natural periodicity d_0 farther from the surface. The proposed mechanism is a tilting of the lamellae. We show that the resulting pattern have similar characteristics to bulk domain walls (tilt grain boundaries).

ACKNOWLEDGMENTS

We would like to thank M. Muthukumar, R. Netz, G. Reiter, T. Russell, M. Schick and U. Steiner for useful comments and discussions. Partial support from the U.S.-Israel Binational Foundation (B.S.F.) under grant No. 98-00429 and the Israel Science Foundation funded by the Israel Academy of Sciences and Humanities — centers of excellence program is gratefully acknowledged.

References

[1] F. S. Bates and G. H. Fredrickson, *Annu. Rev. Phys. Chem.* **41**, 525 (1990).

[2] K. Ohta and K. Kawasaki, *Macromolecules* **19**, 2621 (1986).

[3] L. Leibler, *Macromolecules* **13**, 1602 (1980).

[4] K. Binder, H. L. Frisch and S. Stepanow, *J. Phys. II* **7**, 1353 (1997).

[5] G.H. Fredrickson and E. Helfand, *J. Chem. Phys.* **87**, 697 (1987).

[6] M. W. Matsen and M. Schick, *Phys. Rev. Lett.* **72**, 2660 (1994); M. W. Matsen and F. Bates, *Macromolecules* **29**, 7641 (1996).

[7] G. H. Fredrickson, *Macromolecules* **20**, 2535 (1987).

[8] H. Tang and K. F. Freed, *J. Chem. Phys.* **97**, 4496 (1992).

[9] K. R. Shull, *Macromolecules* **25**, 2122 (1992).

[10] M. S. Turner, *Phys. Rev. Lett.* **69**, 1788 (1992).

[11] M. S. Turner, M. R. Rubinstein and C. M. Marques, *Macromolecules* **27**, 4986 (1994).

[12] M. W. Matsen, *J. Chem. Phys.* **106**, 7781 (1997).

[13] S. H. Anastasiadis, T. P. Russell, S. K. Satija and C. F. Majkrzak, *Phys. Rev. Lett.* **62**, 1852 (1989).

[14] A. Menelle, T. P. Russell, S. H. Anastasiadis, S. K. Satija and C. F. Majkrzak, *Phys. Rev. Lett.* **68**, 67 (1992).

[15] D. G. Walton, G. J. Kellogg, A. M. Mayes, P. Lambooy and T. P. Russell, *Macromolecules* **27**, 6225 (1994).

[16] G. J. Kellogg, D. G. Walton, A. M. Mayes, P. Lambooy, T. P. Russell, P. D. Gallagher and S. K. Satija, *Phys. Rev. Lett.* **76**, 2503 (1996).

[17] P. Mansky, T. P. Russell, C. J. Hawker, J. Mayes, D. C. Cook and S. K. Satija, *Phys. Rev. Lett.* **79**, 237 (1997).

[18] S. T. Milner and D. C. Morse, *Phys. Rev. E* **54**, 3793 (1996).

[19] G. T. Pickett and A. C. Balazs, *Macromolecules* **30**, 3097 (1997).

[20] D. Petera and M. Muthukumar, *J. Chem. Phys.* **107**, 9640 (1997); D. Petera and M. Muthukumar, *J. Chem. Phys.* **109**, 5101 (1998).

[21] T. Geisinger, M. Mueller and K. Binder, *J. Chem. Phys.* **111**, 5241 (2000).

[22] G. G. Pereira and D. R. M. Williams, *Phys. Rev. E* **60**, 5841 (1999); G. G. Pereira and D. R. M. Williams, *Phys. Rev. Lett.* **80**, 2849 (1998).

[23] Y. Fink, J. N. Winn, S. Fan, C. Chen, J. Michael, J. D. Joannopoulos and E. L. Thomas, *Science* **282**, 1679 (1998).

[24] S. Walheim, E. Schäffer, J. Mlynek and U. Steiner, *Science* **283**, 520 (1999).

[25] M. Park, C. Harrison, P. M. Chaikin, R. A. Register and D. H. Adamson, *Science* **276**, 5317 (1997).

[26] Y. Tsori, D. Andelman and M. Schick, *Phys. Rev. E.* **61**, 2848 (2000).

[27] R. R. Netz, D. Andelman and M. Schick, *Phys. Rev. Lett.* **79**, 1058 (1997); S. Villain-Guillot, R. R. Netz, D. Andelman and M. Schick, *Physica A* **249**, 285 (1998); S. Villain-Guillot and D. Andelman, *Eur. Phys. J. B* **4**, 95 (1998).

[28] J. Swift and P. C. Hohenberg, *Phys. Rev. A* **15**, 319 (1977).

[29] M. Seul and D. Andelman, *Science* **267**, 476 (1995).

[30] A. E. Jacobs, D. Mukamel and D. W. Allender, *Phys. Rev. E* **61**, 2753 (2000).

[31] G. Gompper and M. Schick, *Phys. Rev. Lett.* **65**, 1116 (1990); F. Schmid and M. Schick, *Phys. Rev. E* **48**, 1882 (1993).

[32] G. Gompper and S. Zschocke, *Phys. Rev. A* **46**, 1836 (1992).

[33] D. Andelman, F. Brochard and J.-F. Joanny, *J. Chem. Phys.* **86**, 3673 (1987).

[34] T. Garel and S. Doniach, *Phys. Rev. B* **26**, 325 (1982).

[35] Y. Tsori and D. Andelman, *Macromolecules*, in press (2001).

[36] Y. Tsori and D. Andelman, *Europhys. Lett.*, to be published (2001).

[37] Y. Tsori and D. Andelman, *unpublished*.

Mat. Res. Soc. Symp. Proc. Vol. 651 © 2001 Materials Research Society

Anomalous Surface Conformation for Polymeric Gas-Hydrate-Crystal Inhibitors

H.E. King, Jr.; Jeffrey L. Hutter[1]; Min Y. Lin[2]; and Thomas Sun[3]
ExxonMobil Research and Engineering Company
Annandale, NJ 08801
[1] Present address: Department of Physics and Astronomy, University of Western Ontario,
London, ON N6A 3K7
[2] Also at the National Institute of Standards and Technology, Gaithersburg, MD 20899.
[3] Present address: Exxon Chemical Company, Baytown, TX 77522.

ABSTRACT

We have used both conventional small-angle neutron scattering and contrast variation techniques to characterize the polymer conformations of two non-ionic water soluble polymers: poly(ethylene oxide) and poly(N-vinyl-2-pyrollidone). The second of these is able to kinetically suppress hydrate crystallization, and an objective of these studies is to obtain a understanding of this inhibition mechanism. The dilute-solution polymer conformation in a hydrate-forming tetrahydrofuran/water fluid shows only a small difference between the polymers. The single-chain characteristics are unperturbed, but the hydrate inhibitor polymer seems to show an enhanced tendency to form aggregates in solution. This is evidenced by excess low-q scattering following a $q^{-2.5}$ power law. Much more evident is the strong perturbation in the conformation of the inhibitor polymer upon crystallization of the hydrate. We utilize contrast variation methods to resolve the scattering of the polymer on the hydrate surface. Unlike expectations from polymer scaling laws, the resulting layer is considerably thicker (550Å) than the single-chain radius of gyration, 80Å. Also surprising, only 2 percent of the available crystal surface is covered. Within the covered areas, the polymer concentration is significantly enhanced over that in the surrounding solution, about $2.5c^*$ (where c^* is the overlap concentration). This suggests a coverage of clumps of polymer widely separated from one another. Consideration of the concentration and implied spacing of such clumps on the hydrate surface suggests that they can effectively inhibit crystal growth.

INTRODUCTION

Gas hydrates are ice-like crystals consisting of water cages surrounding small molecules such as propane or methane.[1] Enclathration of the small molecules is key to the crystals' stability, and depending upon the gas pressure their melting point can extend to temperatures well above that of ice. Because of this, they crystallize in many settings where ordinary ice is not stable. For example there are many natural-gas hydrate deposits on the sea floor associated with gas seepage into cold ocean water[2]. Similar conditions can be found in oil and gas transport pipelines. Under deep-sea conditions, inside such pipelines one finds a ready supply of water and natural gas at temperatures of ~5°C and pressures of ~100 MPa. Thus, hydrate formation within pipelines is a significant long-standing problem for the oil industry.[3] The consequences of hydrate formation can be very serious. Blockage of the pipeline, with consequent loss of production is one scenario. Such consequences have led to treatments to prevent hydrate formation. A common approach is to add an antifreeze compound, typically methanol, to suppress the formation temperature.[4] Although effective, this approach has drawbacks such as the requirement for high volume fractions of methanol (up to 50 volume percent) with the potential for environmental impact due to accidental spillage. Also, there are considerable costs associated with supply and recovery of the methanol.

As with many other crystallization problems, there is the possibility of kinetic inhibition rather than thermodynamic suppression. Recently it has been found that certain water-soluble polymers can kinetically inhibit hydrate formation.[5] Experiments show that at the low concentrations of use, less than one weight percent, there is little or no shift in the equilibrium hydrate formation point. Yet they inhibit hydrate formation by several degrees Celsius during continuous cooling and can suppress crystallization for long time periods at low levels of subcooling.

Despite these technical achievements, the mechanism of inhibition remains unclear.[6-8] From the time of their discovery, there has been an ongoing debate as to whether the polymers act on the surface of an already existing crystal or whether they associate in some manner with hydrate-forming constituents while still in solution. In the present work we show that polymeric hydrate inhibitors have conformations in solution very similar to that for non-inhibitor polymers. However, we also show that upon hydrate crystal formation a significant change in conformation occurs for the hydrate inhibitors but not the non-inhibitor polymers. Through use of contrast-variation neutron scattering this conformational change is shown to occur due to polymer adsorption on the hydrate crystal surfaces. This polymer layer is characterized by having limited coverage of the surface, but with a thickness several times the characteristic size of a single chain, the polymer radius of gyration. The coverage of these polymer clumps on the surface are consistent with a growth-arrestor functionality for the inhibitor polymer.

EXPERIMENTAL

Sample Preparation

For this work we studied solutions containing D_2O, H_2O, deutrated-tetrahydrofuran(TDF), and one of two water-soluble polymers: poly(ethylene oxide), PEO, a non-inhibitor and poly(N-vinyl-2-pyrollidone), PVP, an inhibitor. The Structure II hydrate crystal studied has a cubic lattice with edge length $a=17.24$ Å and an idealized formula $X \cdot 17H_2O$, where X is the enclathrated molecule, TDF. The influence of D_2O on the hydrate phase diagram has been measured by Hanley et al.[9] and is discussed in Hutter et al.[10] In the experiments two types of samples were studied. For studies of the solution behavior we cooled our samples to 7°C, several degrees above the hydrate formation temperature. For studies with hydrate crystals present, we cooled our samples to about 3°C with continuous stirring to generate uniform crystal sizes. All samples had a 1:25 mole ratio of TDF to water which gives an approximately 50:50 crystal:liquid mixture at this temperature. The polymers, see Table I for their characteristics, were added to this solvent at approximately 0.5 weight percent.

Table I: Polymers Used In This Study

Polymer	Monomer Weight (g/mol)	Molecular Weight (g/mol)	M_W/M_N	A_2 (ml mol/g^2) [a]	Concentration (g/cc)	Concentration (volume fraction)
polyethylene oxide	44.0532	46,000	1.1	2.15×10^{-3}	0.00553	0.00491
poly(N-vinyl-2-pyrollidone)	111.1436	49,000	3.2	8.40×10^{-4}	0.00554	0.00438

[a] Values calculated utilizing data as follows: PEO[11]; PVP[12]

Data Collection and Reduction Procedure

The small angle neutron scattering measurements were performed at the NIST Center for Neutron Research, in Gaithersburg, MD. Two SANS instruments, NG3 and NG7, which are almost identical, were utilized. We chose a suitable range of momentum transfer of $0.003 < q < 0.2$ Å within the instrument capabilities, where $q = 4\pi \sin(\theta/2)/\lambda$ (θ is the scattering angle and λ the neutron wavelength).

Contrast Calculations and Measurement

The measured scattering, $d\Sigma(q)/d\Omega$, can be expressed as the sum of terms, each of which is the product of a structure factor that describes a particular structure in the sample and the contrast corresponding to that term. We determine the contrasts from the scattering length densities n_α of each component α in the system. In general n_α is calculated as the total molecular scattering length of component α divided by its volume V_α:

$$n_\alpha = \frac{1}{V_\alpha} \sum_{molecules} b_i , \qquad (1)$$

where b_i is the scattering length of the ith molecule. Thus, knowing the composition and density of each component, we can calculate its scattering length density. We direct the reader to King et al.[13] for a discussion of the values used in this study. The contrast factors then depend on the differences in scattering length densities between components.

For the contrast-variation experiments several D_2O/H_2O mixtures were used. In this case, the relative scattering lengths are of paramount importance and these are most accurately determined experimentally. The process for this is described in Hutter et al.[10]

RESULTS

Polymer Solutions above the Hydrate Formation Temperature

For our two polymers SANS data were taken in TDF/D$_2$O solutions at a temperature above the hydrate formation point, 7 °C. The concentrations utilized are given in Table I. These concentrations are well below the overlap concentration, c^*; therefore these data will exhibit single-chain scattering. We model the overall scattering as

$$I(q) = (n_p - n_s)^2 S_{Beaucage}(q) + Aq^{-2.5} + b,$$ (2)

The first term, $S_{Beaucage}(q)$, is a Debye-like structure factor for single-chain scattering,

$$S_{Beaucage}(q) = G[\exp(-q^2 R_g^2/3) + d_f \Gamma(d_f/2)/(R_g q^*)^{d_f}]$$ (3)

due to Beaucage.[14; 15] Here, $q^* = q/[\text{erf}(kqR_g/\sqrt{6})]^3$ with $k \approx 1.06$, and $\Gamma(n)$ is the gamma function. The three fitting parameters in this structure factor allow us to extract the single-chain characteristics: polymer radius of gyration R_g, fractal dimension of the polymer coils d_f, and the Guinier prefactor G. The second term in Eq. 2 describes a low-q feature with A, an amplitude to be determined from the fits. The last term, b, is a q-independent background intensity.

For both PEO[11] and PVP,[12] previous room temperature studies in pure water have shown that these polymers exhibit "good-solvent" behavior. Therefore if "good-solvent" conformations persist under the conditions introduced here, we can conclude that the hydrate forming constituents within the solution have not significantly altered the polymer's conformation. One test of this is that the fractal dimension should equal the theoretical value for a linear chain in good solvent, 5/3. For PEO we obtain $d_f = 1.9$ and for PVP $d_f = 1.8$, both in reasonable agreement with the ideal value. Another test is whether the polymer chain is expanded to its good-solvent radius of gyration. From the good-solvent molecular-weight scaling relationships for PEO and PVP (e.g. $R_g = KM_W^m$, where K and m are empirically determined numeric values, with $m \approx 1/d_f$), we can estimate the radius of gyration from the experimentally-determined molecular weights in Table I as, respectively, $R_g = 112$ Å and $R_g = 71$ Å. Again these predicted values are in reasonable accord with the observed values, PEO $R_g = 80$ Å and PVP $R_g = 80$ Å.

The low-q rise in scattering is indicative of large structures in the solutions. Such features are a common observation for polymers in aqueous solution, and there is still debate as to whether they are characteristic of such systems[16] or merely the consequence of contaminants. We have chosen to model these aggregates with a $q^{-2.5}$ power law form, which would imply a fractal structure more compact than that of the polymer chains themselves and without a sharp interface with the surrounding media. This is consistent with the results from Sun and King.[12] The resulting amplitudes are PEO, A=1.3 x10^{13} cm^{-1}; and PVP, A=4.8x10^{13} cm^{-1}. It's interesting that the amplitude for these aggregates is larger for the inhibitor polymer than for PEO. In light of these results, it is interesting to note that in the propane/D$_2$O system[13], we find an enhancement in this low-q scattering beyond that seen in the TDF/D$_2$O system. Thus, supramolecular association may be an important feature of hydrate inhibitor polymers, but the single-chain conformations are little affected by the presence of hydrate-forming constituents.

Polymer Solutions Below Hydrate Formation Temperature, Slurry Scattering

PEO

In Fig. 1a, we show the SANS data for the PEO sample in the slurry state. As expected, the low-q scattering is dominated by Porod scattering from the crystal-liquid interfaces. At higher q-values, the polymer scattering is dominant. If the polymer does not interact with the hydrate crystals, we expect the scattering to consist of a linear combination of Porod and polymer solution scattering. The dashed line in Figure 1a is the result of addition of a Porod term to the (scaled) fit from the solution scattering at 7°C. The Porod amplitude as well as the overall background were allowed to vary, but the solution scattering was simply scaled for the increase in concentration due to exclusion of the polymer into the remaining 0.49 fraction of the sample[10] which is still liquid. The resulting scaling constant of 0.64, comes from the influence of the Second Virial Coefficient, A_2, as shown in Eq. 8 of King et al.[13] . This simple model closely approximates the observed scattering. As a further test of its robustness, we allowed the Porod and the polymer scattering variables to simultaneously vary, giving the best-fit result shown by the solid line in Fig. 1a. The most notable change in the polymer scattering is a shift to a smaller radius of gyration, 62 Å. This is essentially identical to the effect observed by experimentally doubling the solution concentration. We conclude from these results that the PEO in coexistence with the hydrate crystals has the same conformation as a polymer solution at double (0.0113 g/cc) the starting concentration. There is no evidence of an adsorbed layer.

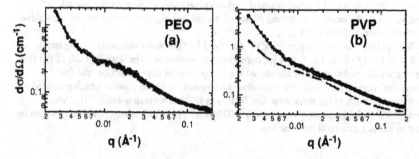

Figure 1. *SANS for PEO (non-inhibitor) and PVP (inhibitor)polymer-crystal-liquid slurries, the concentration of polymer is approximately 0.5 wt. percent and the crystal/liquid ratio is about 50:50. The linear combination of Porod scattering, dominant at low-q, and polymer-solution scattering, dominant at high-q, provides an excellent description for the PEO data (dashed line in (a)). Thus, the polymer conformation for this non-inhibitor is unaffected by the hydrate crystals. The failure of this model for PVP (dashed line (b)), indicates a significant conformation change which is shown to arise from polymer adsorbed on the hydrate crystal surfaces.*

PVP

In Figure 1b, we show the slurry-state scattering from the PVP sample. One again notes a significant low-q scattering. Comparison with Figure 1a shows that the amplitude is significantly enhanced over that from the PEO sample. At higher q values, there is also a higher level of scattering than for PEO. A linear combination of Porod and polymer-solution scattering cannot account for these data, as is evident by the dashed line, which is the scaled (by 0.75 in this case) intensity from the 7°C polymer solution. There is an excess of scattering at all q values, and adding Porod scattering to the solution scattering cannot satisfy the data. If we follow the procedure above of allowing the fit parameters to freely vary, the resulting solid line goes through the data points, but it suggests a physically unrealistic polymer conformation, including a fractal dimension approaching unity due to the slow decline of the intensity at higher q values. This simple two-component scattering model is inadequate to describe the data because it neglects the presence of polymer on the hydrate surfaces as shown in following section.

Contrast Variation Methods Applied to Adsorbed PVP Layer

In the case of an adsorbed polymer layer, the scattered intensity is written as the sum of three structure factors:[10]

$$I(q) = (n_h - n_s)^2 S_{gg}(q) - 2(n_h - n_s)(n_p - n_s) S_{pg}(q) + (n_p - n_s)^2 S_{pp}(q). \tag{4}$$

The first term, S_{gg}, describes the Porod scattering from the crystal-liquid interfaces. The second term, S_{pg}, describes scattering from polymer adsorbed at the crystal-liquid interfaces, and the third term, S_{PP}, contains contributions from both the polymer in the surface layer and that still remaining in the solution. Utilizing contrast-variation methods, where data sets are taken for different values of the neutron scattering power of the solvent, n_s, then combined, one can solve for the three structure factors of Eq. 4.

We applied these techniques for PVP dissolved in TDF/water solutions having molar ratios of D_2O-to-H_2O from 1 to 0.8. Five compositions were used. The data for each of the five samples were collected as detailed above. At the completion of data collection, the five combined sets were utilized in a singular-value decomposition fitting routine which provided the best fit values for each of the three structure factors in Eq. 4 at a given q value. This was repeated for the entire range of q values from 0.003 to 0.2 Å^{-1}. Each structure factor can then be analyzed to extract structural information.

S_{gg}

This structure factor results from Porod scattering, given by

$$S_{Porod} = 2\pi(S/V)q^{-4} \tag{5}$$

Therefore the S_{gg} structure factor values were fit to a q^{-4} power law dependence allowing us to extract a surface area per unit volume term, (S/V). This fit is shown in Figure 2.

As a consistency check for the Porod surface/volume estimate, we computed the estimated increase in surface area due to the polymer-induced change from octahedral to flat-plate hydrate crystal morphology.[10] Using optical microscopy to estimate the length/width ratio for these plates as 40:1, we calculate that the theoretical increase in surface area/unit volume over than of octahedral crystals would be 13-fold. Considering the uncertainties in the scattering contrasts, this seems in reasonable agreement with the observed 7-fold increase.

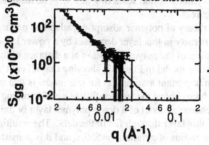

Figure 2. *Structure factor S_{gg} for PVP-in-TDF/water polymer-crystal-liquid slurry extracted via contrast variation techniques. The solid line is a fit to a Porod scattering model. The resulting fit yields a surface/volume ratio of 5000 cm^{-1}, a value 7 times larger than the polymer-free slurry. This is consistent with the polymer-induced crystal morphology change.*

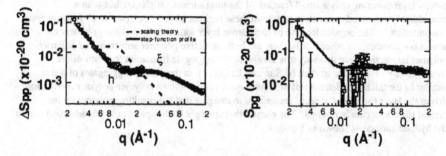

Figure 3. *Structure factors for adsorbed polymer layer in the polymer-crystal-liquid slurry. $\Delta S_{pp} = S_{pp} - S_{poly-soln}$ is the excess scattering above that of the polymer still in solution. The solid line is a fit to an adsorbed layer having a thickness of h=550Å, and an internal mesh size of ξ=24.8Å. The solid line for this model is also shown for the S_{pg} structure factor. The dashed line depicts the expected scattering model if each polymer chain were to contact the surface, with no polymer-polymer adsorption.*

S_{pp} and S_{pg}

As mentioned above, the S_{pp} term contains contributions from both the polymer bound to the hydrate crystal surfaces as well as that still remaining in solution. In previous studies of polymer adsorption by neutron scattering, the excess polymer in solution was eliminated. To do so here would be experimentally impossible, hence we instead subtract this contribution from S_{pp}. The quantity subtracted is the polymer-solution scattering at 7°C. The resulting quantity, ΔS_{pp}, is plotted along with S_{pg} in Figure 3.

According to the scaling theory of polymer adsorption, adsorbed polymer should form a self-similar layer. The polymer density in this layer decreases by a power law out into the solution until reaching the concentration of the polymer solution at a distance of the radius of gyration. We depict the scattering for this model in Figure 3, showing that it is not a reasonable description for our data. An important assumption for this model is that only polymer-substrate binding is important. The chains above the surface follow a self-avoiding random walk. In contrast, to fit our data we choose a model of a dense polymer layer of uniform density. We then obtain the characteristics of this layer through fits to the data. The resulting layer is considerably thicker than the radius of gyration, h=550Å, and it is consistent with a high density of polymer throughout the layer. Consistent with this elevated density, it exhibits blob scattering characteristic of semi-dilute solutions with a mesh size of ξ=25Å. Because the polymer concentration in the layer (ϕ) and the polymer surface coverage (S/V) both enter the ΔS_{pp} and S_{pg} equations as a product, one cannot solve for them independently. However by combining the fits for ΔS_{pp}, which goes as ϕ^2, and S_{pg} which goes as ϕ, we solve independently for these two quantities and obtain, ϕ= 0.07 and (S/V)=80 cm^{-1}. The resulting picture is of a patchy layer covering only a small fraction of the total surface (80/5000), but with a considerably enhanced concentration within these regions, about 2.5c* (where c* is the overlap concentration). The physics leading to such dense layer formation is that the polymers must attract one another. In other words, there are both attractive polymer-substrate and polymer-polymer interactions. In this way it is similar to the aggregate formation for semi-dilute solutions of PVP in water studied by Sun and King.[12] In their study, aggregates of a size similar to the thickness determined here were observed. If such a polymer-polymer interaction drives the layer formation, it seem reasonable to imagine that the resulting polymer on the crystal surface occurs in clumps. The picture that emerges is clumps of polymer adsorbed across the hydrate surface as shown in Figure 4.

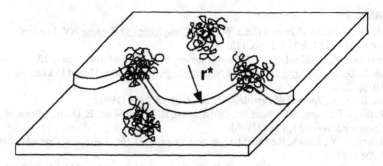

Figure 4. *Suggested mechanism of step pinning by adsorbed polymer clumps. Assuming that the clump size is the same as it's thickness (550Å), the 2% coverage implies a distance between clumps of about l=2300Å. The step growth velocity in this case diminishes towards zero as l-2r*/l, where r* is the critical 2D nucleation radius. The r* value for hydrate crystals is at present unknown, but typical values lie in the range of 100-1000Å. Therefore the polymer clumps shown here could effectively inhibit growth.*

CONCLUSIONS

In conclusion, comparison of the conformations of non-inhibitor and inhibitor polymers in hydrate-forming solutions prior to crystal formation shows only weak evidence for a difference between the two types of polymers. The single-chain characteristics are unperturbed, but the hydrate inhibitor polymers seem to show an enhanced tendency to form aggregates in solution. This is evidenced by excess low-q scattering following a $q^{-2.5}$ power law. Much more evident is the strong perturbation in the conformation of the inhibitor polymer upon crystallization of the hydrate. Contrast variation methods are used to solve for the three structure factors describing this perturbed conformation. This shows that the conformation change arises from polymer adsorbed to the polymer surface. Analysis of this adsorbed layer reveals that the polymer covers only about 2 percent of the total crystal surface. However within the covered areas the polymer is considerably enhanced in concentration above that of the surrounding liquid. This suggests a coverage of clumps of polymer widely separated from one another. Consideration of the concentration and implied spacing of such clumps on the hydrate surface suggests that they can effectively inhibit crystal growth.

ACKNOWLEDGMENTS

We would like to thank the NIST Center for Neutron Research and Exxon PRT for providing beamtime and help during the measurements and the following individuals for their helpful comments: John Huang , Scott Milner, Larry Talley and Dieter Richter.

REFERENCES

[1]Davidson, D. W. In: F. Franks (Ed.), Water: A Comprehensive Treatise NY, London: (Plenum Press, 1973); Vol. 2; pp. 115.

[2]Kvenvolden, K. A.; Ginsburg, G. D. and Soloviev, V. A., *Geo-Marine Letters* **13**, 32 (1993).

[3]Sloan, E. D., Jr., Clathrate Hydrates of Natural Gases, New York: (Marcel Dekker, Inc., 1990), pp. 276.

[4]Sloan, E. D., Jr., *Journal of Petroleum Technology* **43**, 1414 (1991).

[5]Lederhos, J. P.; Long, J. P.; Sum, A.; Christiansen, R. L. and Sloan, E. D., Jr., *Chemical Engineering Science* **51**, 1221 (1996).

[6]Makogon, T. Y.; Larsen, R.; Knight, C. A. and Sloan, E. D., Jr., *Journal of Crystal Growth* **179**, 258 (1997).

[7]Larsen, R.; Knight, C. A. and Sloan, E. D., Jr., *Fluid Phase Equilibria* **150-151**, 353 (1998).

[8]Sloan, E. D., Jr.; Subramanian, S.; Matthews, P. N.; Lederhos, J. P. and Kokhar, A. A., *Industrial and Engineering Chemistry Research* **37**, 3124 (1998).

[9]Hanley, H. J. M.; Meyers, G. J.; White, J. W. and Sloan, E. D., Jr., *International Journal of Thermophysics* **10**, 903 (1989).

[10]Hutter, J.; King, H. E., Jr. and Min, L. Y., *Macromolecules* **33**, 2670 (2000).

[11]Devanand, K. and Selser, J. C., *Macromolecules* **24**, 5943 (1991).

[12]Sun, T. and King, H. E., Jr., *Macromolecules* **29**, 3175 (1995).

[13]King, H. E., Jr.; Hutter, J.; Lin, M. Y. and Sun, T., *Journal of Chemical Physics* **112**, 2523 (1999).

[14]Beaucage, G., *Journal of Applied Crystallography* **29**, 134 (1996).

[15]Beaucage, G., *Journal of Applied Crystallography* **28**, 717 (1995).

[16]de Gennes, P. G., *Comptes Rendus de L'Académie des Sciences (Paris), Serie II* **313**, 1117 (1991).

Mat. Res. Soc. Symp. Proc. Vol. 651 © 2001 Materials Research Society

Spatial fluctuations and the flipping of the genetic switch in a cellular system

Ralf Metzler

Department of Physics, Massachusetts Institute of Technology

77 Massachusetts Ave., Cambridge, MA 02139

Electronic address: metz@mit.edu

Abstract – It has been realised that noise plays an important rôle in cellular processes where fluctuation induced number fluctuations of certain messenger molecules become non–negligible, due to the small total number of these molecules within one cell. In the following, it is argued that spatial fluctuations of such molecules and their impact on genetic switches should be considered as well.

Bacteriophage T4, the so-called λ phage, and its parasitosis with the bacterium *Escherichia coli* has been studied as a paradigm for genetic switches for over half a century. The corkscrew shaped phage, see Fig. 1, injects its DNA into *E.coli* where it fuses with the host DNA. From this moment on, the phage's genetic information is copied each time *E.coli* reproduces itself through cell division, if the parasite-host system is dormant (*lysogeny*). If not, λ uses its host as a miniature chemical plant to reproduce itself, until so many copies of λ are produced that the host cell virtually bursts, and thus a swarm of new λ phages is released (*lysis*). The decision between either route within this if-then loop depends on whether certain antagonist messenger molecules, repressor R or cro, bind to the active operator site on the λ DNA and thus allow for the transcription of the corresponding genes. The interplay between messenger molecules and the subsequent decision which one of two genes is transcribed, i.e., which blueprint is copied and applied, is called the genetic switch [1].

The basic mode of operation of a genetic switch is shown in Fig. 2.

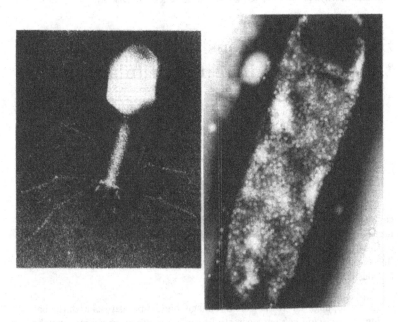

Figure 1: *Left:* Bacteriophage T4, the λ phage. The head contains the DNA which is injected into the host cell *E.coli* through the corkscrew-like pipe. *Right:* Like a cannoli coated with demerara sugar—the *E.coli* cell under attack. Each tiny spot represents a λ phage.

One can, in essence, distinguish between co-operative and non–co-operative switches. In a non–co-operative switch, a bound R molecule dissociates with some rate constant k^{nc}_{diss}. If it is not replaced by another R, but a cro molecule binds to the adjacent operator instead, the process of lysis occurs. In the co-operative scenario, in contrast, on average two repressor units bind to the gene; one R binds much stronger and facilitates the binding of the other R. Now, one R dissociates, but as long as the other R is still present, only new R can bind to its vacant operator site. Cro binding, and therefore lysis, can only occur, if *both* R dissociate, due to the co-operative bond formation a much less likely event. This is how nature created extremely robust control mechanisms for sensitive issues such as reproduction [1].

The chemical processes of transcription, production and degradation of the messenger molecules are a priori noisy. As the total number of these

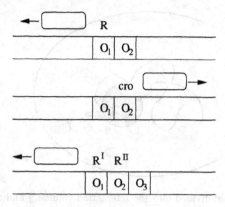

Figure 2: Idealised picture of a genetic switch, controlled by the messenger molecules R and cro. Upper two panels: non–co-operative case. If R dissociates and is eventually replaced by a cro at the adjacent operator site, the switch is flipped, i.e., in the λ example, lysis would occur. Lower panel: Co-operative switch. Here, both R have to be replaced. As both act co-operatively, the replacement by cro is significantly suppressed. In both cases, R or cro enable the transcription of the genes corresponding to lysis or lysogeny [1].

molecules is usually fairly small, a few up to a hundred, noise-induced fluctuations of these numbers may have non-negligible effects on the accuracy of the associated genetic switch. Recently, numerous studies, both analytically and numerically, have corroborated that the co-operative switch is far superior and suppresses fluctuation-based errors of the switch to extremely small rates as typically encountered for reproductive biological systems [3-9]. These models, however, concentrate on the purely time-dependent description of the molecule number $\mathbf{n}(t)$, in our example the number of R and cro, $(n_R(t), n_{cro}(t))^T$. Here, we are going to argue that spatial fluctuations cannot be neglected and should be considered as part and parcel of the process, in order to obtain a further understanding of genetic switches, especially in respect to the study of arrays of switches. Moreover, we show that the superior quality of co-operativity can already be revealed in a purely diffusive model. As we do not consider chemical kinetics number fluctuations in the following, we call the diffusive model the naked switch. The result can in turn be used to efficiently include spatial fluctuations in future modelling and simulations.

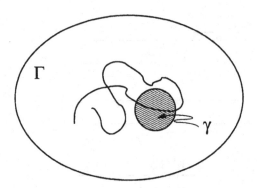

Figure 3: The cell is subdivided into the interaction volume γ and the free volume Γ. The ratio γ/Γ determines the probability for a given molecule to be within γ or not. γ is centred around the decisive λ part of the DNA strand which is depicted in Fig. 2.

Messenger molecules are fairly small in comparison to the whole cell volume, as can be anticipated from Fig. 1 where the difference in size between $E.coli$ and the λ phage is shown. Moreover, it can be safely assumed that (i) the messenger molecules are not actively transported within the cell and (ii) that they diffusive freely. A molecule located on the fringe of the cell cannot possibly contribute to the switch unless it diffuses towards the operator first. As diffusion is fairly efficient, $K \sim 2 \cdot 10^{-6} \mathrm{cm}^2/\mathrm{sec}$, the usual assumption of well-stirredness seems reasonable. However, if one compares the size of an effective interaction volume γ around the operator sites depicted in Fig. 2 with the free cell volume Γ, the probability $p = \gamma/\Gamma$ for any given messenger molecule to be inside γ becomes actually extremely small, compare Fig. 3. Choosing a small multiple of the van der Waals radius ($\propto 10^{-6}\mathrm{cm}$, corresponding to some tens of the molecular radius of R) for the size of γ, this probability is $p \sim 0.0005$. Thus, even the probability that none out of 100 molecules is in γ, reaches 0.9512!

Using the diffusion picture, we find the diffusion approximation $\delta t \sim 10^{-4}\mathrm{sec}$ for the mean time it takes a molecule to cross the interaction volume. This time can be viewed as a stroboscopic time scale after which the system may be regarded renewed. That is, during this time, a molecule can either enter or exit the interaction volume. With this basic model assumption, the division of the free cell volume into γ and Γ creates an effective description which is independent of the details of the volume Γ and, to some extent, even of the assumption of well-stirredness. It allows for an efficient implementation in Monte Carlo simulations that include the spatial dependence.

The basic probabilities of interest for the switching process are composed of the probabilities that no R molecule is in γ, Π_0, and its complement, $\bar{\Pi}_0 = 1 - \Pi_0$, and similarly χ_0 and $\bar{\chi}_0$ for cro. From these, four different realisations can be distinguished,

$$p_1 = \Pi_0 \chi_0; \quad p_2 = \bar{\Pi}_0 \chi_0; \quad p_3 = \Pi_0 \bar{\chi}_0; \quad \bar{\Pi}_0 \bar{\chi}_0.$$

Straightforward calculations lead to a typical time $\langle \delta t \rangle = \delta t / (\chi_0 (\Pi_0 + q\bar{\Pi}_0))$ spent during a switching process *without* taking the lifetimes of the bonds into account. q is the probability that cro binds instead of R if both species are present in γ. Doing so, one obtains the characteristic lysis time

$$T_{nc}^{lys} = \langle \tau_{diss}^{nc} \rangle + \langle \delta t \rangle$$

for the non-co-operative case. Conversely, for the co-operative arrangement, one finds the lysis time

$$T_c^{lys} = \frac{\langle \tau_{diss}^{I\&II} \rangle (p_2 + (1-q)p_3)}{p_{lys} p_{inhib}} + \frac{\delta t}{p_{lys}}$$

with the mean time $\langle \tau_{diss}^{I\&II} \rangle = \langle i \rangle_\eta \tau_\eta + (\langle i \rangle_\eta + 1) \left(k_{diss}^{cI} \right)^{-1}$ consumed by superprocesses η of mean duration τ_η, describing the prolonged vacancy of O_1 such that the R molecule from O_2 can actually dissociate, compare Ref. [6]. Moreover, we use the abbreviation $\tau_{diss}^{cII} = \left(k_{diss}^{cII} \right)^{-1}$ for the dissociation time of this second R molecule. Thus, instead of having an urn model where R and cro are selected randomly, one should necessarily consider a "dynamical Cinderella experiment" where the good and the bad molecules appear and disappear spontaneously, i.e., the "urn" γ is connected to the bath Γ.

With the rough numbers assumed above, one obtains non-co-operative lysis time of the order of seconds, whereas one is up to hundreds of orders of magnitude above this in the co-operative setup. Accordingly, even the naked switch model that purely bases on a diffusive renewal picture, results in an extremely efficient separation of co-operative and non-co-operative schemes. We believe therefore that spatial fluctuations should come into consideration in the modelling of genetic switches. We also believe that this ingredient becomes even more important for switch cascades or networks.

Concluding, nature has found a very robust way of coping with fluctuations for small population processes running off at room temperature or above. We have presented a simple argument that spatial fluctuations indeed come into play in such cellular systems, and being to comparable order of magnitude effect, should necessarily be incorporated in the existing models. Due to the effective interaction between the two volumes γ and Γ on the renewal time scale δt, the implementation of these additional assumptions should be fairly straightforward, and they should not consume too much computation time in the associated simulation routines.

* * *

RM thanks Peter Wolynes for discussions and his hospitality at the University of Illinois where part of this study was carried out. RM also thanks Yossi Klafter for discussions and comments. Financial support from the DFG within the Emmy Noether programme is acknowledged as well.

References

[1] M. Ptashne, *A Genetic Switch: Phage λ and Higher Organisms* (Cell Press/Blackwell, Cambridge, MA, 1992); B. Alberts et al., *The molecular biology of the cell* (Garland, New York, 1994); S. R. Bolsover et al., *From genes to cells* (Wiley, New York, 1997); H. Linder, *Biologie* (Metzlersche Verlagsbuchhandlung, Stuttgart, 1985)

[2] M. A. Shea and G. K. Ackers, J. Mol. Biol. **181**, 211 (1985); E. Aurell, S. Brown, J. Johanson and K. Sneppen, cond-mat/0010286

[3] M. S. H. Ko, H. Nakauchi and N. Takahashi, EMBO J. **9**, 2835 (1990); M. S. H. Ko, J. Theor. Biol. **153**, 181 (1991); BioEssays **14**, 341 (1992); J. Peccoud and B. Ycart, Theor. Popul. Biol. **48**, 222 (1995); D. L. Cook, L. N. Gerber and S. J. Tapscott, Proc. Nat. Acad. Sci. USA **95**, 6750 (1998)

[4] A. Arkin, J. Ross and H. H. McAdams, Genetics **149**, 1633 (1998); H. H. McAdams and A. Arkin, Ann. Rev. Biophys. Biomol. **27**, 199 (1998); H. H. McAdams and A. Arkin, Trends Genet. **15**, 65 (1999); Proc. Nat. Acad. Sci. USA **94**, 814 (1997)

[5] W. Bialek, cond-mat/0005235

[6] R. Metzler, submitted for publication

Mat. Res. Soc. Symp. Proc. Vol. 651 © 2001 Materials Research Society

Rock Wetting Condition Inferred From Dielectric Response

Yani Carolina Araujo, Mariela Araujo and Hernán Guzmán
Reservoir Department, PDVSA Intevep, Caracas, Venezuela.

ABSTRACT

Wettability is a manifestation of rock-fluid interactions associated with fluid distribution in porous media. Conventional wettability evaluation is performed by a sequence of spontaneous and forced displacements of different fluids into a porous sample, a method which is costly and time consuming. A new attractive approach is to estimate this quantity from dielectric measurements, since they can be done rapidly and economically.

The dielectric frequency response of several rock samples of known wettability condition was studied in the range from 10 Hz to 100 MHz. Samples were saturated with brine and oil. The results confirm the strong influence of wetting condition on dielectric response. Water wet samples have significantly higher values of ε' and ε'' (real and imaginary parts of generalized complex permitivity) than oil wet samples. In particular, the high frequency behavior of ε'' is most affected. Different regimes are identified as a function of frequency. They correspond to zones where different polarization effects are manifested. We quantify this effect and find a correlation with the modified Amott wettability index. Based on these findings, we propose an experimental protocol for the indirect measurement of wettability at laboratory scale.

INTRODUCTION

The oil industry is devoted to the recovery of the highest possible amount of hydrocarbons from reservoirs, which typically are at deep locations in the ground, under variable temperature and pressure conditions. For this objective, an interesting and useful information about the reservoir is the knowledge of the spatial distribution of fluids, and their affinity for the mineral surface, a quantity described by the wettability condition [1].

Several direct and indirect methods have been used in the industry to learn about the behavior of multiphase flow in porous reservoirs, which is a challenging task due to the opaque nature of natural porous media. Among those we find the use of a variety of tracers, including radioactive materials, high magnetic susceptibility materials, x-ray absorbing tracers, neutron diffraction techniques and nuclear magnetic resonance which has been used to get saturation on cores [2,3].

Dielectric measurements have been used in oil industry to characterize and investigate physical properties of rock samples by non destructive, rapid, economic and accurate techniques. Dielectric properties of reservoir rocks depend on the permittivity and conductivity, density of the fluids and their saturation, and how they are distributed inside the porous structure. In the literature, it has been reported that the dielectric constant is a very useful parameter to measure the level of water saturation in rocks [4,5]. For saturated samples this parameter depends strongly on frequency.

Wettability is one of the petrophysical properties that most affects the dielectric response of reservoir rocks. Wettability plays an important role in the determination of the location and distribution of reservoir fluids and thus, it influences the oil recovery efficiency from a reservoir [1]. Wettability effects on rock electric properties can be split in two groups: connectivity effects at low frequencies (< 1 KHz) and shape effects from intermediate to high frequencies (> 10 KHz).

The wetting condition of a reservoir rock is commonly classified as homogeneous when the entire rock surface has a uniform affinity for a fluid either water or oil, and heterogeneous, when there are regions in the sample that exhibit different affinities for oil or water. In general, the wetting fluid has the tendency to occupy the small pores of the sample and to contact the majority of the rock surface as thin films, while the non wetting fluid occupies the largest pores commonly as isolated packets.

Since water and oil have different dielectric properties we expect considerable differences in the dielectric response of reservoir samples with different wetting condition. Brines and crude oils differ electrically in two well-known, fundamental ways. Brines are conducting and hydrocarbons not. Water has a large relative dielectric constant (80 at room temperature), while crude relative dielectric constant is between 2 and 2.5. These distinctions are basic and exist from zero frequency up to high frequencies (40 MHz). The electrical behavior of a reservoir rock saturated with fluids (oil and water) can be described in terms of a complex relative dielectric constant, $\varepsilon(\omega)$:

$$\varepsilon(\omega) = \varepsilon_p - \varepsilon_c \qquad (1)$$

where ω is the angular frequency of the electric field applied to the rock, ε_p $(\varepsilon_p(\omega) = \varepsilon_p'(\omega) - i\varepsilon_p''(\omega))$ is the relative dielectric constant that results from all polarization effects (electronic, atomic, dipole and interfacial or Maxwell-Wagner polarization), and ε_c represents the contribution of free ions responsible for the d.c. water conductivity ($\varepsilon_c = i\sigma_{dc}/\varepsilon_o\omega$). For rocks saturated with brine and oil, the different polarizations mechanisms (electronic, atomic and dipolar polarization) [6] provide a constant contribution to $\varepsilon_p(\omega)$ in the range from 100 Hz to 40 MHz. The interfacial polarization is activated by an electric field in a disordered system, and is frequency sensitive in the above mentioned range [4,7].

To quantify the relationship between dielectric response and rock wettability Bona et al. [7] performed measurements in model porous media, sinterized glass and Berea samples. They found that the dielectric parameters were more sensitive to wettability in the frequency range between 100 Hz and 40 MHz. A relevant feature is that from all the petrophysical properties of the studied samples (saturation, capillary pressure, permeability, wettability, etc.) the wetting condition showed the strongest dependence with respect to the dielectric constant. For example, in the frequency range mentioned above, the saturation condition and the porosity of the sample were the less affected. This result suggests that dielectric measurements can be used to determine the wetting condition of reservoir samples quickly and at low cost,.

In this paper we present results from an experimental study of the dielectric response of rock samples from producing wells from Venezuelan reservoirs of different wettability condition, with the objective of assessing the potential use of this indirect technique for the inference of this rock-fluid interaction property.

EXPERIMENTAL PROCEDURE

Five consolidated samples were used to make the experimental analysis presented here. The samples were selected to cover a wide range of wetting conditions ranging from strongly water-wet to oil-wet. Core samples were cleaned with the traditional distillation-extraction method known as Dean Stark. Toluene and Methanol were used as cleaning fluids. After cleaning, the samples were placed inside an oven at 80°C for 16 hours and storaged in a dissicator until their final use. Conventional tests were performed on the samples. Porosity (ϕ) and gas permeability (k) were measured with a Boyle type conventional cell and with a Profile Permeameter PDPK400 from CoreLab respectively.

Wettability condition was estimated from Amott-Harvey test which consists of a series of spontaneous and forced fluid displacements [8,9] by which two indexes (water and oil index) are obtained. If the water index is close to one, the sample is strongly water wet and similarly for the oil phase. The properties of the samples are summarized in Table 1.

Table 1. Petrophysical characteristics of the analyzed rock samples.

Sample	ϕ (%)	k (md)	Amott Index Water	Amott Index Oil	Wetting Condition
WW	17.1	19.6	0.95	0.00	Strongly Water Wet
WWW	10.4	38.4	0.75	0.44	Weakly Water Wet
IW	15.9	42.6	0.63	0.50	Intermediate Wet
WOW	18.9	13.1	0.54	0.76	Weakly Oil Wet
OW	22.3	217.2	0.00	0.75	Strongly Oil Wet

Samples were saturated with brine (13.85 g/L NaCl, 1.76 g/L NaHCO₃, 0.44 g/L SO₄Na.H₂0, 0.39 g/L Na₂CO₃, 0.26 g/L CaCl₂), and medium crude oil (density: 0.81 gr/cc, viscosity: 29.24 cP and static dielectric constant: 2.2) from a Venezuelan reservoir to perform the measurements. To saturate the samples the rock was placed in a pressurized core holder and fluid was injected using a constant rate pump.

Complex dielectric measurements (dielectric permittivity and conductivity) were made on the selected samples. The electrical parameters were evaluated using an HP 4194A Impedance Analyzer in the frequency range 100 Hz to 13 MHz. The temperature was kept constant at 25 °C. To measure the impedance, the samples were placed into a special cell between two parallel plates that form a golden plate capacitor specially designed for the measurements. A two contact method was used. The total impedance Z of the equivalent circuit, that results from the sum of two contributions in series, one the impedance of the capacitor, which accounts for the capacitance of the cell and the wires, and the impedance of the interconnecting system. The complex admittance $(Y=Z)^{-1}$ is calculated taking into account the capacitance of the empty cell and any spurious capacitance from other elements in the system. The dielectric permittivity of the sample was obtained by using:

$$\varepsilon(\omega) = 1 + \frac{Y_C(\omega) - Y_0(\omega)}{i\omega C_0} \tag{2}$$

where C_o is the capacitance and $Y_o(\omega)$ the admittance of empty cell, and $Y_c(\omega)$ the admittance of the cell with the sample.

RESULTS AND DISCUSSION

The saturation dependence of the complex dielectric constant for two types of systems was studied (air-brine and oil-brine). In the first case, we started from a fully saturated (with brine) sample, measured the dielectric response in frequency and then, we reduced the saturation by drying the sample. In the sequence, sample saturation is calculated by using the changes in the sample weight. In the second case, we started with a sample saturated with brine and when the brine saturation is reduced, the pore space is filled with oil. Figure 1 shows the dependence of the real and imaginary parts of the dielectric constant for three of the samples with a clear wettability contrast, samples WW, IW and OW for high water content, and figure 2 shows the real and imaginary parts for a low water content situation (10% of water).

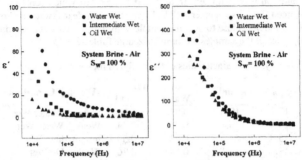

Figure 1. Real and imaginary parts of the dielectric constant at high water content for the air-brine system.

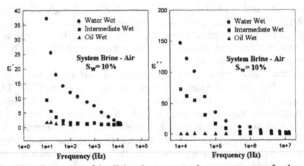

Figure 2. Real and imaginary parts of the dielectric constant at low water content for the air-brine system.

Despite the fact the samples have the same brine saturation, it is found that the water-wet samples have higher values of ε' and ε'' than the oil wet and intermediate wet samples in the frequency range studied. In the low frequency range, ε' and ε'' decrease following a power-law behavior. This behavior is characteristic of charge transport polarization and is related to the fact that for the water-wet sample the wetting phase remains connected, whereas the non-wetting phase does not. At higher frequencies, ε' and ε'' exhibit a change in the behavior reflected as a change of slope in the frequency dependence. It is known that in this regime that Maxwell-Wagner polarization is observable [7].

Figure 3 summarizes the saturation behavior of the real and imaginary parts of the complex dielectric constant for 15 KHz and the wettability contrast. As it is clearly seen, the frequency response is sensitive to fluid distribution in the samples for a given saturation level and there is a good contrast according to the wetting condition.

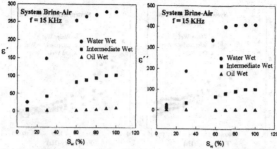

Figure 3. Saturation dependence of the dielectric constant for a constant frequency in air-brine system.

The dielectric constant shows a large increase with increasing saturation at low levels of saturation and then a more gradual, linear increase with saturation at higher saturation levels. At high saturations the change in the dielectric constant can be related to the change in the volume of water. At low saturations the dielectric constant is controlled by bound and free water in the pores of the sample.

Figures 4 and 5 show the behavior for the oil-brine case. Here the most relevant information is obtained from the imaginary part of the dielectric constant. For high water content (around 85%) at high frequencies there is almost no difference between oil wet and intermediate wet samples in terms of the real part of the dielectric constant, the imaginary part is clearly sensitive to the wetting condition of porous rock Thus, the technique is sensitive to the saturation value and type of fluid present in the sample.

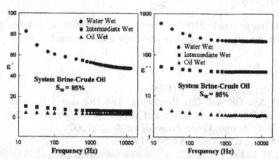

Figure 4. Real and imaginary parts of the dielectric constant at high water content for the oil-brine system.

In the limit of low brine saturation, there is almost no difference in the dielectric response of the water wet and the oil wet samples. The rapid decrease observed in ε' and ε'' can be attributed to polarization mechanisms. However, as soon as the water saturation level increases to 5-10% (which are characteristic values of an oil reservoir at irreducible conditions) there is a clear difference between the rock responses.

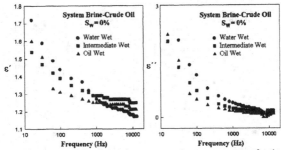

Figure 5. Real and imaginary part of the dielectric constant at low water content for the oil-brine system.

CONCLUSIONS

The dielectric constant has been found to be a strongly frequency-dependent quantity in saturated samples, both the real and imaginary part decrease with increasing frequency.

Dielectric measurements of reservoir samples demonstrate that the frequency response can be used to determine the wettability condition of reservoir rocks. The imaginary part of the dielectric response is higher (in a factor 10^2-10^3) for the water wet sample. At low water content, the real part is sufficiently enough to distinguish the wetting surface condition. The experimental method involved is more simple, reliable and of low cost in terms of equipment and time, when compared to displacement experiments commonly used to evaluate the wetting condition.

ACKNOWLEDGMENTS

The authors would like to thank PDVSA Intevep for permission to publish this paper. The experimental support of Marian Giampaoli is gratefully acknowlegded.

REFERENCES

1. N. R. Morrow, *JPT*, December, 1476 (1990).
2. L. Cuiec, North Sea Oil and Gas Reservoirs, The Norwegian Institute of Technology, *Wettability and Oil Reservoirs*, (Graham and Trotman, 1987) pp. 193-207.
3. M. Araujo, and Y. C. Araujo, *Vision Tecnológica*, **8**, 19 (2000).
4. R. Knight, and A. Nur, *Geophysics*, **52**, 644 (1987).
5. M. R. Taherian, W. E. Kenyon, and K. A. Safinya, *Geophysics*, **55**, 1530 (1990).
6. C. J. F. Bottcher, and P. Bordewijk, Theory of Electric Polarization, Volume II, Elsevier Scientific Publishing, Amsterdam (1978).
7. N. Bona, E. Rossi, C. Venturini, S. Capaccioli, M. Lucchesi, and P. A. Rolla, *Revue de L'Institu Francais du Pétrole*, **53**, 771 (1998).
8. E. Amott, *Trans. AIME*, **26**, 156 (1959).
9. L. E. Treiber, D. L. Archer, and W. W. Owens, *JPT*, December, 531 (1972).

Mat. Res. Soc. Symp. Proc. Vol. 651 © 2001 Materials Research Society

Propagation Dynamics of a Particle Phase in a Single-File Pore.

A.M.Lacasta[1], J.M.Sancho[2], F.Sagues[3] and G.Oshanin[4]

[1] Departament de Fisica Aplicada, Universitat Politècnica de Catalunya, E-08028 Barcelona, Spain

[2] Departament d'Estructura i Constituents de la Matèria, Universitat de Barcelona, E-08028 Barcelona, Spain

[3] Departament de Quimica Fisica, Universitat de Barcelona, E-08028 Barcelona, Spain

[4] Laboratoire de Physique Théorique des Liquides, Université Paris 6, 75252 Paris, France

Abstract

We study propagation dynamics of a particle phase in a single-file pore connected to a reservoir of particles (bulk liquid phase). We show that the total mass $M(t)$ of particles entering the pore up to time t grows as $M(t) = 2m(J, \rho_F)\sqrt{D_0 t}$, where D_0 is the "bare" diffusion coefficient and the prefactor $m(J, \rho_F)$ is a non-trivial function of the reservoir density ρ_F and the amplitude J of attractive particle-particle interactions. Behavior of the dynamic density profiles is also discussed.

Introduction

Particles transport across microscopic pores is an important step in a vast variety of biological, chemical engineering and industrial processes, including drug release, catalyst preparation and operation, separation technologies, especially biological and biochemical, tertiary oil recovery, drying and chromatography [1, 2, 3].

Man-made or naturally occuring porous materials contain a wide range of pore sizes, from meso- to micro- or even nanoscales, in which case the pore diameter is comparable to the molecular size. Such molecular sized channels, of order of a few Angstroms only, appear, for instance, in biological membranes, and are specific to water and ion transport which participate in hydrostatic or osmotic pressure controlled cellular volume regulation [4]. Carbon nanotubes or zeolites, such as, Mordenite, L, AlPO$_4$-5, ZSM-12, may also contain many channels of nearly molecular diameter and can selectively absorb fluids serving as remarkable molecular sieves [5]. For example, AlPO$_4$-5 is composed of nonintersecting and approximately cylindrical pores of nominal diameter 7.3 Angstroms.

A salient feature of transport in molecularly sized pores is that there is a dramatic difference between the diffusion of adsorbates whose size is much smaller than the pore diameter, and those whose size is comparable to it. If the diameter of the diffusing guest molecules exceeds the pore radii, but is still less than its diameter, the particles are able to enter the pore but not able to bypass each other such that initial given order is striktly

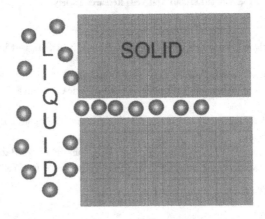

Figure 1: Solid with a single-file pore in contact with a liquid phase.

maintained (see Fig.1). Here, diffusion obeys the so-called "single-file" behavior; the mean-square displacements of tracer molecules do follow $X^2(t) = 2Ft^{1/2}$ [6, 7, 8], where F is referred to as the "single-file" mobility; evidently, such a behavior is remarkably different from the conventional diffusive law $X^2(t) = 2Dt$, observed in situations when the guest molecules are able to bypass each other. Single-file diffusion was evidenced by pulsed field gradient NMR measurements with a variety of zeolites and guest species [6, 7]; in particular, the law $X^2(t) = 2Ft^{1/2}$ has been identified experimentally for C_2H_6 [6] and CF_4 [9] in AlPO$_4$-5.

Diffusion of absorbed particles in single-file pores has also been the issue of a considerable theoretical interest recently. Several approaches have been proposed, based mostly on the lattice-gas-type models, and such properties as concentration profiles or steady-state particle currents have been evaluated [10, 11, 12]. As well, a great deal of Monte Carlo and Molecular Dynamics simulations has been devoted to the problem [13, 14, 15], providing a deeper understanding of the transport mechanisms in single-file pores. On the other hand, still little is known about non-stationary behavior in single-file systems; in particular, how fast do the particle phase propagates within the single-file pores or how does the total mass (or number) of particles entering the pore up to time t grows with time?

In the present paper we focus on the challenging question of the particle phase propagation dynamics in the single-file pores. We report here some preliminary results; a detailed account will be published elsewhere [16]. More specifically, we consider a single-file pore in contact with a reservoir of particles (bulk liquid phase), which maintains a fixed particle density at the entrance to the pore. The pore is modelled, in a usual fashion, as a one-dimensional regular lattice whose sites support, at most, a single occupancy; the particles interaction potential consists of an abrupt, hard-core repulsive part, which insures single-occupancy, and is attractive, with an amplitude $J \geq 0$, for the nearest-neighboring particles only. Introducing then a standard, interacting lattice-gas dynamic rules (see e.g. [17]), we derive evolution equation for the local variables describing mean occupation (density) of the lattice sites, which is analysed both analytically and numerically. Within our appoach, we define the evolution of the total number $M(t)$ of particles entering the single-file pore up to time t and also discuss the dynamics of the density distribution function within the pore. We show that the growth of $M(t)$ is described by $M(t) = 2m(J, \rho_F)\sqrt{D_0 t}$, which can be thought of as the microscopic analog of the Washburn equation, where D_0 denotes the "bare" diffusion coefficient, while the prefactor $m(J, \rho_F)$ is a non-trivial function of the attractive interactions amplitude J and the density ρ_F at the entrance to the pore.

The model

Following earlier works [10, 11, 12], as well as a conceptually close analysis of an upward creep dynamics of ultrathin liquid films in the capillary rise geometries [18], which appears to be well-adapted to the single-file dynamics, we model the single-file system under study (Fig.1) as a semi-infinite linear chain of equidistantly placed sites X (with spacing σ), attached to a reservoir of particles maintained at a constant chemical potential μ (see Fig.2). Each pair of sites is separated by a potential barrier of height E_B, which sets the typical time scale τ_B. Note that the spacing σ can be defined, in case of hard solids, as the interwell distance of a periodic potential describing the interactions of the particles with the solid atoms and τ_B is related to E_B and the reciprocal temperature $\beta = 1/T$ through the Arrhenius formula. For soft solids, in which case the dominant dissipation channel is due to mutual particle-particle interactions, σ can be thought of as the typical distance travelled by particles before successive collisions; here, τ_B is just the ballistic travel time.

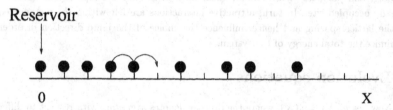

Figure 2: Effective model for transport in a single-file pore.

Further on, we define the particles interaction potential $U(X)$ as:

$$U(X) = \begin{cases} 0, & X > \sigma, \\ -J, & X = \sigma, \\ +\infty, & X < \sigma, \end{cases} \qquad (1)$$

i.e. we suppose that the interaction potential between particle is a hard-core exclusion, which prevents multiple occupancy of any site, and attraction with an amplitude J, $(J \geq 0)$, between the nearest-neighboring particles only. Occupation of the site X at time moment t for a given realization of the process will be described then by the Boolean variable $\eta_t(X)$, such that

$$\eta_t(X) = \begin{cases} 1, \text{the site } X \text{ is occupied,} \\ 0, \text{otherwise.} \end{cases} \qquad (2)$$

Consequently, the interaction energy $U_t(X)$ of the particle occupying at time t the site X for a given realization of the dynamical process is

$$U_t(X) = -J\Big(\eta_t(X + \sigma) + \eta_t(X - \sigma)\Big). \qquad (3)$$

Lastly, we define the particle dynamics (see [17] for more details). We suppose that at time moment t any particle occupying site X waits an exponential time with mean τ_B and then selects a jump direction with the probabilty

$$p(X|X') = Z^{-1} \exp\Big[\frac{\beta}{2}\Big(U_t(X) - U_t(X')\Big)\Big], \quad \sum_{X'} p(X|X') = 1, \qquad (4)$$

where Z is the normalization, $X'(= X \pm \sigma)$ denotes here the target neighboring site and the sum over X' means the sum over all nearest neighbors of the site X. As soon as the target site is chosen, the particle attempts to hop onto it; the hop is instantaneously fulfilled if the target site is empty; otherwise, the particle remains at its position. Physically, it means that repulsive interactions are very short-ranged - much shorter than the lattice spacing, and particles "learn" about them only when they attempt to land onto some already occupied site. In turn, attractive interactions are felt within the distance equal to the lattice spacing and hence, influence the choice of the jump direction in order to minimize the total energy of the system.

Evolution equations

Now, let $\rho_t(X) = \overline{\eta_t(X)}$, where the overbar denotes averaging with respect to different realizations of the process. Assuming local equilibrium, we find then that the time evolution of $\rho_t(X)$ is governed by the following balance equation [16]:

$$\tau_B \dot{\rho}_t(X) = \left(1 - \rho_t(X)\right)\left[\rho_t(X - \sigma)p(X - \sigma|\sigma) + \rho_t(X + \sigma)p(X + \sigma|X)\right] -$$
$$- \rho_t(X)\left[(1 - \rho_t(X + \sigma))p(X|X + \sigma) + (1 - \rho_t(X - \sigma))p(X|X - \sigma)\right], \quad (5)$$

where the average transition rate $p(X|X')$ obeys Eq.(4) with $\eta_t(X)$ replaced by $\rho_t(X)$. Equation (5) accounts for the fact that a particle may appear at time moment t on an *empty* site X by hopping from the *occupied* sites $X \pm \sigma$ with corresponding transition probabilities dependent on the interaction energy of the system; and may leave the *occupied* site X for *unoccupied* sites $X \pm \sigma$.

Next, we turn to the so-called diffusion limit, assuming that τ_B scales as σ^2; we thus suppose that $\sigma \to 0$, $\tau_B \to 0$, but the ratio $\sigma^2/\tau_B = const = 2D_0$, where D_0 is the diffusion coefficient describing motion of an individual, isolated particle. Expanding $\rho_t(X \pm \sigma)$ and $p(X|X \pm \sigma)$ in the Taylor series up to the second order in powers of the lattice spacing σ, we arrive at the desired dynamical equation of the form

$$\dot{\rho}_t(X) = D_0 \frac{\partial}{\partial X}\left[\frac{\partial}{\partial X} + \beta \rho_t(X)(1 - \rho_t(X))\frac{\partial U_t(X)}{\partial X}\right], \quad (6)$$

where $U_t(X)$ is the interaction energy at point X defined by Eq.(3).

Now, a few comments on Eq.(6) are in order. Note first that Eq.(6) is a Burgers-type equation with an environment dependent force,

$$\frac{\partial U_t(X)}{\partial X} \approx -2J\frac{\partial \rho_t(X)}{\partial X} - \sigma^2 J\frac{\partial^2 U_t(X)}{\partial X^2}. \quad (7)$$

When only the first term on the rhs of Eq.(7) is taken into acount, we get from Eq.(6) a one-dimensional diffusion equation

$$\dot{\rho}_t(X) = \frac{\partial}{\partial X}D(\rho_t(X))\frac{\partial}{\partial X}\rho_t(X), \quad (8)$$

with a field-dependent diffusion coefficient

$$D(\rho_t(X)) = D_0\left(1 - 2\beta J \rho_t(X)(1 - \rho_t(X))\right), \quad (9)$$

which is precisely the equation derived earlier by Lebowitz et al [19] and describing hydrodynamic limit dynamics of a system of mutual interacting particles undergoing ballistic motion. On the other hand, if we keep the second term on the rhs of Eq.(7), (which is appropriate if we consider some steady-state solutions [16]), we will obtain the customary equation of the form

$$\dot{\rho}_t(X) = \frac{\partial}{\partial X}M(\rho_t(X))\frac{\partial}{\partial X}\frac{\delta \mathcal{F}(\rho_t(X))}{\delta \rho_t(X)}, \quad (10)$$

where the mobility $M(\rho_t(X))$ is given by

$$M(\rho_t(X)) = \rho_t(X)(1 - \rho_t(X)), \tag{11}$$

while local free energy $\mathcal{F}(\rho)$ obeys

$$\mathcal{F}(\rho) = \int dX \left(f(\rho) + \frac{\sigma^2 \beta J}{2} \left(\frac{\partial \rho}{\partial X} \right)^2 \right), \tag{12}$$

with

$$f(\rho) = \rho \ln \rho + (1 - \rho) \ln (1 - \rho) + \beta J \rho (1 - \rho). \tag{13}$$

Curiously enough, $f(\rho)$, which has been derived in our work starting from a microscopic dynamical model obeying the detailed balance condition, has exactly the same form as the phenomenological Flory-Huggins-de Gennes local free energy density [20, 21]. Note also that for $\beta J \geq 2$, the local free energy $f(\rho)$ in Eq.(13) has a double-well structure whose minima approach 0 and 1 as βJ increases. This implies that the Onsager mobility in Eq.(11) never reaches negative values, contrary to the behavior predicted by Eq.(8) for which one has $D(\rho) < 0$ when $\rho_{c,-} < \rho < \rho_{c,+}$, with

$$\rho_{c,\pm} = \frac{1}{2}(1 \pm \sqrt{1 - 2/\beta J}). \tag{14}$$

Finally, we define the appropriate boundary conditions. As a matter of fact, any particle, in order to enter to the nanopore from the liquid phase, has to surmount an additional barrier E_M related to the enthalpic energy difference between the particle within the pore and in the bulk liquid phase [11, 12]. Supposing that the reservoir (bulk liquid phase) is in equilibrium with the particle phase within the single-file pore, we thus stipulate, following a similar analysis in [18], that the reservoir maintains a constant density ρ_F (see [11, 12] for relation between ρ_F and energetic parameters) at the entrance of the pore (site $X = 0$ in Fig.2). Second boundary condition is rather evident, we just suppose that $\rho_t(X)$ vanishes as $X \to \infty$ at fixed t. Consequently, Eq.(6) (or Eq.(10)) is to be solved subject to the conditions

$$\rho_t(X = 0) = \rho_F, \quad \rho_t(X \to \infty) = 0. \tag{15}$$

Below we discuss solutions of Eqs.(6) and (10) obeying these two boundary conditions.

Results

We focus here on the time-evolution of the total mass of particles $M(t)$, having entered the single-file pore up to time t, i.e.,

$$M(t) = \int_0^\infty dX \, \rho_t(X) \tag{16}$$

To define the time-dependence of $M(t)$, it is expedient to turn to the scaled variable $\omega = X/2\sqrt{D_0 t}$. In terms of this variable Eq.(10) attains the form

$$\frac{d^2\rho(\omega)}{d\omega^2} + 2\omega \frac{d\rho(\omega)}{d\omega} - 2\beta J \frac{d}{d\omega}\left[\rho(\omega)\left(1-\rho(\omega)\right)\frac{d\rho(\omega)}{d\omega}\right] -$$
$$- \beta J\left(\frac{\sigma^2}{4D_0 t}\right)\frac{d}{d\omega}\left[\rho(\omega)\left(1-\rho(\omega)\right)\frac{d^3\rho(\omega)}{d\omega^3}\right] = 0, \tag{17}$$

while the boundary conditions in Eq.(15) become $\rho(\omega = 0) = \rho_F$ and $\rho(\omega \to \infty) = 0$.

Note now that the term in the second line on the rhs of Eq.(17), associated with the second term in the expansion of the interaction energy in the Taylor series, Eq.(7), is irrelevant to the dynamics as $t \to \infty$, since it is multiplied by a vanishing function of time. Hence, the propagation dynamics can be adequately described by Eq.(8), which is not sufficient, however, for description of the steady-state characteristics, such as, e.g. steady-state particle current through a finite pore [16].

Now, in terms of the scaled variable ω, the total mass of particles $M(t)$ reads

$$M(t) = 2 \, m(J, \rho_F) \sqrt{D_0 t}, \tag{18}$$

where the prefactor $m(J, \rho_F)$ is determined by

$$m(J, \rho_F) = \int_0^\infty d\omega \, \rho(\omega) \tag{19}$$

Consequently, Eq.(18), which can be thought of as the microscopic analog of the Washburn equation, signifies that the mass of particles grows in proportion to the square-root of time. Note that similar result has been obtained in [18] for $J \equiv 0$, in which case $m(J, \rho_F) = \rho_F/\sqrt{\pi}$.

Before we discuss behavior of the prefactor $m(J, \rho_F)$, it might be instructive to understand what are the physical processes underlying the $M(t) \sim \sqrt{t}$ behavior (see also [18] for more detailed discussion). To do this, let us first recollect that the boundary condition $\rho_t(X = 0) = \rho_F = const$ is tantamount to the assumption that the reservoir is in equilibrium with the particle phase in the pore. Turning next to the model depicted in Fig.2, we notice that a jump of the rightmost particle of the particle phase away of the reservoir, leads to creation of a "vacancy". When this vacancy manages to reach diffusively, due to redistribution of particles, the entrance of the pore, it perturbs the equilibrium and gets filled by a particle from the reservoir. Hence, the mass of particles within the pore $M(t)$

is proportional to the current $\mathcal{J}$ of vacancies from the front of the propagating phase[1], $M(t) \sim \mathcal{J}$, where the current $\mathcal{J} \sim 1/L$, L being the distance travelled by the rightmost particle away from the entrance of the pore. Consequently, $M(t) \sim L \sim 1/L$, which yields eventually the $M(t) \sim \sqrt{t}$ law. Note also that, from the viewpoint of the underlying physics, the dynamical process under study is intrinsically related to such phenomena as directional solidification, freezing, limited by diffusive motion of the latent heat, or Stefan problem.

To define the prefactor $m(J, \rho_F)$ we have solved Eq.(17) (with the last term set equal to zero) numerically, for various values of the system's parameters ρ_F and J. These results are summarized in Fig.3 where we depict $m(J, \rho_F)$ as a function of ρ_F for several different values of βJ.

Now, one notices that for any values of ρ_F and βJ the prefactor $m(J, \rho_F) > 0$, which implies that, as one can expect on intuitive grounds, there is no transition in the one-dimensional system under study and the particle phase propagates into the pore as soon as $\rho_F > 0$. On the other hand, the prefactor $m(J, \rho_F)$ depends on the system's parameters in a quite non-trivial fashion. While for relatively small βJ the prefactor $m(J, \rho_F)$ varies with ρ_F almost linearly, for large βJ some saturation effect occurs, followed by, for larger βJ, a non-monotoneous ρ_F-dependence, and eventually, after a cusp-like variation, with a rapid growth.

We begin with a small βJ limit, in which case some perturbative analytical calculations are possible. To do this, let us represent $\rho(\omega)$ in the form of the series

$$\rho(\omega) = \sum_{n=0}^{\infty} (2\beta J)^n \rho_n(\omega), \qquad (20)$$

(which is related to expansion in powers of the reservoir density $\rho_F < 1$, as we will see in what follows) and try to calculate explicitly several first terms in such an expansion, constraining ourselves to the quadratic in βJ approximation. After some rather cumbersome but straightforward calculations, we find eventually, that in the quadratic with respect to the parameter $2\beta J$ approximation, the prefactor $m(J, \rho_F)$ obeys:

$$
\begin{aligned}
m(J, \rho_F) = \; & \frac{\rho_F}{\sqrt{\pi}} - (2\,\beta\,J) \left[0.18\,\rho_F^2 - 0.134\,\rho_F^3 \right] \\
& - (2\,\beta\,J)^2 \left[0.025\,\rho_F^3 - 0.047\,\rho_F^4 + 0.018\,\rho_F^5 \right]
\end{aligned}
\qquad (21)
$$

[1]Note, however, that there are some subtleties concerning propagation of the rightmost particle of the phase growing in the single-file pore. As a matter of fact, it has been shown in [18] that in a similar model without attractive interactions (i.e. $J = 0$) the mean displacement of the rightmost particle follows $X(t) \sim \sqrt{t} \ln(t)$, i.e. grows at a faster rate than the mass $M(t)$. Similar behavior is observed for the front of the pahse, propagating in the single-file pore, also in the interacting case, i.e. for arbitrary J [16].

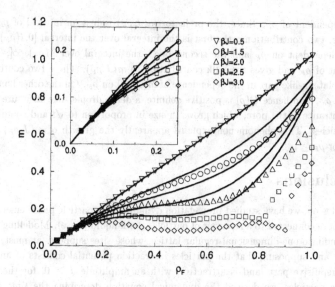

Figure 3: The prefactor $m = m(J, \rho_F)$ versus the density ρ_F for different values of the amplitude J of the attractive particle-particle interactions. Symbols denote the results of numerical solution of Eq.(17), while the solid lines correspond to the analytical result in Eq.(21). The inset displays the behavior for small ρ_F.

This dependence, as one notices, agrees quite well with the numerical solution for relatively low values of βJ over the entire domain of variation of ρ_F, or, for larger βJ, for progressively smaller values of ρ_F.

To get some understanding of the intricate, non-trivial behavior of $m(J, \rho_F)$, observed for larger values of βJ, we analyse dynamical density profiles defined by Eq.(8) versus X and $X/M(t)$ for $\beta J = 3$ and different reservoir densities ρ_F, (see Fig.4). Note now that the form of the density profiles is rather complex and depends largely on whether ρ_F is less than or exceeds $\rho_{c,\pm}$. When $\rho_F \leq \rho_{c,-}$, the form of the density profile is well described by $\rho_t(X) = \rho_F \mathrm{erfc}(X/M(t))$, where $\mathrm{erfc}(X/M(t))$ is the error function. This behavior is essentially the same as the one predicted for non-interacting lattice gas in [18]. On the other hand, when ρ_F exceeds $\rho_{c,-}$, but is less than $\rho_{c,+}$ (for $\beta J = 3$, $\rho_{c,\pm}$ are equal to 0.79 and 0.21, respectively) we have two different regimes: the density rapidly, within the small constant distance $l(\rho_F)$, drops to the value $\rho_{c,-}$ and then evolves as $\rho_t(X) = \rho_{c,-}\mathrm{erfc}(X/M(t))$, where

$\rho_{c,-}$ is independent of ρ_F. Since, the prefactor $m(J, \rho_F)$ is just an integral of $\rho_t(X)$, we have, hence, two contributions: the first is the integral over the interval $[0, l(\rho_F)]$, which is weakly dependent on ρ_F and the second one - the integral over $[l(\rho_F), \infty[$, which is independent of ρ_F and gives the bulk contribution to $m(J, \rho_F)$. These two contributions define the plateau-like part in the dependence of $m(J, \rho_F)$ on ρ_F. On the other hand, when ρ_F exceeds $\rho_{c,+}$ and hence $D(\rho)$ is positive definite, a dense droplet-like structure emerges near the entrance of the pore, which grows in size in proportion to $\sqrt{t}$ and contains most of the particles. This phenomenon explains apparently the growth of $m(J, \rho_F)$ with ρ_F observed for $\rho_F \geq \rho_{c,+}$.

Conclusions

In conclusion, we have studied propagation dynamics of a particle phase emerging in a single-file pore connected to a reservoir of particles (bulk liquid phase). Modelling the pore as a semi-infinite one-dimensional regular lattice, whose sites support, at most, a single occupancy, and supposing that the particles interaction potential consists of an abrupt, hard-core repulsive part, and is attractive, with an amplitude $J \geq 0$, for the nearest-neighboring particles, we derived the dynamical equation describing the time-evolution for the local particle density. This equation has been analysed both analytically and numerically. Within our appoach, we defined the evolution of the total number $M(t)$ of particles entering the single-file pore up to time t and also discussed the dynamics of the density distribution function within the pore. We have shown that the growth of $M(t)$ is described by $M(t) = 2m(J, \rho_F)\sqrt{D_0 t}$, which can be thought of as the microscopic analog of the Washburn equation, where D_0 denotes the "bare" diffusion coefficient, while the prefactor $m(J, \rho_F)$ is a non-trivial function of the attractive interactions amplitude J and the density ρ_F at the entrance to the pore. We have discussed a peculiar behavior of $m(J, \rho_F)$ and explained it through the analysis of the dynamical density distribution within the pore.

References

[1] *Access in Nanoporous Materials*, eds. T.J.Pinnavaia and M.F.Thorpe (Plenum, New York, 1995)

[2] M.Sahimi, *Flow and Transport in Porous Media and Fractured Rock*, (VCH, Weinheim, Germany, 1995)

[3] N.Y.Chen, J.T.F.Degnan and C.M.Smith, *Molecular Transport and Reaction in Zeolites*, (VCH, New York, 1994)

[4] B.Alberts, D.Bray and J.Lewis, *Molecular Biology of the Cell*, (Garland, New York, 1994)

[5] W.M.Meier and D.H.Olson, *Atlas of Zeolite Structure Types*, (Butterworths, London, 1987)

[6] V.Gupta et al., Chem. Phys. Lett. **247**, 596 (1995)

[7] V.Kukla et al., Science **272**, 702 (1996)

[8] D.Keffer, A.V.McCormick and H.T.Davis, Mol. Phys. **87**, 367 (1996)

[9] K.Hahn, J.Kärger and V.Kukla, Phys. Rev. Lett. **76**, 2762 (1996)

[10] C.Rödenbeck, J.Kärger and K.Hahn, Phys. Rev. E **55**, 5697 (1997)

[11] T.Chou, Phys. Rev. Lett. **80**, 85 (1998)

[12] T.Chou, J. Chem. Phys. **110**, 606 (1999)

[13] S.Murad, P.Ravi and J.G.Powles, J. Chem. Phys. **98**, 9771 (1993)

[14] D.S.Sholl and K.A.Fichthorn, J. Chem. Phys. **107**, 4384 (1997)

[15] L.Xu et al., Phys. Rev. Lett. **80**, 3511 (1998)

[16] A.M.Lacasta, J.M.Sancho, F.Sagues and G.Oshanin, in preparation

[17] G.Giacomin and J.L.Lebowitz, Phys. Rev. Lett. **76**, 1094 (1996)

[18] S.F.Burlatsky, G.Oshanin, A.M.Cazabat and M.Moreau, Phys. Rev. Lett. **76**, 86 (1996)

[19] J.Lebowitz, E.Orlandi and E.Presutti, J. Stat. Phys. **63**, 933 (1991)

[20] P.G.de Gennes, J. Chem. Phys. **72**, 4756 (1980)

[21] K.Binder, J. Chem. Phys. **79**, 6387 (1983)

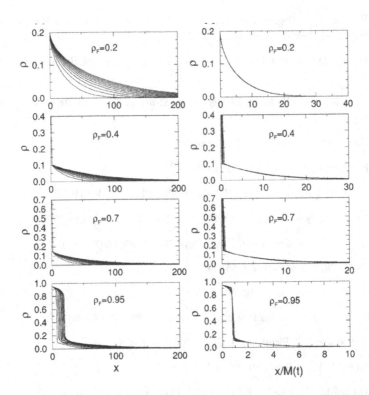

Figure 4: Dynamical density profiles defined by Eq.(8) for $\beta J = 3$ and different values of ρ_F. In the left column $\rho_t(X)$ is plotted versus the space variable X and here different curves on each graph show the dynamics of the density distribution in the single-file pore, Eq.(6), while in the right column we present corresponding plots for $\rho_t(X)$ vs the scaling variable $X/M(t)$.

Mat. Res. Soc. Symp. Proc. Vol. 651 © 2001 Materials Research Society

Relaxation Method Simulation of Confined Polymer Dispersed Liquid Crystals in an External Field

J. J. Castro[1], R. M. Valladares and A. Calles
Departamento de Física, Facultad de Ciencias, UNAM
Apdo. Postal 70-646, 04510 México, D. F.
[1]On sabbatical leave from Departamento de Física, CINVESTAV del IPN
Apdo. Postal 14-740, 07300 México, D.F.

ABSTRACT

Polymer dispersed liquid crystals (PDLC) are materials formed by nematic liquid crystals droplets with radii of a few hundred Å embedded in a polymer matrix. We discuss the use of relaxation methods for the study of the response of the director of a PDLC under the switching of an external electric field. We simulate the confining system by considering different boundary conditions at the droplet surface.

INTRODUCTION

Polymer dispersed liquid crystals (PDLC) have provided the technological basis for different kind of displays [1]. PDLC are materials that consist of microscopic droplets of nematic liquid crystals with typical radii varying from a few hundred Ångstroms to above a micron, embedded in a polymer matrix. The PDLC films can be switched between translucent and transparent states by applying an electric voltage. The reason for this change is the optical mismatch between the nematic droplets and the polymer matrix in the absence of electrical voltage. By applying an electric field, the nematic liquid crystal molecules within the droplets are reoriented creating an optical matching with the matrix, producing therefore no light dispersion. The study of PDLC droplets are also interesting because they exhibit topological defects that are realized in some other fields of physics [2]. The details of liquid crystal director configuration within the droplet play a dominant role in determining the PDLC optical properties. In general this configuration depends on its size, shape and the anchoring strength. Depending on the preparation method and the chosen polymer matrix, the most commonly found director configurations are radial or axial [3,4] for homeotropic surface anchoring, and toroidal or bipolar director configuration [3-6] for tangential anchoring. Studies of electro-optical properties of PDLC have shown that a field threshold must be overcome in order to induced reorientation of the nematic molecules within the droplet (Frederiks transition [7]). Although some theoretical models have been proposed in order to study the induced reorientation in PDLC [8,9], the problem is still far from being completely solved. It has been shown experimentally that the threshold depends on a number of factors related to specific physical and chemical properties that has not been easy to incorporate into the models.

Simulations of the behavior of such materials with different kind of anchoring and its response to applied voltages might help in gaining physical insight on the understanding of the response of the molecular orientation within the droplet. In the present work we study the behavior of the director of a nematic droplet with different boundary conditions that simulates particular anchoring situations, with and without external electric fields. This is made by the use of a relaxation method and visualization algorithms that allow to follow the distribution of the director towards equilibrium.

THEORY

We consider a lattice model for a nematic liquid crystal within the droplet, where the molecules are fixed at each lattice site. The model is limited to planar rotations of the molecules. The lattice is constructed in such a way as to maintain constant the distance between nearest neighbor particles

The molecules are described by molecular directors $\mathbf{n}(\mathbf{R})$, where $\mathbf{R}$ determines the lattice site. A pair of molecules with directors $\mathbf{n} = \mathbf{n}(\mathbf{R})$ and $\mathbf{n'} = \mathbf{n}(\mathbf{R'})$ separated by a distance $\mathbf{r} = \mathbf{R'}-\mathbf{R}$ are interacting through the potential

$$V_B (\mathbf{n,n',r}) = -\frac{C}{r^6}\left[\mathbf{n}\cdot\mathbf{n'} -3\,\frac{\varepsilon}{r^2}(\mathbf{n}\cdot\mathbf{r})\,(\mathbf{n'}\cdot\mathbf{r})\right]^2 , \qquad (1)$$

where $C > 0$ is the interaction strength and ε is the anisotropy parameter which can vary between 0 and 1. For $\varepsilon = 0$ we obtain the isotropic Maier-Saupe interaction, whereas for $\varepsilon = 1$ we obtain the induced dipole-induced dipole interaction [10].

For the nematic-polymer interaction we assume the Rapini-Papoular form [11]

$$V_S (\mathbf{n,\Pi,r}) = -\frac{D}{r^6}\left[\mathbf{n}\cdot\mathbf{\Pi}\right]^2 , \qquad (2)$$

where $D > 0$ is the interaction strength and Π is the substrate-induced easy axis.

We take into account all interactions between nematic molecules in the bulk and the interactions between the nematic molecules and the molecules that conform the substrate. The external electric field is assumed uniform within the nematic and we study the response of the director within the droplet as a function of the applied voltage and the surface anchoring.

The total energy of the system V, is composed of the bulk and the surface contributions as

as the energy coming from the external electric field

$$V = V_B + V_S + V_E . \qquad (3)$$

We look for the stabilization of the system trough the use of a relaxation method that allows us to find the minimum of the total energy as a function of the distribution of the director $\mathbf{n} = \mathbf{n}(\mathbf{R})$, that is:

$$\frac{\partial V}{\partial n(R)} = 0 . \qquad (4)$$

The present model permits dealing in a simple way with the response of the molecular orientation within the droplet at external applied electric fields. The anchoring condition at the surface can be easily incorporated and treated in a systematic way. We also have developed the proper software to visualize the director distribution.

RESULTS

In the present work we have determined the director distribution of a PDLC for different anchoring conditions with and without external electric field. In particular we study the effect of radial, tangential and oblique anchoring. In figures 1, 2 and 3 we show the director distribution that stabilized the system for the three different anchoring conditions, which are shown as lines at the boundary surface. As can be seen from those figures the symmetry of the anchoring is preserved in the nematic. This is a consequence of the interaction potential that was used for those cases were the anisotropy factor has been neglected.

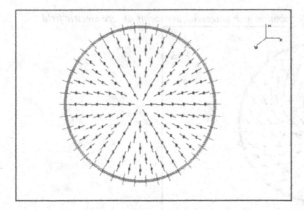

Figure 1. *Molecular director distribution with radial anchoring at zero electric field.*

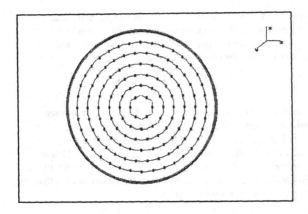

Figure 2. *Molecular director distribution with tangential anchoring at zero electric field.*

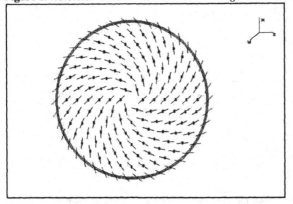

Figure 3. *Molecular director distribution with oblique (45°) anchoring at zero electric field.*

In figure 4 we show the molecular orientation response for the radial distribution to weak external fields (the intensity of the electric field is measured in arbitrary units). In this case we see how the effect of the weak external field (5.0 a.u.) starts to reorient the molecules. As it is shown in figure 5, the situation is completely different when we increase the intensity of the external field. For this case (50.0 a.u.) we have achieved a completely reorientation of the nematic molecules. This can be related to Frederiks transition, were a threshold potential must be overcome in order to induced reorientation of the nematic molecules within the droplet.

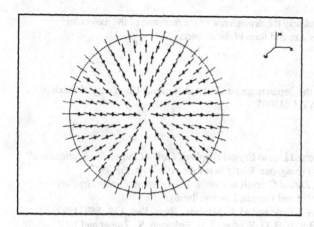

Figure 4. *Molecular director distribution with radial anchoring at a weak electric field (5 a.u.) along the X axes.*

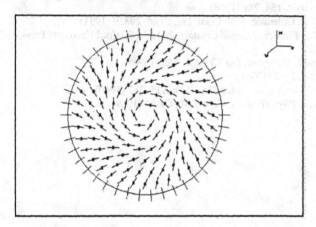

Figure 5. *Molecular director distribution with radial anchoring at a strong electric field (50 a.u.) along the X axes.*

CONCLUSIONS

This work presents a model for the study of the molecular orientation within a droplet of a PDLC and the molecular re-distribution under the influence of an external electric field. We have shown how the use of relaxation methods permits the systematic study of the director behavior within a nematic droplet with general anchoring conditions. The utility of this model lies on the simplicity for treating in a general way the dependence of the response of the molecular orientation to parameters such as size and form of the droplets.

ACKNOWLEDGMENTS

Research partially supported by the Supercomputer Center at National University of Mexico under contract No. UNAM-CRAY-I-941008.

REFERENCES

1. J. W. Doane, Polymer Dispersed Liquid Crystal Displays, *Liquid Crystal-Applications and Uses* , Vol. 1, ed. B. Bahadur (Singapur: World Scientific, 1990) pp.361-395.
2. P. Crawford and S. Žumer, *Liquid Crystals in Complex Geometries Formed by Polymer and Porous Networks*, (Taylor and Francis, London, 1996).
3. A. Golemme, S. Žumer, J. W. Doane and M. E. Neubert, *Phys. Rev.* **A37**, 559 (1988).
4. R. Ondris-Crawford, E. P. Boyco, B. G. Wagner, J. H. Erdmann, S. Žumer and J. W. Doane,
   J. Appl. Phys. **69**, 6380 (1991).
5. P. Drzaic, *Mol. Cryst. Liq. Cryst.* **154**, 289 (1988).
6. R. Aloe, G. Chidichimo and A. Golemme, *Mol. Cryst. Liq. Cryst.* **203**, 9 (1991).
7. P. G. De Gennes, J. Prost, *The Physics of Liquid Crystals*, 2nd ed., (Oxford University Press, New York, 1993).
8. B. G. Wu, J. H. Erdmann and J. W. Doane, *Liq. Cryst.* **5**, 1453 (1989)
9. K. Amundson, *Phys Rev.* **E53**, 2412 (1996) .
10. W. Maier and A. Saupe, *Z. Naturforsch.* **A14**, 882 (1959); **A15**, 287 (1960).
11. A. Rapini and M. Papoular, *J. Phys.* (France), Colloq. **30**, C4-54 (1969).

Mat. Res. Soc. Symp. Proc. Vol. 651 © 2001 Materials Research Society

Transport properties in ionic media

A.-L. Rollet[1], M. Jardat[1], J.-F. Dufrêche[1], P. Turq[1]* and D. Canet[2]

(1) Laboratoire LI2C, Université Pierre et Marie Curie Bat. F
case 51, 4 place Jussieu, 75252 Paris cedex 05, France
fax : 33144273834, *email : pt@ccr.jussieu.fr
(2) Laboratoire de méthodologie RMN, Université Henri Poincaré
BP 239, 54056 Vandoeuvre-les-Nancy, France

Abstract

Transport coefficients in charged media exhibit strong variations, according to the conditions of displacement of the particles. Electrical transport, characterized by the simultaneous displacement of positive and negative charges in opposite directions obeys Ohm's law, but its variation with concentration (non-ideality), depends on several types of interactions, whose time of establishment varies from picosecond to nanosecond. Several diffusion processes can occur: mutual diffusion, where ions move simultaneously in the same direction, keeping local electroneutrality, and self diffusion where individual ionic particles move separately. The variation of diffusion coefficients with concentration depends on non-ideality factors analogous to those occuring in conductance, and their experimental evidence is facilitated by the availability of experimental techniques owing different characteristic times of observation. This phenomenon is particularly noticeable for self-diffsuion coefficients, where the dynamical processes can be observed from the picosecond range (neutron quasi-elastic scattering), to millisecond (NMR) and to hour scale (radiactive tracers). The results are especially enhanced for porous charged media like ion exchanging membranes (nafions).

Those results are be explained here theoretically in the framework of continuous solvent model theories (brownian dynamics) and experimentally in the study of self-diffusion in nafion membranes.

1 Introduction

The transport processes in liquids and solutions depend both on the time and the space conditions. In any case, the short time dynamics of the solute particles is dominated by ion-solvent collisions. Furthermore, in confined media, the boundary conditions of the system bring special features such as particle-wall collisions : this last effect dominates the long-time behaviour of the transport processes. Thus the transport coefficients and the dynamics depend on the time and scale windows related to the process under consideration. Another factor closely related to the present analysis is the establishment time of the different long-ranged interactions in liquids.

In ionic solutions, two effects are significant :
- the characteristic time of establishment of hydrodynamic interactions
- the time duration of charge fluctuations (departures from electroneutrality : Debye time).

In this case, the spatial fluctuation is also associated to a characteristic length which is the Debye length. All these quantities have very simple classical expressions for diluted solutions, but a precise and simple description is not obvious in complex media. In confined media, for example, the question of existence of separated characteristic times or of continuous multimodal diffusion remains yet open.

In the present paper, we will discuss the characteristic times related to the self-diffusion and electrical conductance of electrolyte solutions in bulk and confined media. In the first part, we will deal with the establishment of different forces by a mesoscopic approach. This mesoscopic Brownian dynamics allows to obtain long time scales (several ns),that are necessary for the undergoing of the different relaxation processes. Some experimental consequences of those effects will be presented in the following part. In this second section, we will present multiscale variations of the counterion self-diffusion coefficient in a confined charged porous medium, the nafion.

2 Theoretical aspects in bulk electrolytes : Brownian dynamics simulation

For times t larger than the Brownian motion time τ_B which is less than one picosecond for aqueous solutions, the solvent can be treated as a continuum. For times greater than the velocity relaxation time of ions which is about one picosecond for simple ions, the velocities of the solute are always in equilibrium with the solvent. The motion of particles are then described in the position space by a stochastic equation of motion. This is the base of Brownian dynamics simulation.

To compute collective transport coefficients like the electrical conductivity, one has to consider time scales exceeding significantly the electrostatic relaxation time τ_D, which is about 100 ps for simple 1-1 electrolyte solutions in the molar range. Recently, an efficient Brownian dynamics simulation method has been proposed [3,4] which permits to calculate long trajectories for ions and to compute the transport coefficients in the molar range. This method includes the calculation of hydrodynamic interactions between solute particles. We recall here the results obtained in the case of KCl aqueous solutions at 298 K.

The transport coefficients of KCl solutions are compared to experimental data in Figure 1 and 2. The conductance is strongly decreased by hydrodynamic interactions. When they are neglected, the conductivity is simply given by the relaxation effect. The latter lowers the conductance with respect to the value at infinite dilution. However this effect is small and shows almost no variation with concentration. When hydrodynamic interactions are taken into account, the simulations are in excellent agreement with the experimental data.

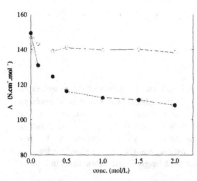

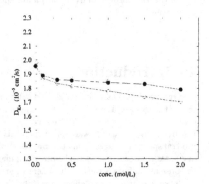

Figure 1: Equivalent conductance of aqueous KCl solutions at 298 K : experimental data ($\Diamond$), Brownian dynamics simulations without ($\circ$) and with ($\bullet$) hydrodynamic interactions.

Figure 2: K$^+$ self diffusion coefficient of aqueous KCl solutions at 298 K : experimental data ($\Diamond$), Brownian dynamics simulations without ($\circ$) and with ($\bullet$) hydrodynamic interactions.

On the opposite, the self-diffusion coefficient of K^+ is increased by hydrodynamic interactions. In this case both effects have opposite influences on the transport coefficient, as the relaxation effect decreases the self-diffusion.

3   Experimental example of porous charged medium: Nafion membrane

The Nafion 117 membrane [8,9] is a porous and charged medium composed by a linear polymer of fluorocarbon with an anionic ending : a sulfonate group. These anionic groups build inverted micelles and thus form, in the hydrophobic polymer matrix, hydrophilic cavities in which aqueous solutions can penetrate. The diameter of these cavities is about 40 Å. They are connected by small pores whose size has been estimated at 10 Å length and 10 Å diameter. Another interesting characteristic of the Nafion membranes is the anionic site concentration in the hydrophilic domains : it is about 4 mol.l^{-1}.

In this experimental section, we deal with the determination of the self-diffusion coefficient of the tetramethyl ammonium ion $N(CH_3)_4^+$ by different experimental techniques. These measurements at different times scales allow to get a complete picture of the ionic transport processes in the membrane. The times range extends from the picosecond with Neutron Quasi-Elastic Scattering experiments to millisecond with NMR experiments and to minute with radiotracers experiments.

3.1   Experimental results

3.1.1   NQES

The NQES experiments were performed with the time-of-flight spectrometer mibemol at Laboratoire Léon Brillouin (CEA Saclay, France). The studied samples were : Nafion membranes equilibrated with solutions of $N(CH_3)_4^+$ in D_2O at 0.5 and 2 mol/kg ; $N(CH_3)_4Cl$ in D_2O solutions at 1, 3.5, 4.5 and 8 mol/kg. The elastic and quasi-elastic peaks were extracted from the scattered intensity using the "fitmib" program (LLB).

For a simple diffusion the incoherent scattering function $S_{inc}(q,\omega)$ can be written in the following manner [10] :

$$S_{inc}(q,\omega) = \frac{1}{\pi} \frac{y(q)}{\omega^2 + (y(q))^2} \tag{1}$$

with $\hbar\omega$ the energy transfer and $y(q)$ the half width at half maximum of $S_{inc}(q)$.

In the limit of small q, $y(q)$ becomes :

$$\lim_{q \to 0} y(q) = Dq^2 \tag{2}$$

where D is the self-diffusion coefficient.

The self-diffusion coefficients of $N(CH_3)_4^+$ ions are thus obtained from a linear regression on a plot of $y(q)$ as a function of q^2. The resolution 20 μeV corresponds approximately to an observation time of $9 \ 10^{-11}$ s and, with a self-diffusion coefficient about 10^{-5}cm^2.s^{-1}, to a displacement of 3 Å. This length is quite small as compared with Nafion cavity size, thus allowing one to use the same model for determining the self-diffusion coefficient of $N(CH_3)_4^+$ ions in solutions and in Nafion membrane from NQES spectra. In order to compare the self-diffusion in solution and within a Nafion membrane, the values of D are plotted as a function of the molality of $N(CH_3)_4^+$ on Figure 3. One can notice that the self-diffusion coefficients in Nafion membrane are placed on the curve formed by the values of the self-diffusion coefficient in solution. It appears therefore that the self-diffusion process is similar, within this time and space scales, to that in non-confined solution.

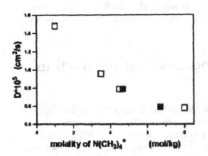

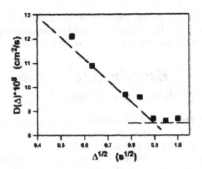

Figure 3: Variation of self-diffusion coefficient of $N(CH_3)_4^+$ in Nafion membrane (■) and in non-confined solution (□) versus $N(CH_3)_4Cl$ molality.

Figure 4: Variation of the apparent self-diffusion coefficient of $N(CH_3)_4^+$ in Nafion membrane (■) versus the square root of the diffusion interval Δ in NMR sequence. The two straight lines are representative of approximations valid at short and long times, respectively. Example of Nafion membrane equilibrated with 0.5 mol/kg.

The decrease of self-diffusion coefficient when the electrolyte concentration is increased is due to viscosity effects.

3.1.2 NMR

The NMR measurements have been performed on Nafion membranes equilibrated with solutions of 0.1, 0.5, 1 and 2 mol/kg $N(CH_3)_4^+$ in D_2O. The NMR method uses in this study employs gradients of the radio-frequency field B_1, instead of usual gradients of B_0 [11]. More details on this technique are given elsewhere [12,13]. The experimental results show a decrease in the measured self-diffusion coefficient (apparent D) when the diffusion interval Δ is increased except for high values of Δ (> 1 s) (see Figure 4).

Such a behaviour indicates that the observed diffusion process is confined in space, at NMR scale. It seems then that the distribution of cavities is not homogeneous. Mitra et al.[14,15] have developed a model which described such diffusion. In this work, the following limiting cases have been used :

$$D(\Delta) = D_0 \left[1 - \frac{4}{9\sqrt{\pi}} \frac{S}{V} \sqrt{D_0 \Delta}\right] \quad \text{(short times)} \qquad (3)$$

Where S/V is the ratio of the surface and volume of the pores and D_0 the value of the self-diffusion for a non-confined medium.

$$D(\Delta) = D_0 \left(\frac{1}{\alpha}\right) \quad \text{(long times)} \qquad (4)$$

Where α is the tortuosity of the medium.

The values obtained from NMR are two orders of magnitude lower than those from NQES (see figure 8). It seems then that passing through the connecting pores is a limiting factor of the diffusion process.

In the same time, the effect of electrolyte concentration has been investigated. On Figure 5 the values of the self-diffusion coefficient of $N(CH_3)_4^+$ have been plotted as a function of the external molality, for short times represented by the limiting values D_0 and for long times D_{inf} that is when $D(\Delta)$ does not depend anymore on Δ. It must be noticed that the diffusion is sensitive to electrolyte concentration only at short times.

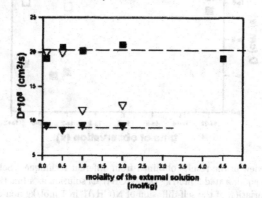

Figure 5: Variation of the self-diffusion coefficient of $N(CH_3)_4^+$ in Nafion versus the molality of external $N(CH_3)_4Cl$ solution. Values obtained by the radiotracer method (■). Values obtained by NMR : extrapolated values D_0 (∇) and for long time Δ D_{inf} (▼).

3.1.3 Radiotracers

A precise description of this method has been given in a previous publication [16].

The studied systems were Nafion membranes equilibrated with solutions of 0.1, 0.5, 1 and 2 mol/kg $N(CH_3)_4Cl$ in H_2O.

With such a radiotracer method, one follows the transfer of $N(CH_3)_4^+$ ions through the whole membrane. On Figure 5 are represented the values of the self-diffusion coefficient as a function of the molality of the external solution. No variation is noticed. These results confirm the NMR results for long Δ. This difference probably arises of from $N(CH_3)_4^+$ size.

References

[1] Turq P., Lantelme F., Friedman H.L. Brownian dynamics : its application to ionic solutions. J. Chem. Phys., 1977, vol 66, No 7, p 3039-3044.

[2] Allen M.P., Tildesley D.J. Computer Simulation of Liquids. Oxford Science Publications, 1987.

[3] Jardat M., Bernard O., Turq P., Kneller G.R. J. Chem. Phys., 1999, vol 110, No 16, p 7993-7999.

[4] Jardat M., Durand-Vidal S., Turq P., Kneller G.R. Brownian dynamics J. Mol. Liq., 2000, vol 85, p 45-55.

[5] Ermak D.L. J. Chem. Phys., 1975, vol 62, p 4189.

[6] Rossky P.J., Doll J.D., Friedman H.L. J. Chem. Phys., 1978, vol 54, p 1086.

[7] Durand-Vidal S., Simonin J.P., Turq P., Bernard O. J. Phys. Chem., 1995, vol 94, p 6733.

[8] Gebel G., Loppinet B. Ionomers : characterisation, theory and applications. Schlick S. Ed., CRC Press : Boca Raton, FL, 1996, chapitre 5.

[9] Hsu W.Y., Gierke T.D., J. Membrane Sci., 1983, vol 13, p 307.

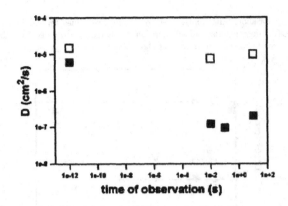

Figure 6: An overall view of the results obtained from different techniques: Self-diffusion coefficient of $N(CH_3)_4^+$ in Nafion equilibrated 2 mol/kg ($\blacksquare$) $N(CH_3)_4Cl$ solution as a function of the times scale. For comparison the variation of the self-diffusion of $N(CH_3)_4^+$ in 1 mol/kg non-confined solution ($\square$) has also been plotted.

[10] Bée M. Quasi-elastic neutron scattering : principles and application in solid state chemistry, biology and materials science. Adam Hilger Ed., Bristol and Philadelphia, 1988.

[11] Johnson C.S.Jr, Encyclopaedia of NMR, Harris G. Ed., Wiley Press.

[12] Canet et al. J. Magn. Reson., 1989, vol 81, p 1.

[13] Dupeyre et al. J. Magn. Reson., 1991, vol 95, p 581.

[14] Mitra P.P., Sen P.N., Schwartz L.M., Le Doussal P. Phys. Rev. Lett., 1992, vol 68(24), p 3555.

[15] Latour L.L., Mitra P.P., Kleinberg R.L., Sotak C.H. J. Magn. Reson., 1993, vol 101(A), p 342-346.

[16] Rollet A.-L., Simonin J.-P., Turq P. Phys. Chem. Chem. Phys., 2000, vol 2(5), p 1029.

[17] Rollet A.-L., Simonin J.-P., Turq P. Gebel G. submitted to J. Polym. Sci. Polym. Phys. Ed.

AUTHOR INDEX

SUBJECT INDEX